Fluid Mechanics and Its Applications

Founding Editor

René Moreau

Volume 135

Series Editor

André Thess, German Aerospace Center, Institute of Engineering Thermodynamics, Stuttgart, Germany

The purpose of this series is to focus on subjects in which fluid mechanics plays a fundamental role. As well as the more traditional applications of aeronautics, hydraulics, heat and mass transfer etc., books will be published dealing with topics, which are currently in a state of rapid development, such as turbulence, suspensions and multiphase fluids, super and hypersonic flows and numerical modelling techniques. It is a widely held view that it is the interdisciplinary subjects that will receive intense scientific attention, bringing them to the forefront of technological advancement. Fluids have the ability to transport matter and its properties as well as transmit force, therefore fluid mechanics is a subject that is particulary open to cross fertilisation with other sciences and disciplines of engineering. The subject of fluid mechanics will be highly relevant in such domains as chemical, metallurgical, biological and ecological engineering. This series is particularly open to such new multidisciplinary domains. The median level of presentation is the first year graduate student. Some texts are monographs defining the current state of a field; others are accessible to final year undergraduates; but essentially the emphasis is on readability and clarity.

Springer and Professor Thess welcome book ideas from authors. Potential authors who wish to submit a book proposal should contact Dr. Mayra Castro, Senior Editor, Springer Heidelberg, e-mail: mayra.castro@springer.com

Indexed by SCOPUS, EBSCO Discovery Service, OCLC, ProQuest Summon, Google Scholar and SpringerLink

Clovis R. Maliska

Fundamentals of Computational Fluid Dynamics

The Finite Volume Method

Clovis R. Maliska
SINMEC/CFD Laboratory
Mechanical Engineering Department
Federal University of Santa Catarina
Florianópolis, Santa Catarina, Brazil

ISSN 0926-5112   ISSN 2215-0056 (electronic)
Fluid Mechanics and Its Applications
ISBN 978-3-031-18234-1   ISBN 978-3-031-18235-8 (eBook)
https://doi.org/10.1007/978-3-031-18235-8

© The Editor(s) (if applicable) and The Author(s), under exclusive license to Springer Nature Switzerland AG 2023, corrected publication 2023
This work is subject to copyright. All rights are solely and exclusively licensed by the Publisher, whether the whole or part of the material is concerned, specifically the rights of translation, reprinting, reuse of illustrations, recitation, broadcasting, reproduction on microfilms or in any other physical way, and transmission or information storage and retrieval, electronic adaptation, computer software, or by similar or dissimilar methodology now known or hereafter developed.
The use of general descriptive names, registered names, trademarks, service marks, etc. in this publication does not imply, even in the absence of a specific statement, that such names are exempt from the relevant protective laws and regulations and therefore free for general use.
The publisher, the authors, and the editors are safe to assume that the advice and information in this book are believed to be true and accurate at the date of publication. Neither the publisher nor the authors or the editors give a warranty, expressed or implied, with respect to the material contained herein or for any errors or omissions that may have been made. The publisher remains neutral with regard to jurisdictional claims in published maps and institutional affiliations.

This Springer imprint is published by the registered company Springer Nature Switzerland AG
The registered company address is: Gewerbestrasse 11, 6330 Cham, Switzerland

This book is dedicated to my grandson and friend Augusto (13) which in all opportunities he met me during the writing he remembered to ask me how is the book going, a keen perception in recognizing that I was involved in an important matter, and also to my three lovely granddaughters Mariana (9), Isabela (7) and Cecilia (5). Always when they visited me, as they knew I was in the office working on the book, they used to bring me a cup of coffee or milk and some cookies or dessert. Beautiful and enjoyable moments. I hope now I will have more time for them all. In Memory of my Parents Delphina and Antônio

Foreword

In the Preface, the author describes how he knocked on my office door looking for a third supervisor after his first and second had tragically passed away. Based on his excellent academic record and wonderful sense of humor, it was difficult to say no; yet, there was lingering doubt about how saying yes might affect my own longevity. As it turned out saying yes left me doubly blessed: I gained both a very talented research associate and a treasured friend.

As the years passed, I have admired how he established a remarkable university CFD research facility and how he drove the practical application of CFD not only in Brazil but in all of South America. Later, he was also instrumental in founding a commercial CFD company through which his expertise has helped clients around the world.

And now we should be grateful that he has written this book. Students learning CFD will benefit by working their way through topics starting with the simple iterative solution of the conservation equation for a single variable and ending with methods to solve the full 3D equations of fluid flow and heat transfer using unstructured grids and multigrid solvers. Along this journey, concepts are carefully explained using simple 1D examples and by applying physical insight. The reader is effectively prepared for important developments by often giving a preview before the full explanation pages or chapters later. There is even helpful advice on how to go about debugging a computer code. And once CFD is being applied with confidence, this book will remain a valuable resource on the office bookshelf.

I thank the author for his time and energy to preserve his knowledge and advice for all of us in the pages of this book.

Waterloo, ON, Canada George D. Raithby

The original version of the book was revised: The symbol used for the v-velocity component in the original version of this book has been updated due to a printing issue. The correction to the book is available at https://doi.org/10.1007/978-3-031-18235-8_16

Preface

I was introduced to the field of computational fluid dynamics in my Ph.D. program at the University of Waterloo, Canada, under the supervision of Prof. George D. Raithby. He is one of the world's renowned scientists in fundamentals of CFD, pioneer in delivering seminal works to the scientific community. How this all started deserves few words. When preparing myself to go to Canada, I met Dr. Julio Militzer in a conference in my hometown Florianopolis. He had just arrived from his doctorate in Waterloo. Knowing that I was going to Waterloo, he advised me to try working with George Raithby, since "he is the best in the department", he said.

I went to Canada on May 1978 with George's name in my mind. Due to scientific collaborations among universities, my supervisors were already defined: Prof. Nicoll and Prof. Alpay. It is unbelievable, but both of them died few months after my arrival. So, I ended up knocking on Prof. Raithby's door looking for a supervisor, following Julio's advice. At that time George was full of students under supervision, since he was in the most important moment of creative numerical developments. He kindly invited me for a cup of coffee in the lounge. There, he told me about the possible research lines such as pressure–velocity coupling, grid generation, among many others. Some of them, I fully understood a year or more later.

George told me he would have to look on my profile and would see me the day after. Back to George's office, the next day he told me: As you are a faculty member on a leave for your Ph.D. studies, and in a difficult situation right now, I am going to supervise you. So, let's work. From now on, I will call you Clovis and, please, call me George. That day decided my professional life.

Back to my Department of Mechanical Engineering of the Federal University of Santa Catarina, in 1981, I created the courses Computational Fluid Dynamics I and II, taught at the graduate level in mechanical engineering, and started an active life in teaching heat transfer, fluid mechanics and numerical fluid flow, supervising master and Ph.D. students in the development of numerical techniques. Up to now, around 50 M.Sc. and 25 Ph.D. students worked with me. Perhaps, one the most significative results for having entered in the CFD field was the creation of the startup company ESSS, leaded by Clovis Jr. and Marcus Vinicius of my CFD Group at the university.

The company is now present in several countries and leader in South America on CFD and scientific software developments.

In parallel to my academic activities, I was always involved in leading collaborative research projects with major Brazilian companies, mainly in aerodynamics and petroleum reservoir simulation. In those areas, fundamental topics, always based on numerical developments, were studied along more than three decades. These engineering consulting works helped me to understand the linkage among numerical techniques and the real engineering world of simulation and to judge up to where in complexity the numerical techniques should go on the solution.

In 1995, I wrote my book on numerical fluid flow with a 2nd Edition in 2005. I wrote in Portuguese such that undergraduate students not mastering properly the English language, and not affording buy an international textbook, could benefit from the text in a moment in which CFD was growing and wide spreading in the academy and industries. It was a pioneer book written in Portuguese and well accepted by the scientific community. Following the trends in my teaching activities, recently, I made available at the SINMEC YouTube channel, two of my graduate courses, recorded in classroom, convection heat transfer and numerical fluid flow, of 30 h each, spoken in Portuguese. A third set of lectures, of about 10 h on numerical methods for unstructured grids, this one spoken in English, is also available on YouTube.

However, it was always in my plans to write a book for the international community, sharing with students, CFD professionals and analysts around the world what I have learned along those years in the field, particularly in the way of teaching numerical fluid flow. The day-to-day at the University, added to the research projects, always hindered this initiative, until I decide to have 6 months leave of absence from the University and write it. Here it is. I am very happy to have accomplished the goal.

As an engineer, when teaching numerical fluid flow, I always try to be with one foot on the physics (engineering) and the other one on the basics of the numerical methods. These two pillars are complementary, and they do not survive without each other. I believe this is the recipe for an efficient learning and for successful engineering simulations. This belief is fully immersed in this book, trying to offer to students and professionals a clean text which, hopefully, may be of enjoyable reading too. The goal is a compromise among simplicity and the required fundamental deepness of the subjects, creating the foundations for further advancements.

Florianopolis, SC, Brazil
July 2022

Clovis R. Maliska
clovis.maliska@ufsc.br
clovis.maliska@gmail.com

Acknowledgements

The efforts of writing a book are never fully foreseen when you start to write it. With this one was not different. It took me seven months working all day long. But this is not what really counts. The contents of a book don't drop instantaneously in your mind, as it is the result of many, many years of learning, digesting information and discussing with many people. I can't name the contribution of all them, as it would be a long list and I would commit injustices. I just want to say sincere thanks to all my students and collaborators for the fruitful and enjoyable time I spent with them during all these years. I am closing by now 44 years of activity in CFD and my students deserve a warm and heartily thanks. I have no words to express my gratitude.

I want to name and say thanks to one of my colleagues, Prof. António Fábio, who was with me along my career, helping me supervising students and working on research and industrial projects. A sharp mind with whom it was always a pleasure to discuss physics and numerical techniques. With him, my doctorate student Aideé and Dr. Honório, read part of the manuscript. Thanks to them for this great help. I am grateful to Vitória who, with dedication and patience, drew the figures for the book.

With this book I want to homage Prof. Raithby. He is the starting point of all these years which I am working on CFD. It was a privilege to work and learn with him. He is a brilliant researcher, with unbounded enthusiasm and an extremely kind human being. We visited him three years ago, he is in good health and, believe or not, I found him working on pressure-velocity coupling. Just amazing!!! Perhaps, he is not satisfied with today's numerical techniques for handling this issue, and more clever insights are still needed.

Finally, I would like to thank my family, Clovis Jr., Karina, Luise and Rui for always promoting the family union, and especially my wife Ana Maria for her continuous help and support during all my career, since the beginning when we went to Canada.

Contents

1	**Introduction**	1
	1.1 Preliminaries	1
	1.2 Available Tools for the Engineer	2
	1.3 Classes of Numerical Methods Available	6
	1.4 Objectives and Scope of This Book	9
	1.5 Applications of Computational Fluid Dynamics	11
	Reference	12
2	**Conservation Equations—Physical and Mathematical Aspects**	13
	2.1 Introduction	13
	2.2 Models Formulation Levels	13
	2.3 Conservation Equations	15
	2.3.1 Mass Conservation Equation	15
	2.3.2 Linear Momentum Conservation Equations	17
	2.3.3 Energy Conservation Equation	23
	2.4 Elliptic, Parabolic and Hyperbolic Problems	31
	2.4.1 Preliminaries	31
	2.4.2 Parabolic and Hyperbolic Problems	31
	2.4.3 Elliptic Problems	33
	2.5 True and Distorted Transient	35
	2.6 Conclusions	37
	2.7 Exercises	37
	References	39
3	**The Finite Volume Method**	41
	3.1 Introduction	41
	3.2 The Task of a Numerical Method	42
	3.3 Why Finite Volume Methods is a Good Choice	43
	3.4 Few Words About the Conservative Property	46
	3.5 Cell-Center and Cell-Vertex Methods	48
	3.6 One Dimensional Transient Heat Diffusion	50
	3.7 Explicit, Implicit and Fully Implicit Formulations	53

		3.7.1	Explicit Formulation	53
		3.7.2	Fully Implicit Formulation	57
		3.7.3	Implicit Formulation	60
	3.8	Linearization of the Source Term		62
	3.9	Boundary Conditions		64
		3.9.1	Balances for the Boundary Volumes	64
		3.9.2	Using Fictitious Volumes	65
		3.9.3	About Boundary Conditions in Cell-Vertex	67
	3.10	Discretization of the 3D Diffusion Equation		68
	3.11	Structure of the Matrix of Coefficients		70
	3.12	Handling Non-linearities		72
	3.13	Relevant Issues When Discretizing the Equations		73
		3.13.1	Positivity of Coefficients	73
		3.13.2	Fluxes Continuity at Interfaces	73
		3.13.3	Linearization of Source Term with S_P negative	76
		3.13.4	Truncation Errors	76
		3.13.5	Consistency, Stability and Convergence	78
	3.14	Conclusions		79
	3.15	Exercises		80
	References			83
4	**Solution of the Linear System**			85
	4.1	Introduction		85
	4.2	Iterative Methods		86
		4.2.1	Jacobi	86
		4.2.2	Gauss-Seidel	87
		4.2.3	SOR-Successive Over Relaxation	87
		4.2.4	Alternating Direction Implicit Methods	88
		4.2.5	Incomplete LU Decomposition	91
		4.2.6	A Note on Convergence of Iterative Methods	94
		4.2.7	Multigrid Method	95
	4.3	Conclusions		107
	4.4	Exercises		107
	References			111
5	**Advection and Diffusion—Interpolation Functions**			113
	5.1	Introduction		113
	5.2	The General Equation		114
	5.3	The Difficulty of the Advective-Dominant Problem		115
	5.4	Interpolation Functions for ϕ		119
		5.4.1	The Physics Behind the Interpolation Functions	120
		5.4.2	One Dimensional Interpolation Functions	121
		5.4.3	Numerical or False Diffusion	129
		5.4.4	Two and Three-Dimensional Interpolation Functions	136
	5.5	Conclusions		140

	5.6	Exercises	141
	References		142
6	**Three-Dimensional Advection/Diffusion of ϕ**		**145**
	6.1	Introduction	145
	6.2	Integration of the 3D Equation for ϕ	145
	6.3	Explicit Formulation	149
		6.3.1 True Transient	150
		6.3.2 Distorted Transient	150
	6.4	Fully Implicit Formulation	152
	6.5	Conclusions	153
	6.6	Exercises	154
7	**Finding the Velocity Field—Pressure/Velocity Couplings**		**157**
	7.1	Introduction	157
	7.2	System of Equations	157
		7.2.1 About Segregated and Simultaneous Solution	159
	7.3	Segregated Formulation. Incompressibility	161
	7.4	Variable Arrangement on the Grid	163
		7.4.1 Co-located Grid Arrangement	164
		7.4.2 Staggered Grid Arrangement	166
	7.5	Co-located PV Coupling (CPVC) Methods	167
		7.5.1 Rhie and Chow-Like Methods	168
		7.5.2 PIS—Physical Influence Scheme	173
	7.6	Segregated PV Coupling (SPVC) Methods	175
		7.6.1 Chorin's Method	177
		7.6.2 SIMPLE—Semi Implicit Linked Equations	178
		7.6.3 SIMPLER—Simple-Revisited	183
		7.6.4 PRIME—Pressure Implicit Momentum Explicit	185
		7.6.5 SIMPLEC—Simple Consistent	187
		7.6.6 PISO—Pressure Implicit with Split Operator	188
		7.6.7 SIMPLEC for Co-located Grids	189
		7.6.8 PRIME for Co-located Grids	191
	7.7	Boundary Conditions for p and p'	192
	7.8	Simultaneous Solution and the Couplings	194
	7.9	A Note on Boundary Conditions	198
		7.9.1 Impermeable Boundary—ϕ Prescribed	199
		7.9.2 Impermeable Boundary—Flux of ϕ Prescribed	199
		7.9.3 Inflow and Outflow Boundary Conditions	200
		7.9.4 General Comments About Boundary Conditions	201
		7.9.5 Incompressible Flows	201
		7.9.6 Compressible Flows	202
	7.10	Conclusions	203
	7.11	Exercises	203
	References		204

Contents

8 All Speed Flows Calculation—Coupling $P \rightarrow [V - \rho]$ 207
   8.1 Introduction 207
   8.2 Pressure–Velocity and Pressure-Density Coupling 208
       8.2.1 Linearization of the Mass Flow 209
   8.3 Two-Dimensional All Speed Flow Discretization 210
       8.3.1 Velocity Relations as Function of p'- SIMPLEC 213
       8.3.2 Density Relations as Function of p'- SIMPLEC 213
       8.3.3 Velocity/Density Relations as Function of p-PRIME 215
   8.4 Conclusions 217
   8.5 Exercises 217
   References 218

9 Two and Three-Dimensional Parabolic Flows 221
   9.1 Introduction 221
   9.2 Two-Dimensional Parabolic Flows 222
       9.2.1 External Two-Dimensional Parabolic Flows 223
       9.2.2 Internal Two-Dimensional Parabolic Flows 224
   9.3 Three-dimensional Parabolic Flows 224
       9.3.1 External Three-Dimensional Parabolic Flows 224
       9.3.2 Internal Three-Dimensional Parabolic Flows 227
   9.4 Conclusions 233
   9.5 Exercises 233
   References 233

10 General Recommendations for Conceiving and Testing Your Code 235
   10.1 Introduction 235
   10.2 Writing Your Code 236
       10.2.1 Generalities 236
       10.2.2 Coding Languages 237
       10.2.3 Tools to Aid the Development 238
   10.3 Running Your Application 239
       10.3.1 Compiling 240
       10.3.2 Size of the Mesh 240
       10.3.3 Convergence Criteria 240
   10.4 Choosing Test Problems—Finding Errors 242
       10.4.1 Heat Conduction—2D Steady State 242
       10.4.2 Transient Heat Conduction—One Dimensional 243
       10.4.3 One Dimensional Advection/Diffusion 244
       10.4.4 Two-Dimensional Advection/Diffusion 244
       10.4.5 Entrance Flow Between Parallel Plates 246
   10.5 Observing Details of the Solution 246
       10.5.1 Symmetry of the Solution 247
       10.5.2 The Coefficients 247
       10.5.3 Testing the Solver of the Linear System 248

	10.6	Conclusions	250
		References	251
11	**Introducing General Grids Discretization**	253
	11.1	Introduction	253
	11.2	Structured and Non-structured Grids	254
	11.3	The Concept of Element	256
	11.4	Construction of the Control Volume	258
	11.5	Conclusions	261
12	**Coordinate Transformation—General Curvilinear Coordinate Systems**	263
	12.1	Introduction	263
	12.2	Global Coordinate Transformation	264
		12.2.1 General	264
		12.2.2 Length Along a Coordinate Axis	268
		12.2.3 Areas (or Volumes) in the Curvilinear System	269
		12.2.4 Basis Vectors	271
		12.2.5 Vector Representation in the Curvilinear System	276
		12.2.6 Mass Flow Calculation	280
		12.2.7 Example of a Nonorthogonal Transformation	284
		12.2.8 Calculation of the Metrics of a Transformation	288
	12.3	Nature of the Discrete Transformation	290
		12.3.1 Preliminaries	290
		12.3.2 The Nature of the Transformation	292
	12.4	Equations Written in the Curvilinear System	301
	12.5	Discretization of the Transformed Equations	311
	12.6	Comments on the Solution of the Equation System	318
		12.6.1 Simultaneous Solution	318
		12.6.2 Segregated Solution	321
	12.7	Boundary Conditions	322
		12.7.1 No-Flow Boundary ($\rho U = 0$). ϕ Prescribed	323
		12.7.2 No-Flow Boundary ($\rho U = 0$). Flux of ϕ Prescribed	324
		12.7.3 Bounday With Mass Flow ($\rho U \neq 0$). Mass Entering With ρU Known	324
		12.7.4 Boundary With Mass Flow ($\rho U \neq 0$). Mass Leaving With ρU Unknown	324
	12.8	Conclusions	325
	12.9	Exercises	326
		References	330
13	**Unstructured Grids**	333
	13.1	Introduction	333
	13.2	Cell-Center Methods	336
		13.2.1 Conventional Finite Volume Method	336

		13.2.2	Voronoi Diagrams	344
	13.3	EbFVM—Element-based Finite Volume Method		354
		13.3.1	Geometrical Entities	356
		13.3.2	Local Coordinates. Shape Functions	357
		13.3.3	Determination of $(\nabla\phi)_{ip}$	361
		13.3.4	Determination of ϕ_{ip}	364
		13.3.5	Family of Positive Advection Schemes	367
		13.3.6	Integration of the Conservation Equations	371
		13.3.7	Assembling Strategies	376
		13.3.8	Boundary Conditions	378
	13.4	Conclusions		380
	13.5	Exercises		381
	References			385
14	**Pressure Instabilities: From Navier–Stokes to Poroelasticity**			387
	14.1	Introduction		387
	14.2	Pressure Instabilities		388
		14.2.1	Remedy 1	389
		14.2.2	Remedy 2	396
	14.3	Conclusions		401
	References			404
15	**Applications**			407
	15.1	Introduction		407
	15.2	Aerodynamics		407
		15.2.1	All Speed Flow Over a Blunt Body	407
		15.2.2	Ice Accretion on Aerodynamic Profiles	411
	15.3	Porous Media Flows		413
	15.4	Conclusions		420
	References			422
Correction to: Fundamentals of Computational Fluid Dynamics				C1
Index				425

Nomenclature

A	Coefficients of discretized equations
$\dot{m}$	Mass flow rate
V	Volume
M	Mass inside the control volume
V	Velocity vector
n	Normal vector
T	Temperature
c_P	Specific heat at constant pressure
m	Mass
$\dot{g}$	General generation term in Table (2.2)
F	Force vector
u, v, w	Cartesian velocity components
p	Pressure
V^1, V^2, V^3	Contravariant velocity components
V_1, V_2, V_3	Covariant velocity components
B	Body forces; independent term in the linear system
S	Source term
q'''	Heat generation by time and volume
q''	Heat flux (density)
E	Energy
e	Energy/mass
i	Internal energy/mass
h	Enthalpy/mass
k	Thermal conductivity
h	Convection heat transfer coefficient
f	General function
$\dot{Q}, \dot{W}$	Heat and work by unit of time, Eq. (2.39)
$D(\ )/Dt$	Total derivative
t	Time coordinate (also time level)
Δt	Time step
x, y, z	Spatial Cartesian coordinates

ξ, η, γ Spatial curvilinear coordinates

Superscripts

o	Old time level
I	Agglomerated grid (multigrid method)
i	Previous grid (multigrid method)
$*$	Approximated field

Subscripts

in	Entering the control volume
out	Leaving the control volume
CV	Control volume
R	Relative velocity
diff	Diffusive
gen	Generation
P	Central coefficient
ip	Integration point
e, w, n, s, b, f	Integration points (Cartesian system)
f	Boundary of the control volume
NB	Neighbor coefficients
∞	Environmental conditions (convection)

Greek Letters

μ	Viscosity
ρ	Density
λ	Second coefficient of viscosity, Eq. (2.25)
Γ^ϕ	General diffusion coefficient
τ	Stress tensor
ϕ	Generic variable
θ	Parameter in the explicit, implicit formulations
Φ	Dissipation function, Eq. (2.58)
β	Expansion coefficient
α	Diffusivity or Biot's coefficient

Chapter 1
Introduction

1.1 Preliminaries

The use of numerical techniques to solve complex problems in all areas of engineering and physics is nowadays a reality, thanks to the rapid development of high-speed computers with large storage capacity. Due to this computational availability, the development of algorithms and methods to solve the most diverse problems has received enormous attention of numerical analysts and engineers, which makes numerical techniques a topic of increasing interest for the scientific and industrial community. Computational Fluid Dynamics (CFD) is a topic which experienced an enormous growth and became a mandatory subject in undergraduate level and in industry, following the same path already observed in graduate studies and research. Moreover, the versatility and generality of numerical methods for the simulation of engineering fluid flow problems, and the relative simplicity of application of these techniques, are additional motivating factors for their use. All modern industry technologies, like IoT, Digital Twins, online processes control, to mention some of them, rely strongly on numerical simulation. The continuous growth of the available computational capacity is the main driver for those technologies. Simulation is mandatory in the design process of equipments, reducing dramatically costs and development time.

To have a glance on the rapid growth of the computational capacity it is enough to remember that in the 1960/70s a super-computer costing millions of dollars to run CFD applications, which today can be run in personal computers, was necessary. In summary, it is becoming easier and easier to use numerical techniques to solve engineering problems, both in the academic/scientific and industrial environment, as the costs for the acquisition of the necessary hardware are getting cheaper and cheaper. This book deals with the fundamentals of the finite volume method, the method which is the engine of the most used packages for solving fluid flow problems combined with heat transfer, multiphase and turbulence.

1.2 Available Tools for the Engineer

To appreciate the power of numerical simulation for solving engineering problems, let us comment about the capabilities of the available tools for the engineer in designing a product or understanding a specific physical phenomenon. Engineer have at their disposal, fundamentally, three tools:

- Analytical methods;
- Numerical methods (numerical experiments); and
- Laboratory experiments.

Analytical and numerical approaches belong to the class of theoretical methods, since both aim at solving the partial differential equations represented by the mathematical model. The difference between them resides in the errors embodied in the solution and in their capabilities. Analytical solutions contain only errors due to machine calculation, the so-called machine errors. Unfortunately, analytical methods are applicable only to problems whose simplifying assumptions take them too far away from the real physical phenomenon. They are usually applied to simple geometries with simple boundary conditions. Obviously, analytical solutions should not be discarded, and one of their important applications is precisely to validate limiting cases of numerical methods, and to assist in the development of more robust numerical methods. A significant advantage is that the solution can be obtained in closed form, requiring very low computational times, with no other errors besides the machine errors. If an analytical method is good enough to solve the problem of interest, such that the solution contains the answer for the engineering problem, it should be preferred. It is a basic rule in engineering practice to use the appropriate tool for what you expect from the solution.

On the other hand, numerical solutions, supported by the powerful high-speed computers, is the choice in modern engineering, since complex problems, represented by systems of partial differential equations, can be attacked with great facility. Numerical solutions, however, are contaminated with discretization errors which are dependent of the size of the grid. Discretization errors are of several types, being the two most important the numerical diffusion and numerical oscillation. We will discuss them in detail, since they originate when interpolation functions are introduced in the discretization.

Numerical simulation (also called numerical experiments) has, virtually, no restrictions, being able to solve problems with complex mathematical models with general boundary conditions, defined in complex geometries and releasing results in a time frame required in the development of new equipment or analysis. The time and cost of designing new equipment can be significantly reduced with the use of numerical simulation, compared with laboratory experiments. Currently, CFD tools are starting to be integrated with other numerical tools, creating an interactive work environment, where the final design is practically achieved through computational tools, leaving the final adjustments for the laboratory experiments. There are situations, for example, when the mathematical model is already known and fully

1.2 Available Tools for the Engineer

validated, such that laboratory experiments are no longer needed, and the problem can be reliably solved through computers simulations, thus reducing time and costs.

Regarding laboratory experiments, their great advantage is the fact that they can deal with the real configuration and real physics. It is, however, very expensive, and often cannot be applied for safety reasons, as is the case of heat transfer in the core of nuclear reactors, or because of the difficulty of reproducing the real conditions, as, for example, in high supersonic flows or in the simulation of oil reservoirs, in order to mention a few situations. In the absence of established mathematical models, it is often the only alternative available to the engineer or physicist. One important task of laboratory experiments is to help in the validation of numerical solutions. At the same time, complex numerical simulation push for the development of new experimental techniques able to measure what is simulated. It must be clear to the reader that besides the strong potential of numerical solutions, they do not fully replace laboratory experiments. They are complementary. The help of simulation during the product development allows to have the final product close to the ideal one, but before putting it in operation or launch it to the market, laboratory tests must be performed.

The trend is to have increasingly sophisticated laboratory experiments in order to use the results to corroborate mathematical and numerical models, when investigating and understanding new phenomena that still need to be mathematically modeled, and for a final evaluation of a given product. The laboratory will, certainly, no longer perform the repetitive task, which will be left to the computer, generating data for parametric analysis.

Therefore, what should be practiced in engineering is the proper association of numerical simulation with selected laboratory experiments. The combination of these techniques will result in a better and cheaper design. However, the ability in doing this requires well trained people in physics and numerical techniques. There is no doubt that this is the way for practicing modern engineering, in which numerical simulation is increasingly playing a decisive role, walking side-by-side with laboratory experiments. As mentioned, Industry 4.0, Digital Twins and Internet of Things (IoT) rely strongly on numerical simulation. Few words about the quality of the numerical solution is worth to be mentioned at this point.

Considering the engineering aspects of a numerical solution, there are two distinct types of errors that can make the numerical solution depart from the reality of a physical problem. In the first level there are the numerical errors themselves, resulting from the poor solution of the mathematical model (differential equations), represented by discretization errors, therefore, related to the numerical method as a whole. To detect them, the results should be compared with other solutions, analytical or numerical, of the same mathematical model to verify if the differential equation was correctly solved. This process, that we refer in this book as numerical validation, is also known as verification in the international literature, and attests the quality of the numerical method.

In a second level, there are the errors resulting from the use of differential equations (mathematical model) that do not represent with fidelity the physical phenomenon, called herein physical validation, also known as validation. Physical validation, therefore, is concerned with the fidelity of the mathematical model to be adherent to the physics of the problem. This requires the comparison of the numerical results with the real world. Therefore, a numerical tool is adequate and reliable for helping in solving an engineering problem when one has in hands a numerical method that correctly solves the differential equations, and that, faithfully, represents the physical phenomenon. It is worth remembering that it does not help, from an engineering point of view, to have an excellent numerical method if the mathematical model (i.e., the chosen differential equations) does not represent the phenomenon. Nor does it matter to have a good mathematical model if the numerical method cannot deliver to the user an accurate solution of the system of equations.

Figure 1.1 details the two levels of validation (numerical and physical validation). Comparison of the numerical results with analytical results, if any, or with other validated numerical results characterizes numerical validation. On the other hand, the comparison of the numerical results with the experimental results identifies the physical validation. Thus, since the errors can be of different origins, whenever errors are detected, the procedures listed in the square boxes in this figure should be checked.

A few words about verification and validation, two denominations largely used for checking a broad class of systems and used in numerical simulation are here appropriate. A brief review of the literature brings to us several articles trying to explain the differences in these procedures, since both words have similar meaning in many languages, and they allow margin for misunderstanding. In software development, for example, it is said that verification is the procedure of checking if the software complies with the imposed specifications. And what are the imposed specifications of a software for solving partial differential equations? It is nothing more than the requirement of having a good solution of the equations. And this is, precisely, the numerical validation, since it checks if the numerical algorithm (the kernel of the software) is correctly solving the mathematical model. Validation, on the other hand, it is said, is the procedure of checking if the software complies with the customer expectation. And what is the customer expectation of a software (numerical method), but to guarantee that the mathematical model embodied in the software gives physically consistent results? This is exactly physical validation. Therefore, in this book we prefer to call **numerical validation** and **physical validation** instead of verification and validation, since the meaning is directly understood by the words employed, with no ambiguity, especially in the specific case of numerical solution. If these two words would let clear its meaning in numerical simulation, it wouldn't need to have so many papers in the literature trying to explain them.

In this book, attention is focused on modeling problems involving fluid flow with or without heat transfer, which is applicable for multiphase and turbulent flows. The solution of these problems requires the handling of the highly nonlinear Navier–Stokes equations coupled with the mass conservation equation through a very intricated coupling. At the same time, whenever possible, we emphasize the importance of not losing sight of the physics of the phenomenon being modeled, since the physics can help building the numerical strategies. The methods that will be studied can be applied for any physical problem which involves fluid flow. We will also briefly

1.2 Available Tools for the Engineer

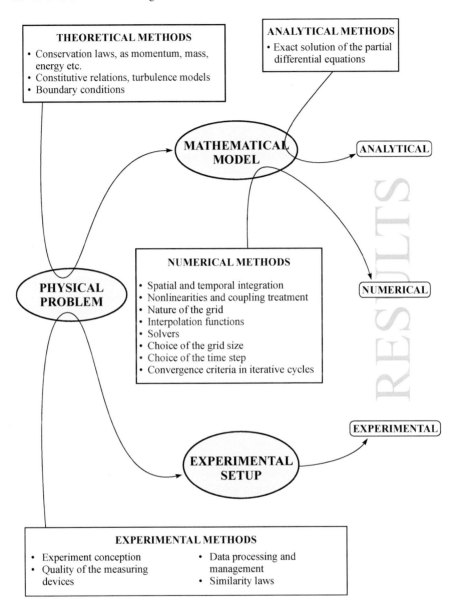

Fig. 1.1 Available tools for the engineer

discuss the use of finite volume techniques in solid mechanics, an area in which those methods are not frequently applied, more because of cultural factors than for its numerical capabilities, what has been demonstrated in several publications that it can solve solid mechanics problems with no numerical difficulties and with quality [1].

The numerical methods available for treating these equations and other partial differential equations are briefly discussed.

1.3 Classes of Numerical Methods Available

The traditional methods for the numerical solution of partial differential equations are the Finite Difference Method (FDM), Finite Volume Method (FVM), and Finite Element Method (FEM). With the great development of numerical methods and their penetration into engineering, it is common to find controversial discussions about the effectiveness and generality of each method and comparisons between them. Many statements about these methods stem from the lack of knowledge of their nature and, therefore, a few words on this subject are important at this point.

Historically, up to the 70 s, finite differences were always used in the fluid mechanics area, while finite elements were used in the structural area to solve, mainly, elasticity problems. These problems, from the physical point of view, are different, since fluid flow problems are highly nonlinear with advection terms present in the Navier–Stokes equations, while elasticity problems resemble purely diffusive heat transfer problems (elliptic), in general with linear characteristics. It was therefore natural that the researchers of finite differences concentrated their efforts on trying to master the nonlinearities of the advective terms and the difficult coupling between pressure and velocity, conditions not encountered in elliptic linear problems.

For this reason, the problem of treating irregular geometries was left behind in the finite differences area, and the method had its entire development mainly based on orthogonal coordinate systems, such as Cartesian, cylindrical and spherical coordinate systems. As a consequence, many people mistakenly link the finite differences method to orthogonal coordinate systems, when, in fact, it can be applied to any type of mesh, even unstructured ones, with the drawback, of course, of computing numerical derivatives along a coordinate axes when the mesh points are not aligned with these axes. Therefore, it is obvious, that it would be cumbersome to use it for irregular unstructured meshes. Let's recall that the basics of finite differences method is just to replace the continuous operators in a partial differential equation by their discrete counterpart. Finite differences, as a full method for solving partial differential equations is no longer, or rarely used. The researchers working with finite differences for fluid flow problems in the late 70 s, due to its familiarity with this kind of physics, migrated to the finite volume approach. A major advancement in numerical fluid flow took place in the mid-1970s, when orthogonal coordinate systems were replaced by generalized coordinate systems coincident with the boundaries of the domain, what allowed finite volume methods to solve problems in irregular geometries. Until around the 2000s it was the dominant discretization in important commercial packages available on the market for solving heat transfer and fluid flow problems.

Finite element methods, by its way, are derived by applying a weighted residual statement to the elements, circumventing the requirement of having a variational

1.3 Classes of Numerical Methods Available

formulation of the partial differential equation. The use of unstructured meshes was always the strongest point of the method.

The picture we had until the early 1970s, therefore, was the finite volume (finite differences) with great experience in the area of fluids, but without the ability to treat irregular geometries, and finite elements skilled in the treatment of irregular geometries, but with difficulties to treat the advective terms present in the equations of motion. Early attempts to use the conventional Galerkin method for problems with strong advection were unsuccessful, since the classical Galerkin method with first order elements is only suitable for purely diffusive problems. This would be equivalent to the use of central differences in finite differences, which is known to produce oscillations in problems with dominant advection.

In the context of finite elements, there were important improvements, employing other interpolation functions to allow for the adequate treatment of advective terms. Interpolation functions, weighting the importance of diffusive and advective effects, like the hybrid schemes employed in finite volumes, enabled a significant advance in finite element in the area of fluid flows. Later, formulations in which these functions are developed along the streamline were also devised, equivalent to the skew schemes used in finite volumes, allowing finite elements to deal with fluid problems minimizing the effects of numerical diffusion.

Finite volume methods, on the other hand, create the approximated equations by a balance in a control volume or, what is equivalent, integrating the partial differential equation in conservative form. This means that the discretized equations are the discrete counterpart of the differential equation, therefore, satisfying the conservation principles for any size of the grid. The possibility of linking the physical interpretation to the mathematics has had a considerable influence on the development of more robust numerical algorithms. Just to mention, it can be shown that the finite volume method is also a weighted residual approach method, the sub-domain method, in which the weighting function is equal to 1 in the control volume in consideration and zero in all others. If the equation is in its conservative form, the integration results in a balance of the property in the control volume.

In the context of commercial packages for fluid flows, the finite volume method is still the method employed in all those with industrial penetration, with exception of specific niches. The preference is based on robustness, due to the conservative characteristics of the method, since, in fluid flow, it is very important to satisfy mass, momentum, energy and other properties at the discrete level. Since what is sought with a numerical method is the solution of a set of partial differential equation, which is the representation of the conservation of properties at the point level, it seems logical that the approximated equations should represent the conservation at the finite volume level. Obeying this, there will be no possibility of generation/disappearance of quantities, such as mass, momentum, and energy, within the domain of computation. If the conservation is satisfied only via boundary conditions, there may be non-physical generation/disappearance of properties within the domain, which will change the solution locally, or even preclude the determination of the solution.

Developments in finite element technology, like mixed-finite element and discontinuous Galerkin can assure conservation principles within the formalism of finite elements. Special elements are constructed such that it resembles the classic staggered grid arrangement of variables, a routine practiced in the finite volume methods since the early 1970s. The drawback of these extended finite element method is that, usually, they increase the degree of freedom, requiring additional storage and increasing computational time and implementation complexity.

In the context of finite volume methods there are important variants, since any numerical procedure that obtains its approximate equations through balances in a control volume is a finite volume method. Cell-center and cell-vertex methods will be discussed, independent of the grid type, if curvilinear or unstructured. One important variation of the finite volume methods is the cell-vertex method using unstructured grids, in which the control volumes are constructed using part of the elements created by the grid generator. This method deals with elements and control volumes, giving rise to a strong method with high grid flexibility. Within this class is the method whose control volumes are created by the method of medians named CVFEM-Control Volume-based Finite Element Method.

This denomination, however, is ambiguous, since it conveys the reader the idea that it is a finite element method that uses control volumes. In fact, it is a finite volume method which uses the element, as used in finite element, as an entity of the method. In addition, it uses the same shape functions employed in finite element methods for interpolation and the usual element-by-element sweep to assemble the system of equations. Thus, perhaps, a more appropriate name for this class of methods would be EbFVM - Element-based Finite Volume Method, which correctly defines a finite volume method which uses the element to construct the control volumes.

Still in the class of FVM, we have those whose control volumes are Voronoi diagrams, obtained from a Delaunay triangulation. In this case, the discretization is locally orthogonal and only two mesh points are required for the correct determination of the fluxes. It is an interesting method, but its difficulty lies on the grid generation. This method will be briefly discussed in this text.

Another method that has gained space in the past decade is the Boundary Element Method (BEM). Its advantage is the possibility of dealing with the discretization of the boundary only, without the need to discretize the internal domain. The method can be applied when it is possible to transfer the influence of the operator from the domain to the boundary. Although attractive, it is a method that is still far from meeting the demands of the complex problems solved by the other methods. Undoubtedly, it is an area of research that deserves efforts and it could fit the needs for certain classes of problems.

Nowadays it is observed that both methods, finite volume and finite element, are solving all kind of physical problems using general unstructured grids. If we look from a mathematical point of view, it couldn't be different, since all numerical methods can be derived from the weighted residuals approach, employing different weighting functions. One strong point in favor of finite volumes is its conservative property for grids of any size, since independence of the grid size for conservation is mandatory. The physics must be always obeyed, not only when the grid size is very fine, that is, it must also be conservative when the solution contains truncation errors. We will return to the topic of conservation in Chap. 3.

As a final comment about the methods, both finite elements and finite differences do not have the control volume as an entity in the formulation, therefore they are not conservative at discrete level. Finite volume techniques, on the other hand, are conservative at discrete level. It is worthwhile to mention that any numerical method, if consistent, will produce the same results when the grid size goes to zero, that is, at point level, situation in which all discretization errors must go to zero for all methods. Since one is always working with discrete grids, the conservation at discrete level is of utmost importance.

1.4 Objectives and Scope of This Book

With the enormous growth of the field of Computational Fluid Dynamics (CFD), many books have been published in the field. Most of them are, however, addressed to the researcher or the student already familiar with numerical methodologies. Some of them try to cover, besides the finite volume basics, other topics, such as turbulence and multiphase flows, for example, and the essentials of the finite volume method get lost in a diversity of topics. Some of them uses heavy mathematical notation and the meaning of the conservation equations and of the algorithms are lost. Physics, and even mathematics becomes hidden in the notations. Our view of the area along decades of teaching engineering subjects reveals that this discourage the students. There is still a big lack of textbooks that present the developments of the finite volume method in a growing complexity, in a clean structure, from the fundamentals of the method and reaching the latest approaches in the practice of engineering simulation, avoiding the need of jumping chapters and searching topics among lots of information, as is frequent in several textbooks in the area of CFD.

This text aims precisely at filling this gap, allowing the learning process to evolve gradually up to the point of interest, writing simple computer codes to fix in the student's memory the numerical topics addressed in the book. The goal is to provide a text that the reader can master the key points of the finite volume method, touching the fundamentals and felt instigated to think about them, and not just grasp an overview of the methodology. And, at the same time, to be an enjoyable reading.

The text is divided in two parts: the first one is concerned with the basic concepts of the finite volume method, using problems represented by simple partial differential equations, but still containing all the required ingredients to understand the procedure. The second part presents the formulation of the finite volume method for general curvilinear coordinate systems and unstructured grids. Curvilinear grids are becoming less used because it is not as flexible as unstructured grids, however, we will dedicate some space because of its historical importance in computational fluid dynamics and its role in global and local transformation of coordinates, as auxiliary tools in many numerical developments. Regarding unstructured grids, it is the standard discretization used nowadays in fluid flow simulation, and the conventional finite volume method using cell-center, for general unstructured grids, and Voronoi diagrams will be considered. In the cell-vertex category, the Element-based Finite Volume Method (EbFVM) will be discussed in reasonable depth.

In the first part of the text, for simplicity and for improving learning, the developments will be carried out using the Cartesian coordinate system, without prejudice to the complete understanding, since all the basic formulation serves for any coordinate system. Therefore, if a beginner wants to learn the fundamentals of the method it is enough to read the first part. The material fits for a high-level course in undergraduate programs and normal one in graduate programs. The second part extends all knowledge of the first part to curvilinear and unstructured grids and can be a second course in the graduate level. This second part is dedicated to more advanced students and for researchers and engineers working with numerical simulation of fluid flows. The book is organized as follows:

Chap. 1 discusses the importance of the numerical simulation in modern engineering, and the methods currently available for solving problems in Fluid Mechanics and Heat Transfer. Chap. 2 discusses the physical and mathematical aspects of the conservation equations, presenting a didactic way to derive the conservation equations. Important issues in the derivation of the conservation equations are discussed, which definitely helps the reader to clearly understand the nonlinearities of the Navier–Stokes equations, an issue which normally remains unclear for the beginners in fluid mechanics. The derivation of the conservation equations in most textbooks are done grounded on mathematics only, and this hinders the intimacy of the reader with the equations.

Chapter 3 presents the basic formulation of the finite volume method using a one-dimensional heat conduction problem, always trying to show that the concepts acquired with this simple problem are general and can be easily extended to 2D and 3D discretization. Chap. 4 brings to the reader a brief presentation of the methods for solving the linear system of algebraic equations, with emphasis in one efficient multigrid method for accelerating the iterative solution of linear systems.

In Chap. 5, the important problem of interpolation functions and the concepts of numerical diffusion and numerical oscillation are discussed. We will try to put lights to the understanding of these two important errors due to the interpolation functions used. In Chap. 6, for the sake of completeness, the discretization of a general three-dimensional equation is performed, considering the velocity field as known. How to calculate the velocity field is reserved for Chap. 7, in which the key details of the pressure–velocity couplings will be discussed. The connection between segregated and simultaneous solution and these couplings will be presented. Chapter 8 presents a methodology for solving problems for any flow regime, while Chap. 9 discusses three-dimensional parabolic problems. Chapter 10 gives recommendations for those who are developing their codes, especially beginners and students, for testing and debugging, suggesting analytical solutions for comparisons, convergence criteria etc. Chapter 11 makes the bridge from the previous chapters to the chapters to come, where curvilinear and unstructured grids will be treated.

Chapter 12 deals with methods in generalized curvilinear coordinate systems. Notions of coordinate transformation, metric tensor, metrics of the transformation, Jacobian, contravariant, covariant and physical velocities, geometrical interpretation of the mathematical quantities, among other concepts, are given. Global and local coordinate systems are briefly discussed, the former, as the basis for generalized coordinate methodologies, and the latter for unstructured grid and element-based methods.

Chapter 13 is reserved for finite volume methods for unstructured meshes of cell-center and cell-vertex types. In the cell-center class of methods Voronoi diagrams are included, while in the cell-vertex category emphasis is given on the EbFVM, probably the most neat and general numerical technology for treating problems with hybrid unstructured meshes.

Chapter 14 is devoted for exploring an interesting similarity of the pressure–velocity coupling in Navier–Stokes equations with the pressure–displacement coupling in poroelasticity. At the same time, it is demonstrated that finite volume techniques can be applied with success in solid mechanics problems keeping its conservative properties.

Chapter 15 presents few results of the work performed by the author and his students and colleagues using generalized coordinates and unstructured meshes. The methodologies that will be presented in this text can be employed in a series of problems of practical interest. The next section seeks to present part of this universe.

1.5 Applications of Computational Fluid Dynamics

Fluid flow, with or without heat transfer is involved in virtually all energy production processes, environmental phenomena, thermal equipment, aeronautical and aerospace engineering, reactor engineering, bioengineering, clean energy generation etc. And the numerical simulation of these flows plays a key role in understanding and quantifying the data for engineering calculations.

Nowadays, industries are already using computers in large scale, even revolutionizing the design in features that would be impossible with the use of the wind tunnel measurements only, since numerical simulation allows to execute many numerical experiments quickly and with a low cost. In the design of automobiles and their components, in rotating machinery, fire-fighting systems in large rooms, in the determination of position and size of the insufflation and ventilation in air conditioning environments, in the design of refrigeration equipment, CFD simulation is always present.

Application in environmental engineering, as in the prediction of pollutant dispersion in the atmosphere and by the discharge of pollutants in rivers, lakes and soil, in the solution of numerous problems of multiphase flows found in the oil industry, in the design of combustors, boilers etc., numerical simulation is of great help to the analyst. The list is endless, and it just can be said that where there is an engineering problem, simulation is there. We may say that all processes and phenomena will be simulated in the future, and CFD will play an important role in this environment. The reader is invited to add other engineering problems to this list.

Reference

1. Cardiff P, Demirdzic I (2021) Thirty years of the finite volume method for solid mechanics. Archives of Comp. Meth. in Eng. 28:3721–3780

Chapter 2
Conservation Equations—Physical and Mathematical Aspects

2.1 Introduction

To succeed in a numerical simulation, it is mandatory to know the physics of the problem under analysis. And, in the case of fluid flows, understanding and knowing how to interpret the terms of the conservation equations is fundamental for correctly modelling the problem. In most textbooks of fluid mechanics, for example, when deriving the momentum conservation equations, the very important non-linear advective terms of these equations are hidden in the total derivative of momentum when applying Newton's 2nd Law. This approach does not make clear, from a physical point of view, how the nonlinearities appear, since the physical interpretation of the total derivative is not always mastered by the beginners in fluid mechanics.

This chapter is devoted for deriving and presenting the conservation equations using a strong physical appealing, seeking to show the dual role played by velocity in those equations. Besides that, the nature of the equation will be given, again based on physical grounds, instead of just classify them in a mathematical way. The knowledge gained in this chapter helps considerably the creation of the mathematical model which represents the physics. In this textbook, in spite of having a numerical scope, the physics and the geometric interpretation of the terms in the equations will be always emphasized.

2.2 Models Formulation Levels

Obtaining the numerical solution of any physical problem requires, initially, the ability to create the corresponding mathematical model. The mathematical model should be such that it can be solved in non-prohibitive computational times with results satisfying the engineering needs. Achieving this goal is not an easy task. Firstly, it should be decided the level at which the conservation balances should be done. To mention the extremes, conservation balances can be done either at the molecular level, originating an equation for each molecule, or over control volumes that may even, in certain directions, be coincident with the solution domain. At these extremes, the complexity of the numerical method suitable for each situation varies enormously.

Table 2.1 shows the various possible levels of the model formulation. With the current computational availability, solving engineering problems within level 1 is impractical. Level 2 is also impractical, as it requires the solution of the differential equations for time intervals in the order of the turbulent fluctuations. At this level, the turbulent fluctuations are part of the true transient, and if such refined spatial and temporal meshes were possible, all the turbulent fluctuations in all spatial and time scales would be captured. Turbulence models would be no longer needed. The spatial meshes required, for example, for problems with Reynolds numbers of the order of 10^5–10^7 are of the order of 3.10^{13}–7.10^{17}, respectively [1]. The attempts to attack the problem at this level, is known as DNS (Direct Numerical Simulation). These techniques, however, are still at the research level and are not part of the computational tools for the routine solution of engineering problems. Level 3 is the level in which the models that solve fluid flows with heat and mass transfer for practical interest are accommodated today. Turbulence, on these levels, is treated by adopting turbulence models as closures of the spatial and time-averaged Navier–Stokes equations, the RANS (Reynolds Averaged Navier–Stokes Equations) approach. Among the most widespread methods in this class, which use two partial differential equations, are (k − ϵ) and its variants, (k − ω) and SST (Shear Stress Transport) [1]. Turbulence models are not in the scope of this book and the reader is referred to an existing extent literature on the subject.

Table 2.1 Levels of model formulation

Level in which balances are made	Information required	Type of resulting equation
Cons. for each molecule $L \ll L_m$ $t \ll t_m$	Molecular mass, laws of molecular momentum exchange, force fields, magnetic fields etc.	One equation for each molecule (level 1)
Balances in which $t_m \ll t \ll t_t$ $L_m \ll L \ll L_t$	Properties reflecting the molecular behavior, as μ, k, ρ etc.	Set of partial differential equations (level 2)
Balances in which $t_t \ll t$ $L_t \ll L$	Properties reflecting the molecular behavior, as μ, k, ρ etc. Turbulence models required	Set of partial differential equations (level 3)

t_m: average time of molecular collisions, t_t: turbulent time scale
t: average time in which balances are made, L: average length in which balances are made, L_m: molecular mean free path, L_t turbulent length scale

2.3 Conservation Equations

Following the basic idea of [2], the derivation of the equations for conservation of mass, momentum, and energy is now presented, adding important physical interpretation that helps the understanding of the equations. Our goal is to make the understanding of the non-linearities of the momentum equations as clear as possible.

2.3 Conservation Equations

2.3.1 Mass Conservation Equation

Consider a fluid flow with velocity $\mathbf{V}$ and a system, shown in Fig. 2.1 at times t and $t + \Delta t$, with a constant velocity $\mathbf{V}_A$ of the area element dA. The change of the mass in the system in this time interval is given by,

$$m_1(t + \Delta t) + m_2(t + \Delta t) - m_1(t) - m_3(t) = \Delta m|_{\text{system}} \quad (2.1)$$

Dividing by Δt,

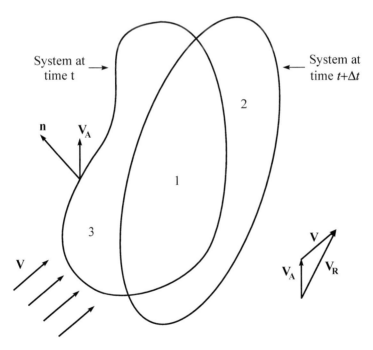

Fig. 2.1 System for deriving the conservation equations

$$\frac{m_1(t+\Delta t) - m_1(t)}{\Delta t} - \frac{m_3(t)}{\Delta t} + \frac{m_2(t+\Delta t)}{\Delta t} = \left.\frac{\Delta m}{\Delta t}\right|_{system} \quad (2.2)$$

Recognizing from Fig. 2.1 that the regions 3 and 2 represent, respectively, the mass entering and leaving the control volume, and region 1 becomes the control volume when $\Delta t \to 0$, Eq. (2.2) can be written as

$$\dot{m}_{in} - \dot{m}_{out} + \left.\frac{\Delta m}{\Delta t}\right|_{system} = \left.\frac{\Delta m}{\Delta t}\right|_{CV} \quad (2.3)$$

In a system there is no flow crossing its boundaries, therefore, the total mass inside a system is constant in time, or, in other words, its variation in time is zero, that is, $(\Delta m/\Delta t)_{system} = 0$, and Eq. (2.3) results,

$$\dot{m}_{in} - \dot{m}_{out} = \left.\frac{\Delta m}{\Delta t}\right|_{CV}, \quad (2.4)$$

which is exactly the mass conservation equation for a discrete control volume, that is, the equation used in the finite volume method. We will be back to this topic in Chap. 3.

Equation (2.4) can be written in its integral form, as

$$\frac{\partial}{\partial t}\int_V \rho \, dV = -\int_A \rho(\mathbf{V}_R.\mathbf{n}) dA, \quad (2.5)$$

in which $\mathbf{V}_R$ is the relative velocity, and $\rho(\mathbf{V}_R.\mathbf{n})dA$ is the mass flow crossing the area element dA. The surface integral gives the net flow in the control volume. Using the divergence theorem, the surface integral is transformed in a volume integral, and considering a control volume of fixed shape in time, and infinitesimal, in the sense that the property may be considered constant in the interior of the control volume, the differential form of the mass conservation equation is obtained,

$$\frac{\partial \rho}{\partial t} + \frac{\partial}{\partial x}(\rho u_R) + \frac{\partial}{\partial y}(\rho v_R) + \frac{\partial}{\partial z}(\rho w_R) = 0 \quad (2.6)$$

For a control volume fixed in space, the relative velocity coincides with the flow velocity, and the mass conservation equation results,

$$\frac{\partial \rho}{\partial t} + \text{div}(\rho \mathbf{V}) = 0 \quad (2.7)$$

Equation (2.7) may also be obtained performing a mass balance for a control volume as shown in Fig. 2.2, which implies applying Eq. (2.4) to the control volume.

2.3 Conservation Equations

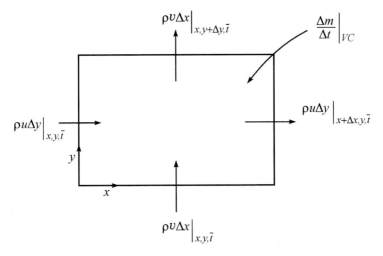

Fig. 2.2 Mass balance for a control volume

This is what is commonly done in the first course in Fluid Mechanics. It is left to the reader this exercise.

For the sake of simplicity, the figures describing the conservation balances in this chapter are drawn in 2D, but extension for 3D is quite simple.

2.3.2 Linear Momentum Conservation Equations

To derive the equation for linear momentum conservation, the same procedure is utilized. The momentum is evaluated at times t and $t + \Delta t$, and the difference is the variation of momentum in the system in the time interval Δt, given by

$$m_1 \mathbf{V}_1(t + \Delta t) + m_2 \mathbf{V}_2(t + \Delta t) - m_1 \mathbf{V}_1(t) - m_3 \mathbf{V}_3(t) = \Delta (m\mathbf{V})_{\text{system}} \quad (2.8)$$

Dividing Eq. (2.8) by Δt, it results

$$\frac{m_1 \mathbf{V}_1(t + \Delta t) - m_1 \mathbf{V}_1(t)}{\Delta t} - \frac{m_3 \mathbf{V}_3(t)}{\Delta t} + \frac{m_2 \mathbf{V}_2(t + \Delta t)}{\Delta t} = \frac{\Delta}{\Delta t}(m\mathbf{V})_{\text{system}} \quad (2.9)$$

For each property in deriving the conservation equation, one needs to invoke its corresponding law for systems. In this case, Newton's Second Law for a system says,

$$\frac{\Delta}{\Delta t}(m\mathbf{V})_{\text{system}} = \sum \mathbf{F}, \quad (2.10)$$

and the momentum conservation equation for a control volume is given by

$$\dot{m}\mathbf{V}|_{in} - \dot{m}\mathbf{V}|_{out} + \sum \mathbf{F} = \frac{\Delta}{\Delta t}(m\mathbf{V})_{CV} \qquad (2.11)$$

Since the idea is to derive all conservation equations using the same physical reasoning, it is didactic to compare Eq. (2.3) with Eq. (2.11). In both equations the third term in the left-hand side comes from a law for systems. In addition, in Eq. (2.11), ($\dot{m}$), the mass that crosses the boundaries of the control volume carries, by advection, the entity $\mathbf{V}$ (momentum/mass) to the interior of the control volume, while in Eq. (2.3) it carries itself, which, by definition is the entity $\mathbf{1}$ (mass/mass).

Recognizing that to determine $\dot{m}$ the velocity field will be needed. It is exactly in the product $\dot{m}\mathbf{V}$ which resides the strong non-linearities of the Navier–Stokes equation.

Expanding this understanding for a general entity, the balance of an entity in a control volume by unit of time can be expressed in the usual literal form, by

$$\left(\frac{entity}{time}\right)_{in} - \left(\frac{entity}{time}\right)_{out} + \left(\frac{entity}{time}\right)_{generated} = \frac{\Delta}{\Delta t}(entity)_{CV}$$

Therefore, the net forces in Eq. (2.11) are equivalent to a generation of momentum, which is quite clear by the Newton's 2nd Law.

It helps to interpret the flow as the carrier of several properties per unit of mass. That is, a flow can transport (advect) momentum, energy, turbulent kinetic energy, dissipation of turbulent kinetic energy, chemical species, entropy etc. In particular, it transports itself, represented by the mass conservation equation. In this context, Eqs. (2.3) and (2.11) can be written for a general variable ϕ (entity advected/unit of mass), as

$$\dot{m}\phi|_{in} - \dot{m}\phi|_{out} + \dot{g}^\phi \Delta V = \frac{\Delta}{\Delta t}(m\phi)_{CV}, \qquad (2.12)$$

in which the term $\dot{g}^\phi \Delta V$ is obtained from the corresponding conservation law for systems for the entity in consideration, as shown in Table 2.2.

This term, $\dot{g}^\phi \Delta V$, for mass conservation is equal to zero, because the total mass inside a system does not change in time. It is $\sum \mathbf{F}/dV$ (force/unit of volume) for the momentum conservation when $\phi = u, v,$ or w. The expression of $\dot{g}^\phi$ for the conservation of energy will be seen later. Equation (2.12) can also be obtained from a balance of ϕ in the control volume shown in Fig. 2.3.

Table 2.2 Parameters for Eq. (2.12)

ϕ	1	u	v	w	e
$\dot{g}\Delta V$	0	$\sum F_x$	$\sum F_y$	$\sum F_z$	$\dot{Q}_{in} - \dot{W}_{out}$

2.3 Conservation Equations

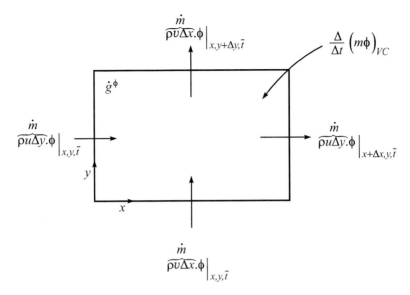

Fig. 2.3 Balance for a generic variable ϕ

This figure helps to clarify one of the common doubts of students in performing a momentum balance in a control volume. For example, when realizing the balance for the scalar u momentum equation, one is tempted to say that no momentum enters by the south face, just because we are dealing with the u momentum equation. Here enters the concept of transporting (or advecting) and transported (or advected) velocity, which helps to dissipate this doubt. It is easy to recognize that as long as there is a v velocity (transporting), there will be u momentum (transported) entering, or leaving the south face, as well as there will be v momentum, and w momentum in a 3D case, crossing this surface. In this figure, it can be observed that the flow, characterized by the mass flow rate $\dot{m}$, causes a net transport of any scalar ϕ by advection into the control volume.

Please, notice that the words entity and property are used, irrespectively, to identify a physical entity. Let's not confuse with the thermophysical properties of a fluid.

The momentum conservation equation is a vector quantity, and the three scalar components are,

$$(\dot{m}u)_{in} - (\dot{m}u)_{out} + \sum F_x = \frac{\Delta}{\Delta t}(mu)_{CV} \qquad (2.13)$$

$$(\dot{m}v)_{in} - (\dot{m}v)_{out} + \sum F_y = \frac{\Delta}{\Delta t}(mv)_{CV} \qquad (2.14)$$

$$(\dot{m}w)_{in} - (\dot{m}w)_{out} + \sum F_z = \frac{\Delta}{\Delta t}(mw)_{CV} \qquad (2.15)$$

Writing Eq. (2.11) in its integral form,

$$-\int_A \mathbf{V}\rho(\mathbf{V}_R.\mathbf{n})\,dA + \sum \mathbf{F} = \frac{\partial}{\partial t}\int_V \rho\mathbf{V}\,dV, \qquad (2.16)$$

it is identified the term, $\rho(\mathbf{V}_R.\mathbf{n})\,dA$, which, as before, represents the mass flow rate through the area element at the boundaries of the control volume.

The relative velocity is the responsible for the mass inflow and outflow of the control volume. If the control volume, for example, moves with the same velocity of the fluid, no mass enter or leaves the control volume. When the control volume is fixed, the relative velocity is equal the flow velocity. Employing again the divergence theorem and considering an infinitesimal control volume, Eq. (2.16) can be written as

$$\frac{\sum \mathbf{F}}{dV} = \rho\left[\frac{\partial \mathbf{V}}{\partial t} + (\mathbf{V}_R.\nabla)\mathbf{V}\right], \qquad (2.17)$$

in which dV is the infinitesimal volume, and $\sum \mathbf{F}$ is still to be found as function of the stress tensor acting on the fluid. It should be remembered that $\sum \mathbf{F}$ is also an infinitesimal quantity related to the control volume dV. Expanding the right-hand side of this equation, and leaving it in a conservative form, one obtains, for example, for the components of the momentum equation,

$$\frac{\sum F_x}{dV} = \frac{\partial}{\partial t}(\rho u) + \frac{\partial}{\partial x}(\rho u_R u) + \frac{\partial}{\partial y}(\rho v_R u) + \frac{\partial}{\partial z}(\rho w_R u)$$

$$\frac{\sum F_y}{dV} = \frac{\partial}{\partial t}(\rho v) + \frac{\partial}{\partial x}(\rho u_R v) + \frac{\partial}{\partial y}(\rho v_R v) + \frac{\partial}{\partial z}(\rho w_R v)$$

$$\frac{\sum F_z}{dV} = \frac{\partial}{\partial t}(\rho w) + \frac{\partial}{\partial x}(\rho u_R w) + \frac{\partial}{\partial y}(\rho v_R w) + \frac{\partial}{\partial z}(\rho w_R w) \qquad (2.18)$$

Notice that the left-hand side of this equation still need to be written in infinitesimal form as the right-hand side is. This is done with the help of Fig. 2.4. The forces, shown in a 2D situation for the sake of simplicity, results in the following balances

$$\frac{\sum F_x}{dV} = \frac{\partial}{\partial x}(\tau_{xx}) + \frac{\partial}{\partial y}(\tau_{yx}) + B_x \qquad (2.19)$$

$$\frac{\sum F_y}{dV} = \frac{\partial}{\partial x}(\tau_{xy}) + \frac{\partial}{\partial y}(\tau_{yy}) + B_y \qquad (2.20)$$

The reader is invited to solve Exercise 2.7 to explore these concepts. Carrying out the force balances for all the coordinate directions, the three components of the momentum conservation equation are obtained. Since momentum is a vector quantity, its scalar equations for the three coordinate directions are given by

2.3 Conservation Equations

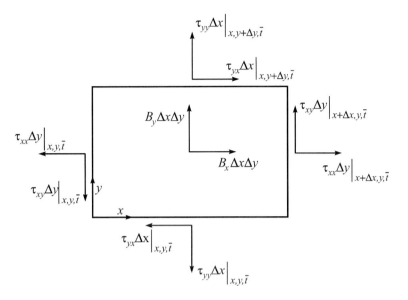

Fig. 2.4 Forces acting on a control volume

$$\frac{\partial}{\partial t}(\rho u) + \frac{\partial}{\partial x}(\rho u_R u) + \frac{\partial}{\partial y}(\rho v_R u) + \frac{\partial}{\partial z}(\rho w_R u)$$
$$= \frac{\partial}{\partial x}(\tau_{xx}) + \frac{\partial}{\partial y}(\tau_{yx}) + \frac{\partial}{\partial z}(\tau_{zx}) + B_x \quad (2.21)$$

$$\frac{\partial}{\partial t}(\rho v) + \frac{\partial}{\partial x}(\rho u_R v) + \frac{\partial}{\partial y}(\rho v_R v) + \frac{\partial}{\partial z}(\rho w_R v)$$
$$= \frac{\partial}{\partial x}(\tau_{xy}) + \frac{\partial}{\partial y}(\tau_{yy}) + \frac{\partial}{\partial z}(\tau_{zy}) + B_y \quad (2.22)$$

$$\frac{\partial}{\partial t}(\rho w) + \frac{\partial}{\partial x}(\rho u_R w) + \frac{\partial}{\partial y}(\rho v_R w) + \frac{\partial}{\partial z}(\rho w_R w)$$
$$= \frac{\partial}{\partial x}(\tau_{xz}) + \frac{\partial}{\partial y}(\tau_{yz}) + \frac{\partial}{\partial z}(\tau_{zz}) + B_z \quad (2.23)$$

Equations (2.21–2.23) are known as the equations of motion and are valid for any fluid. The key question now is to specialize them for different fluids, what is done through the constitutive relations, equating the stress tensor with the deformation rate. To find this relation for all fluids of practical interest is challenging for physicists, chemists and engineers. If no reliable constitutive relation is available, the simulation carried out may not be useful. For Newtonian fluids, extremely abundant in our environment, the constitutive relation is a simple one, relating the stress tensor linearly with the rate of deformation, as

$$\tau_{ij} = -p\delta_{ij} + \mu\left(\frac{\partial u_i}{\partial x_j} + \frac{\partial u_j}{\partial x_i}\right) + \delta_{ij}\lambda \operatorname{div}\mathbf{V} \tag{2.24}$$

Considering for the second coefficient of viscosity the usual relation,

$$\lambda = -\frac{2}{3}\mu, \tag{2.25}$$

the components of the stress tensor for a Newtonian fluid are,

$$\tau_{xx} = -p + \mu\left(\frac{\partial u}{\partial x} + \frac{\partial u}{\partial x} - \frac{2}{3}\operatorname{div}\mathbf{V}\right) \tag{2.26}$$

$$\tau_{yy} = -p + \mu\left(\frac{\partial v}{\partial y} + \frac{\partial v}{\partial y} - \frac{2}{3}\operatorname{div}\mathbf{V}\right) \tag{2.27}$$

$$\tau_{zz} = -p + \mu\left(\frac{\partial w}{\partial z} + \frac{\partial w}{\partial z} - \frac{2}{3}\operatorname{div}\mathbf{V}\right) \tag{2.28}$$

$$\tau_{xy} = \tau_{yx} = \mu\left(\frac{\partial u}{\partial y} + \frac{\partial v}{\partial x}\right) \tag{2.29}$$

$$\tau_{xz} = \tau_{zx} = \mu\left(\frac{\partial u}{\partial z} + \frac{\partial w}{\partial x}\right) \tag{2.30}$$

$$\tau_{zy} = \tau_{yz} = \mu\left(\frac{\partial w}{\partial y} + \frac{\partial v}{\partial z}\right) \tag{2.31}$$

The system comprising the mass conservation equation and the Navier–Stokes equations form the important set of equations for solving isothermal fluid flows. It is,

$$\frac{\partial \rho}{\partial t} + \frac{\partial}{\partial x}(\rho u) + \frac{\partial}{\partial y}(\rho v) + \frac{\partial}{\partial w}(\rho w) = 0 \tag{2.32}$$

$$\begin{aligned}
&\frac{\partial}{\partial t}(\rho u) + \frac{\partial}{\partial x}(\rho uu) + \frac{\partial}{\partial y}(\rho vu) + \frac{\partial}{\partial z}(\rho wu) = -\frac{\partial p}{\partial x} \\
&+ \frac{\partial}{\partial x}\left(\mu\frac{\partial u}{\partial x} - \frac{2}{3}\mu\operatorname{div}\mathbf{V}\right) + \frac{\partial}{\partial y}\left(\mu\frac{\partial v}{\partial x}\right) + \frac{\partial}{\partial z}\left(\mu\frac{\partial w}{\partial x}\right) \\
&+ \frac{\partial}{\partial x}\left(\mu\frac{\partial u}{\partial x}\right) + \frac{\partial}{\partial y}\left(\mu\frac{\partial u}{\partial y}\right) + \frac{\partial}{\partial z}\left(\mu\frac{\partial u}{\partial z}\right) + B_x
\end{aligned} \tag{2.33}$$

$$\begin{aligned}
&\frac{\partial}{\partial t}(\rho v) + \frac{\partial}{\partial x}(\rho uv) + \frac{\partial}{\partial y}(\rho vv) + \frac{\partial}{\partial z}(\rho wv) = -\frac{\partial p}{\partial y} \\
&+ \frac{\partial}{\partial x}\left(\mu\frac{\partial u}{\partial y}\right) + \frac{\partial}{\partial y}\left(\mu\frac{\partial v}{\partial y} - \frac{2}{3}\mu\operatorname{div}\mathbf{V}\right) + \frac{\partial}{\partial z}\left(\mu\frac{\partial w}{\partial y}\right) \\
&+ \frac{\partial}{\partial x}\left(\mu\frac{\partial v}{\partial x}\right) + \frac{\partial}{\partial y}\left(\mu\frac{\partial v}{\partial y}\right) + \frac{\partial}{\partial z}\left(\mu\frac{\partial v}{\partial z}\right) + B_y
\end{aligned} \tag{2.34}$$

2.3 Conservation Equations

$$\frac{\partial}{\partial t}(\rho w) + \frac{\partial}{\partial x}(\rho u w) + \frac{\partial}{\partial y}(\rho v w) + \frac{\partial}{\partial z}(\rho w w) = -\frac{\partial p}{\partial z}$$
$$+ \frac{\partial}{\partial x}\left(\mu \frac{\partial u}{\partial z}\right) + \frac{\partial}{\partial y}\left(\mu \frac{\partial v}{\partial z}\right) + \frac{\partial}{\partial z}\left(\mu \frac{\partial w}{\partial z} - \frac{2}{3}\mu \mathrm{div}\mathbf{V}\right)$$
$$+ \frac{\partial}{\partial x}\left(\mu \frac{\partial w}{\partial x}\right) + \frac{\partial}{\partial y}\left(\mu \frac{\partial w}{\partial y}\right) + \frac{\partial}{\partial z}\left(\mu \frac{\partial w}{\partial z}\right) + B_z \qquad (2.35)$$

In this system of equation resides all difficulties in solving fluid flow problems, represented by the non-linearities and the pressure–velocity coupling, issues that will be discussed in detail in the coming chapters. If μ is constant and $\mathrm{div}\mathbf{V} = 0$, the second line in the three above equations disappear.

2.3.3 Energy Conservation Equation

Referring again to Fig. 2.1, the variation of energy E (kinetical + internal) from time t to $t+\Delta t$ is given by,

$$E_1(t + \Delta t) + E_2(t + \Delta t) - E_1(t) - E_3(t) = \Delta E|_{\text{system}}, \qquad (2.36)$$

or

$$\dot{E}_{in} - \dot{E}_{out} + \left.\frac{\Delta E}{\Delta t}\right|_{\text{system}} = \left.\frac{\Delta E}{\Delta t}\right|_{CV} \qquad (2.37)$$

Again, see the similarity among Eqs. (2.3) and (2.11). For the energy conservation equation, the $\dot{g}^\phi \Delta V$, representing the Law for systems, is obtained from the 1st Law of Thermodynamics, which reads

$$\left.\frac{\Delta E}{\Delta t}\right|_{\text{system}} = \dot{Q}_{in} - \dot{W}_{out} = \dot{g}^e \Delta V, \qquad (2.38)$$

resulting in the energy conservation equation for a control volume, given by

$$\dot{Q}_{in} - \dot{W}_{out} + \dot{E}_{in} - \dot{E}_{out} = \left.\frac{\Delta E}{\Delta t}\right|_{CV} \qquad (2.39)$$

Equation (2.39) is the energy balance for the control volume, with the two first terms derived from the 1st Law of Thermodynamic for systems. The first one represents the heat crossing the boundaries entering in the control volume by diffusion and the heat "generation" inside the control volume. The usually called energy "generation", in fact, does not exist, since energy is not created, but just transformed. Here, the example could be an electric current passing through a resistor transforming

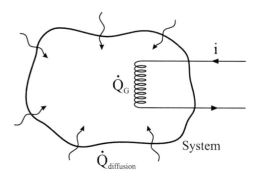

Fig. 2.5 Heat by diffusion + generation

electrical energy into heat, as can be seen in Fig. 2.5. The second term represents the work done/received which, according to the convention adopted here, is positive when the work is performed by the system. In this term the work done by a shaft is included, or the work realized by the viscous forces over the system, among other forms of energy transfer between the system and the exterior. The third and fourth terms represent the net energy advected to the interior of the control volume.

Defining the energy per unit of mass as $e = E/m$, Eq. (2.38) can be written in its integral form, as

$$-\int_A e\rho(\mathbf{V}_R \cdot \mathbf{n}) dA + \dot{Q}_{in} - \dot{W}_{out} = \frac{\partial}{\partial t} \int_V e\rho \, dV \tag{2.40}$$

Following the same steps done for the momentum conservation equation, one obtains

$$\frac{\partial}{\partial t}(\rho e) + \frac{\partial}{\partial x}(\rho u_R e) + \frac{\partial}{\partial y}(\rho v_R e) + \frac{\partial}{\partial z}(\rho w_R e) = \frac{\dot{Q}_{in} - \dot{W}_{out}}{dV} \tag{2.41}$$

The left-hand side of Eq. (2.41) contains the transient and the advection terms written in conservative form and, in the right-hand side the diffusion and source terms, which will be relate to the variables of the problem, like velocity and temperature.

It is nice to see all the similarities in the derivation of the conservation equations. It is a point which helps to understand the rationale behind the derivation and a strategy for seeing them as a balance for a generic variable. Re-write the mass conservation equation and the x component of the momentum equation (it could be the y component) in the same form as Eq. (2.41) to see their similarities. They are,

$$\frac{\partial \rho}{\partial t} + \frac{\partial}{\partial x}(\rho u_R) + \frac{\partial}{\partial y}(\rho v_R) + \frac{\partial}{\partial w}(\rho w_R) = 0 \tag{2.42}$$

$$\frac{\partial}{\partial t}(\rho u) + \frac{\partial}{\partial x}(\rho u_R u) + \frac{\partial}{\partial y}(\rho v_R u) + \frac{\partial}{\partial z}(\rho w_R u) = \frac{\sum F_x}{dV} \tag{2.43}$$

2.3 Conservation Equations

The left-hand side shows the accumulation and the transport by the bulk flow (advection) of the entity in consideration by unit of mass, while the right-hand side is the term coming from the application of the laws for a system. It will give rise to the diffusion and source terms in Eq. (2.43). For the global mass conservation its right-hand side is zero, since mass doesn't change inside the system. Therefore, the above equations can be written in the general following form, in which $\dot{g}^\phi$ can be found in Table 2.2,

$$\frac{\partial}{\partial t}(\rho\phi) + \frac{\partial}{\partial x}(\rho u_R \phi) + \frac{\partial}{\partial y}(\rho v_R \phi) + \frac{\partial}{\partial z}(\rho w_R \phi) = \dot{g}^\phi \qquad (2.44)$$

Back to the job of determining the $(\dot{Q}_{in} - \dot{W}_{out})/dV$ term, considering again Fig. 2.5, the electrical resistance symbolizes any kind of energy transformation, named heat "generation" in our equation. The net heat entering the control volume is,

$$\dot{Q}_{in} = \dot{Q}_{diff} + \dot{Q}_{gen}, \qquad (2.45)$$

with the heat by diffusion given by

$$\dot{Q}_{diff} = -\int_A \mathbf{q}.\mathbf{n} dA = -\int_V \nabla.\mathbf{q} dV \qquad (2.46)$$

Applying the divergence theorem and summing up the heat by diffusion and heat "generated", and considering a very small control volume

$$\frac{\dot{Q}_{in}}{dV} = q''' - \nabla.\mathbf{q} \qquad (2.47)$$

Using Eq. (2.47), the energy equation becomes,

$$\rho \frac{De}{Dt} = \nabla.\mathbf{q} + q''' - \frac{\dot{W}_{out}}{dV} \qquad (2.48)$$

It is still required the determination of $-\dot{W}_{out}/dV$, which is the work by unit of time made by the forces acting on a fluid flow field over the control volume. To find $\dot{W}_{out}$, Fig. 2.6 depicts all terms required for the energy balance to be done. Knowing that $-\dot{W}_{out} = \dot{W}_{in}$, and performing the balance, the result is

$$\begin{aligned}
-\dot{W}_{out} = {} & \tau_{xx} u dy dz|_{x^+} - \tau_{xx} u dy dz|_{x^-} + \tau_{yx} u dx dz|_{y^+} - \tau_{yx} u dx dz|_{y^-} \\
& + \tau_{zx} u dx dy|_{z^+} - \tau_{zx} u dx dy|_{z^-} + \sum_i \tau_{iy} v dA + \sum_i \tau_{iz} w dA \\
& + (B_x + B_y + B_z) dx dy dz,
\end{aligned} \qquad (2.49)$$

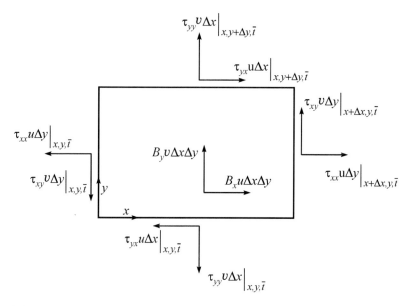

Fig. 2.6 Work done by the stresses

in which the summation of forces times velocity in the y and z direction are represented in a compact form.

Dividing both sides of Eq. (2.49) by $dxdydz$ and recalling the definition of the derivative, it reads,

$$
\begin{aligned}
-\frac{\dot{W}_{out}}{dV} &= \frac{\partial}{\partial x}(\tau_{xx}u) + \frac{\partial}{\partial y}(\tau_{yx}u) + \frac{\partial}{\partial z}(\tau_{zx}u) + B_x u \\
&+ \frac{\partial}{\partial x}(\tau_{xy}v) + \frac{\partial}{\partial y}(\tau_{yy}v) + \frac{\partial}{\partial z}(\tau_{zy}v) + B_y v \\
&+ \frac{\partial}{\partial x}(\tau_{xz}w) + \frac{\partial}{\partial y}(\tau_{yz}w) + \frac{\partial}{\partial z}(\tau_{zz}w) + B_z w
\end{aligned} \quad (2.50)
$$

The derivatives in Eq. (2.50) can be expanded, resulting in

$$
\begin{aligned}
-\frac{\dot{W}_{out}}{dV} &= u\left[\frac{\partial}{\partial x}(\tau_{xx}) + \frac{\partial}{\partial y}(\tau_{yx}) + \frac{\partial}{\partial z}(\tau_{zx})\right] \Rightarrow u\rho\frac{Du}{Dt} \\
&+ v\left[\frac{\partial}{\partial x}(\tau_{xy}) + \frac{\partial}{\partial y}(\tau_{yy}) + \frac{\partial}{\partial z}(\tau_{zy})\right] \Rightarrow v\rho\frac{Dv}{Dt} \\
&+ w\left[\frac{\partial}{\partial x}(\tau_{xz}) + \frac{\partial}{\partial y}(\tau_{yz}) + \frac{\partial}{\partial z}(\tau_{zz})\right] \Rightarrow w\rho\frac{Dw}{Dt} \\
&+ \tau_{xx}\frac{\partial u}{\partial x} + \tau_{yx}\frac{\partial u}{\partial y} + \tau_{zx}\frac{\partial u}{\partial z}
\end{aligned}
$$

2.3 Conservation Equations

$$+ \tau_{xy}\frac{\partial v}{\partial x} + \tau_{yy}\frac{\partial v}{\partial y} + \tau_{zy}\frac{\partial v}{\partial z}$$
$$+ \tau_{xz}\frac{\partial w}{\partial x} + \tau_{yz}\frac{\partial w}{\partial y} + \tau_{zz}\frac{\partial w}{\partial z} \quad (2.51)$$

In Eq. (2.51) the terms in brackets, identified by the arrows, come from the momentum conservation equation and can be written as,

$$u\rho\frac{Du}{Dt} + v\rho\frac{Dv}{Dt} + w\rho\frac{Dw}{Dt} = \rho\frac{D}{Dt}\left[\frac{|\mathbf{V}|^2}{2}\right], \quad (2.52)$$

and Eq. (2.51) takes the following form,

$$-\frac{\dot{W}_{out}}{dV} - \rho\frac{D}{Dt}\left[\frac{|\mathbf{V}|^2}{2}\right] = \tau_{xx}\frac{\partial u}{\partial x} + \tau_{yx}\frac{\partial u}{\partial y} + \tau_{zx}\frac{\partial u}{\partial z}$$
$$+ \tau_{xy}\frac{\partial v}{\partial x} + \tau_{yy}\frac{\partial v}{\partial y} + \tau_{zy}\frac{\partial v}{\partial z}$$
$$+ \tau_{xz}\frac{\partial w}{\partial x} + \tau_{yz}\frac{\partial w}{\partial y} + \tau_{zz}\frac{\partial w}{\partial z} \quad (2.53)$$

Defining the internal energy by unit of mass by

$$i = e - e_{\text{kinetics/mass}}, \quad (2.54)$$

subtracting the kinetic energy in both sides of Eq. (2.48) and using Eq. (2.54), it results

$$\rho\frac{D}{Dt}\left[e - \frac{|\mathbf{V}|^2}{2}\right] = \nabla.\mathbf{q} + q''' - \frac{\dot{W}_{out}}{dV} - \rho\frac{D}{Dt}\left[\frac{|\mathbf{V}|^2}{2}\right] \quad (2.55)$$

The two last terms in Eq. (2.55) can be replaced by the right-hand side of Eq. (2.53), becoming,

$$\rho\frac{Di}{Dt} = -\nabla.\mathbf{q} + q''' + \tau_{xx}\frac{\partial u}{\partial x} + \tau_{yx}\frac{\partial u}{\partial y} + \tau_{zx}\frac{\partial u}{\partial z}$$
$$+ \tau_{xy}\frac{\partial v}{\partial x} + \tau_{yy}\frac{\partial v}{\partial y} + \tau_{zy}\frac{\partial v}{\partial z}$$
$$+ \tau_{xz}\frac{\partial w}{\partial x} + \tau_{yz}\frac{\partial w}{\partial y} + \tau_{zz}\frac{\partial w}{\partial z}, \quad (2.56)$$

or, substituting the stress tensor given by Eqs. (2.26–2.31), it results

$$\rho\frac{Di}{Dt} = -\nabla.\mathbf{q} + q''' - p\nabla.\mathbf{V} + \mu\Phi, \quad (2.57)$$

in which the dissipation function is given by

$$\Phi = 2\left\{\left(\frac{\partial u}{\partial x}\right)^2 + \left(\frac{\partial v}{\partial y}\right)^2 + \left(\frac{\partial w}{\partial z}\right)^2\right\} - \frac{2}{3}\{\nabla.\mathbf{V}\}^2$$
$$+ \left(\frac{\partial u}{\partial y} + \frac{\partial v}{\partial x}\right)^2 + \left(\frac{\partial v}{\partial z} + \frac{\partial w}{\partial y}\right)^2 + \left(\frac{\partial w}{\partial x} + \frac{\partial u}{\partial z}\right)^2 \quad (2.58)$$

In terms of enthalpy the energy equation becomes

$$\rho \frac{Dh}{Dt} = -\nabla.\mathbf{q} + q''' - \frac{Dp}{Dt} + \mu\Phi, \quad (2.59)$$

or, in terms of the temperature and specific heat at constant pressure, c_p, it reads

$$\rho c_p \frac{DT}{Dt} = \nabla(k\nabla T) + q''' + \beta T \frac{Dp}{Dt} + \mu\Phi \quad (2.60)$$

or

$$\frac{\partial}{\partial t}(\rho c_p T) + \frac{\partial}{\partial x}(\rho u c_p T) + \frac{\partial}{\partial y}(\rho v c_p T) + \frac{\partial}{\partial z}(\rho w c_p T)$$
$$= \frac{\partial}{\partial x}\left(k\frac{\partial T}{\partial x}\right) + \frac{\partial}{\partial y}\left(k\frac{\partial T}{\partial y}\right) + \frac{\partial}{\partial z}\left(k\frac{\partial T}{\partial z}\right) + \beta T \frac{Dp}{Dt} + \mu\Phi \quad (2.61)$$

We have used for our developments, up to now, the control volumes in the Cartesian coordinate system. However, all mathematical steps made herein can be applied to any irregular control volume. In Fig. 2.7 it is shown a non-Cartesian control volume, over which a balance of the property ϕ is made. The resulting equation is again Eq. (2.12). The balance gives,

$$\dot{m}_w \phi_w + \dot{m}_s \phi_s - \dot{m}_e \phi_e - \dot{m}_n \phi_n + \dot{g}^\phi \Delta V = \frac{\Delta}{\Delta t}(m\phi)_P, \quad (2.62)$$

in which the symbols e, w, n and s are still being used to denote the interfaces of the control volume centered at P. These points will be called integration points when irregular control volumes with any number of faces will be considered.

Absolutely, there are no conceptual differences in the conservation equations written for control volumes of different shapes. The only difference is in the calculation of the mass flows across the faces of the control volume and in the determination of $\dot{g}^\phi$. In Fig. 2.7, for example, since the faces are not aligned with the Cartesian axes, it will be no longer possible to calculate the mass flows with only one of the Cartesian components u, v or w. All three components will be needed simultaneously in the same face, giving rise to the contravariant components of the velocity vector, which is, in fact, the normal velocity. The use of irregular control volumes will be discussed in Chap. 12 and beyond.

The system of equations comprising Eqs. (2.32–2.35) plus the energy conservation equation written conservative form, as

2.3 Conservation Equations

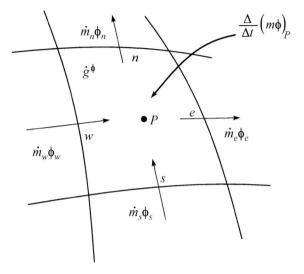

Fig. 2.7 Balance for a non-Cartesian control volume

$$\frac{\partial}{\partial t}(\rho c_p T) + \frac{\partial}{\partial x}(\rho u c_p T) + \frac{\partial}{\partial y}(\rho v c_p T) + \frac{\partial}{\partial z}(\rho w c_p T)$$
$$= \frac{\partial}{\partial x}\left(k\frac{\partial T}{\partial x}\right) + \frac{\partial}{\partial y}\left(k\frac{\partial T}{\partial y}\right) + \frac{\partial}{\partial z}\left(k\frac{\partial T}{\partial z}\right) + \beta T \frac{Dp}{Dt} + \mu \Phi + S^\phi, \quad (2.63)$$

form the system of equation that is largely employed in engineering applications. In Eq. (2.63) β is the compressibility coefficient, which is zero for incompressible fluids. A close view of Eqs. (2.32–2.35) and Eq. (2.63) reveals that all conservation equations have all the same appearance and can be written in a general form as

$$\frac{\partial}{\partial t}(\rho \phi) + \frac{\partial}{\partial x}(\rho u \phi) + \frac{\partial}{\partial y}(\rho v \phi) + \frac{\partial}{\partial z}(\rho w \phi)$$
$$= \frac{\partial}{\partial x}\left(\Gamma^\phi \frac{\partial \phi}{\partial x}\right) + \frac{\partial}{\partial y}\left(\Gamma^\phi \frac{\partial \phi}{\partial y}\right) + \frac{\partial}{\partial z}\left(\Gamma^\phi \frac{\partial \phi}{\partial z}\right) + S^\phi \quad (2.64)$$

However, the energy equation, Eq. (2.63), to be cast in the form of Eq. (2.64) requires that the c_p be removed from the transient term as well as from the advection terms. For certain situations, as for ideal gas and liquids, Eq. (2.63) can be written as

$$\frac{\partial}{\partial t}(\rho T) + \frac{\partial}{\partial x}(\rho u T) + \frac{\partial}{\partial y}(\rho v T) + \frac{\partial}{\partial z}(\rho w T)$$
$$= \frac{\partial}{\partial x}\left(\frac{k}{c_p}\frac{\partial T}{\partial x}\right) + \frac{\partial}{\partial y}\left(\frac{k}{c_p}\frac{\partial T}{\partial y}\right) + \frac{\partial}{\partial z}\left(\frac{k}{c_p}\frac{\partial T}{\partial z}\right) + S^\phi, \quad (2.65)$$

Table 2.3 Parameters for the conservation equations

Entity	ϕ Entity/mass	Transport coefficient Γ^ϕ	Source term S^ϕ
Mass	1	0	0
x momentum	u	μ	According to the problem
y momentum	v	μ	According to the problem
z momentum	w	μ	According to the problem
Energy	T	k/c_P	According to the problem

in which ϕ now stands for T and no longer for total energy e, internal energy (i), enthalpy (h) or $c_p T$. In Eq. (2.64) all terms not appearing explicitly are cast in the source term.

Pay attention in Table 2.3, now, for the energy equation, ϕ stands for T and Γ^T stands for k/c_p. In a fluid flow problem there are other scalars to be solve in the system of partial differential equations and, normally, they are represented by Eq. (2.64).

These equations deserve a deeper analysis to fully understand them. The first term takes care of the transient of the problem which should be, in any numerical simulation, always maintained in the partial differential equation, even if the interest is only the steady state regime. If a distorted transient is employed, the time step serves as a relaxation parameter with physical support. The remaining terms in the left-hand side are the advection terms, responsible for advecting ϕ (property/unit of mass).

Inspecting the following term, representing the advection of ϕ in the x direction,

$$\frac{\partial}{\partial x}(\rho u_R \phi),$$

one sees that ρu_R is the mass flow by unit of area advecting the property per unit of mass in the bulk flow. As discussed, the strong non-linearities of the Navier–Stokes equation appears when ϕ is equal to u, v or w. When the relative velocity is equal to the flow velocity, for the x-momentum equation, for example, this term becomes

$$\frac{\partial}{\partial x}(\rho u u)$$

Of course, the two velocities are the same, but is wise to "see" them as different entities. The first one is the carrier (advecting) velocity and the second one, the carried (advected) velocity. In the linear system, the advecting velocity takes part of the coefficients and the advected is the variable what will receive the interpolation function, that is, the unknown of the problem. Equation (2.64) can also represent other transport properties besides mass, momentum, energy, species, like turbulent kinetic energy and dissipation of turbulent kinetic energy when the turbulence models $k - \varepsilon$ or $k - \omega$, or some of their variants, are used.

For compressible flows the state equation for determining pressure is written as function of density and temperature, as

$$p = f(\rho, T) \qquad (2.66)$$

For a 3D laminar compressible flow with heat transfer, the system of partial differential equations comprises the mass conservation, three components of the momentum conservation, energy conservation and the state equation. If density is a function of temperature only, rigorously the fluid is compressible but, in this case, density is no longer a variable of the equation system but only a physical property dependent on temperature, like viscosity, for example. A problem with a constant density, or a function of temperature only, receives the same numerical treatment.

2.4 Elliptic, Parabolic and Hyperbolic Problems

2.4.1 Preliminaries

The classification of partial differential equations in elliptic, parabolic and hyperbolic is normally done using the relations among the coefficients of the equation. Considering that heat transfer and fluid mechanics problems are described by systems of partial differential equations, this classification is not helpful in connecting the boundary conditions to the type of the equation, since they are always of mixed type. For instance, the conservation equations for compressible flows form a system of equations known as mixed hyperbolic/parabolic, if the transient terms are maintained, and mixed elliptic/hyperbolic, if they are neglected. It seems more didactical and practical not to classify the equation itself, but classify according to the mathematical nature of the equations in each coordinate direction. For example, the equation for a transient fluid flow considering viscous effects in all coordinate directions, is a parabolic equation in time and elliptic in space. Next, following this approach of interpreting the nature of the partial differential equations, examples are given for these classes of problems.

2.4.2 Parabolic and Hyperbolic Problems

From the numerical point of view, it is important to recognize the equations characteristics, in order to take possible computational advantages, such as CPU time and computational storage. Therefore, it is useful to define the problems of fluid flow and heat transfer in problems that allow the solution to be obtained by marching process in a given coordinate (spatial or temporal) and those which do not allow such a procedure.

It can be said that hyperbolic and parabolic problems allow the marching process, whilst the elliptic ones do not. In fluid flows, marching problems are those that need no downstream boundary conditions, that is, they depend solely upon upstream information. The advective terms of the Navier–Stokes equations show this behavior and are parabolic, thus becoming easy to understand that, if no other means of transport of information exists in that direction, it will be impossible that downstream information be transmitted upstream. Therefore, no boundary conditions will be need in the downstream boundary.

Figure 2.8 shows the classical parabolic problem of two-dimensional flow over a flat plate. In this problem, the diffusion effects in the x direction are neglected and, as there are no effects due to pressure, since there are no obstacles to the flow, the x direction is parabolic Therefore, in the x direction, only the advective term remains, and no downstream boundary conditions are needed. The problem is, therefore, solved by marching from the initial spatial conditions and solving a one-dimensional elliptic problem at each station x. The solution marches as far as there is interest on it. The governing x momentum equation for this problem is given by Eq. (2.67) (Exercise 2.3), in which the pressure gradient is known. In this direction only a first derivative exists, and, mathematically, that is why only one boundary condition is required. In the other hand in the y direction the problem is diffusive dominated represented by a second order derivative term, which requires, therefore, two boundary conditions in this direction. It couldn't be different, since solving a partial differential equation is just integrating it in the coordinate directions, originating constants that should be found using the boundary conditions.

Computationally, there is a great advantage in this treatment, since the required storage corresponds to only two stations: the calculation one and the upstream station, whilst, if the treatment were elliptic, global storage would be required. Still more important is the fact that the complete solution is a set of independent one-dimensional solutions, extremely faster to be obtained than the solution involving all mesh points of the domain. The only one boundary condition required in x corresponds to the beginning of the plate.

Therefore, according to the order of the derivatives it is also possible to identify if the problem is parabolic or elliptic in a given direction. The difference between the parabolic and the hyperbolic march is that the first one occurs along a coordinate direction, while the second one, along the characteristics. The difficulty of the hyperbolic march lies in the fact that it is not know the boundary conditions of the elliptic problem in the other direction(s). Therefore, in a complex 3D problem it becomes easier to treat the problem elliptically in all directions.

2.4 Elliptic, Parabolic and Hyperbolic Problems

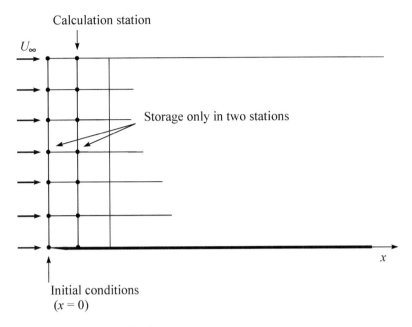

Fig. 2.8 Boundary layer on a flat plate

2.4.3 Elliptic Problems

Elliptic problems are those where the physical information travels in all coordinate directions. Diffusive and pressure effects are elliptic, which, if they are present in the phenomenon, require the specification of boundary conditions in both extremes of the axis in consideration. In a fluid flow these effects travel also in the opposite direction of the velocity, conferring the elliptical characteristics of the flow. Depending on how strong the advection is compared to diffusion, the later can be neglected, changing the mathematical nature of the equation in that direction. This shows the importance of linking the physics with the classification of the equations, since it will dictate the type of boundary conditions required.

It is easy to understand, based on physics, that in a given medium, when an increase in temperature occurs at one point, heat will be diffused in all directions according to the value of the thermal conductivity. Heat diffusion, as well as mass and momentum diffusion, are elliptic phenomena and, therefore, require boundary conditions along the whole boundary of the domain, the so-called boundary value problem. Figure 2.9 shows, for a single axis, the influence on the domain for elliptic, parabolic and hyperbolic problems when a perturbation is introduced at point P in a fluid flow.

Diffusive terms possess second order derivatives, thus requiring boundary conditions at both ends of the solution domain along the axis under consideration. Considering the x coordinate shown, a perturbation at point P affects the domain upstream

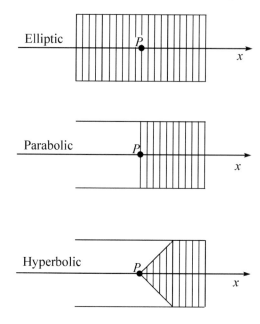

Fig. 2.9 Parabolic, elliptic and hyperbolic problems

and downstream of P, in the elliptic case; only downstream of P in the parabolic case; and only downstream form P and over a certain region (obviously not known), in the hyperbolic case. The first problem does not admit the march procedure, while the other two do.

To give another example and conclude this section, consider a supersonic flow over a blunt body of revolution presented in Fig. 2.10, where only one plane is shown, for the sake of simplicity. The flow region is divided in three parts. In region I the velocity is constant and equal to u_∞ (non-disturbed flow). Region II confines a flow with Mach < 1, subsonic and, thus, elliptical, and region III is supersonic (hyperbolic), consequently admitting a marching procedure for the solution.

Let's begin the comments about the physics of this problem considering Euler's equation, that is, without considering the viscous terms. Region II is an elliptical region since, due to the presence of the blunt body, the pressure wave "travels" contrary to flow with the local sound velocity. When the flow velocity equals the velocity in which the pressure information travels upstream, a shock wave is established and, upstream from the shock, it is identified the region I, where the flow is undisturbed and supersonic. The higher the flow velocity, the nearer to the body the shock will occur. It is easy to understand that, when the flow is subsonic, incompressible, there is no shock formation because the pressure wave travels contrary to the flow with an infinity velocity. In this case, the flow has no condition of preventing the propagation, due to its lower velocity and no shock is formed.

Region III is totally supersonic. Given the inexistence of diffusive effects (inviscid flow) and pressure (the body does not present protuberances in this region), the effects propagate only downstream. This flow nature allows region III be solved by

2.5 True and Distorted Transient

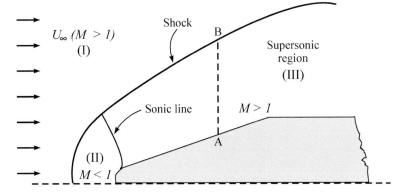

Fig. 2.10 Supersonic flow over a blunt body

a marching process. If viscous effects were considered, region III would exhibit the formation of a boundary layer next to the body. In this region, the problem is, once again, of marching along the body. Observing line A-B, the velocity is known at these points, they are u_∞ (Euler equations), but it is not known the position in space of point B. It would be an elliptic axis if the spatial position of point B was known. It is located on the boundary of the shock and, logically, the position of the shock is what one wants to know solving this problem.

There are two main classes of strategies for solving supersonic flows, shock fitting and shock capturing. In the first one the shock is treated as a discontinuity, while in the former the shock is captured by finding the large gradients in the domain, what will require very fine grids.

Up to this point comments were done on the specific physical characteristics of each flow region, suggesting an adequate numerical model for each of them. Obviously, regions I, II and III may be solved as a single calculation domain, avoiding the separation of problems in regions whose boundaries are unknown. Solving the whole domain will require conditions in all boundaries, and the specific characteristics of the flow will be revealed by the results, including the shock. That is the usual form of tackling complex problems of this nature. For the sake of completeness, Fig. 2.11 shows a typical grid that can be used for this kind of problem.

2.5 True and Distorted Transient

The distorted transient approach, when the interest is the steady-state solution, is compulsory when solving fluid flow problems, since the nonlinearities and the coupling doesn't allow to use large time steps. To discuss this topic let's consider just one nonlinear equation, represented herein by the two-dimensional transient heat conduction on a plate with temperature-dependent thermal conductivity. The understanding can be extended to a system of transient equations. According to our

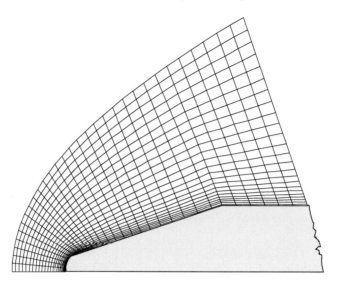

Fig. 2.11 Curvilinear coordinate system for a blunt body

definitions, this problem is parabolic in time and elliptic in the two spatial directions. If we are not interested in the trtue transient solution, but only in the steady state one, two alternatives of tackling the problem may be employed.

The first one is keeping the transient term in the equation, requiring the solution of the two-dimensional spatial problem for each time level until the steady state is reached obeying some specified tolerance. The size of the time step should be as large as to permit convergence, recalculating the matrix coefficients, which are variable due to the varying thermal conductivity, in each time level. The transient results, are distorted, by two reasons; firstly, the linear system is not solved accurately in each time level (for saving computational time) and, secondly, the size of the time step can generate a solution with no grid resolution in time.

The second alternative is to remove the transient term from the equations. However, just as before, the problem must be solved using iterations, updating the matrix coefficients which depends on the temperature to take into account the nonlinearities. These successive iterations are equivalent to performing a distorted march in time (distorted transient) approach, which maintains the time coordinate in the equation. Relaxation is compulsory to obtain convergence, but the relaxation parameter is non-physical. In the other hand, dealing with a distorted transient keeping the time in the equations and playing with the time step is equivalent as having a physical relaxation parameter. It is worthwhile to mention that only few linear and simple problems can be solved using an infinite time step, what is equivalent as removing the time from the equation and solving just one linear system.

The main learning from this discussion is that it is highly recommended to maintain the time coordinate in the numerical scheme, following the true or distorted transient, according to what is sought. Maintaining the time coordinate does not entail any additional numerical complication, and one should take advantage of that, creating numerical methods where the progress in time may be used as a relaxation parameter, as commented. In this way, the relaxation coefficient is kept under the control of the user and may be variable among the equations and among the control volumes.

In a problem where only the steady state solution is sought and the transient term was maintained, the estimation of the variable to start the iterative procedure plays the role of the initial condition.

2.6 Conclusions

Computational Fluid Dynamics (CFD) deals with the conservation equations, like mass and momentum, expressing the physics in a mathematical way. The nature of these equations, their similarities, the non-linearities and coupling of these equations are, unfortunately matters not sufficiently considered when these equations are derived in most fluid mechanics and heat transfer textbooks. Because there are so many ways in deriving them, some of them invoke the basic laws and work mathematically to reach the final equations. Others perform balances, but without differentiating the advected and advecting velocities, of crucial importance for developing CFD algorithms and understanding fluid flows. This chapter tried to derive the conservation equation in a unified form, recognizing easily, in a physically base, how the important nonlinearities of the momentum equations appear. At the same time, whenever possible, it was directed for the use in CFD. The ultimate goal of this chapter was, definitively, eliminate all difficulties in understanding the conservation equations. The chapter can be also used in any introductory course in fluid mechanics.

2.7 Exercises

2.1 To get familiarity with the conservation equations, write the system of equations for the following flow situations

- Viscous incompressible;
- Viscous compressible;
- Inviscid Compressible (Euler equations).

Using Eqs. (2.33–2.35), show that the source terms S^u, S^v and S^w contain only the body forces and pressure gradient if ρ and μ are constant.

2.2 Using the same procedure done for the mass, momentum and energy conservation equations, derive the mass conservation equation for species. Then, consider the incompressible two-dimensional flow, with heat transfer, of two species

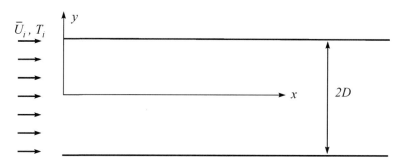

Fig. 2.12 Flow between parallel plates

forming a single phase (dry air + water vapor, for example) with concentrations $C_1(x, y)$ and $C_2(x, y)$. Write the system of partial differential equations that needs to be solved for the determination of all variables, that is u, v, p, T, C_1 and C_2. What are the boundary conditions for these variables for the situation when the flow is of dry air over a water surface (pool), admitting the water surface to be rigid?

2.3 For the problem of incompressible two-dimensional flow between two parallel flat plates, according to Fig. 2.12, write, in Cartesian coordinates, the form of the energy equation which is adequate for the problem. Give the necessary boundary conditions for the cases when the velocity (or the Peclet number, to be more rigorous) tends to zero and to infinity

2.4 The equation is the momentum equation for the flow over a flat plate after the boundary layer hypothesis were applied. According the way elliptic and parabolic coordinates are defined in this book, classify this equation and specify which are the boundary conditions required to solve it. Also, comment on the approach of determining the pressure gradient in such a problem.

$$\frac{\partial}{\partial x}(\rho u u) + \frac{\partial}{\partial y}(\rho v u) = -\frac{d\overline{p}}{dx} + \frac{\partial}{\partial y}\left(\mu \frac{\partial u}{\partial y}\right), \quad (2.67)$$

2.5 For the two-dimensional heat conduction problem shown in Fig. 2.13 with a uniform heat generation $q'''[W/m^3]$, what restriction must be obeyed in order to be possible to apply a prescribed heat flux at the north face? If such restriction is not respected, what are the consequences from the point of view of obtaining the solution?

2.6 The governing equation for the vibrating movement of a spring under tension, fixed by its ends is given by

$$\frac{\partial^2 u}{\partial t^2} = c^2 \frac{\partial^2 u}{\partial x^2}, \quad (2.68)$$

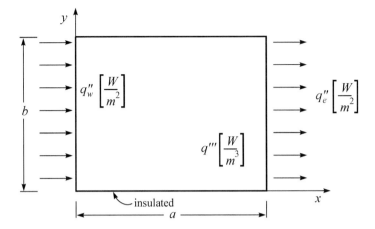

Fig. 2.13 Two-dimensional heat conduction

in which u is the displacement, t is time, x is the spatial coordinate and c is a constant. Classify this equation and give the necessary initial and boundary conditions for this problem.

2.7 Starting from Eqs. (2.13–2.15) obtain Eq. (2.18) using a control volume momentum balance. This exercise is very important to understand the origin of the nonlinearities in the Navier–Stokes equations.

2.8 Using the control volume shown in Fig. 2.4, obtain Eqs. (2.19) and (2.20).

References

1. Volfang V, Thomas E, Florian M (2002) Heat transfer predictions using advanced two-equation turbulence models. CFX Technical Memorandum-CFX-VAL10/0602. www.software.aeat.com/cfx/(2002)
2. Burmeister LC (1983) Convective heat transfer. Wiley

Chapter 3
The Finite Volume Method

3.1 Introduction

The previous chapters introduced the discipline known as CFD-Computational Fluid Dynamics, one of the numerical alternatives for solving engineering problems, and brought to the reader the derivation of the conservation equations based fundamentally on physical knowledge, giving a general treatment for all conservation equations. Relevant physical and mathematical aspects of these equations were discussed.

The objective of this chapter is to introduce the fundamentals of the finite volume methodology for fluid flow problems. The strategy employed is the development of the subject using simple differential equations in a growing level of difficulty, in a block-by-block construction approach, allowing undergraduate students and beginners to be introduced to the field of CFD without difficulties. Finite volume methods have a strong physical appealing and no deep mathematics involved, what makes the learning easy and enjoyable. A motivation for the use of finite volume methods based on the discrete conservation will be addressed, introducing the cell-center and cell-vertex methods.

With this approach in mind, this chapter presents the one-dimensional transient heat conduction problem as a prototype equation for demonstrating the method. With this equation, it will be possible to show all the integration procedure, both temporal and spatial, pertinent to the finite volume method, and to present the application of the boundary conditions for diffusive problems. For the first time interpolation function will be introduced and, for these problems, it can be linear (central differencing) without causing difficulties such as those appearing when advection is present, a subject of Chap. 5. The explicit, implicit, and fully implicit formulations will be presented when using this prototype equation.

Linearization of the source term, structure of the matrix of coefficients, boundary conditions, flux continuity at interfaces, truncation errors and consistency, stability and convergence will be all topics covered in this Chapter. The idea is that after this chapter the student can write a simple code solving a transient heat conduction problem. The solution methods for solving the linear system are seen in Chap. 4.

3.2 The Task of a Numerical Method

The task of a numerical method is to solve one or more differential equations by replacing the existing derivatives with algebraic expressions involving the unknown variables. When an analytical solution is not possible, and we need to rely on a numerical approximation of the differential equation, it is accepted to have the solution for a discrete number of points, with a certain error, hoping that the larger this number of points is, the closer to the exact solution our approximate (or numerical) solution will be. An analytical method with the ability to solve such equations would give us the solution in closed form, and it would then be possible to calculate the values of the dependent variables at the infinitesimal level, that is, for an infinite number of points.

Thus, it is easy to understand that if it is decided to calculate only N values of the variable in the domain, there will be N unknowns, requiring N algebraic equations for closure, forming a system of N equations at N unknowns. If we want to make our calculations more precise there will be an increasing in the number of unknowns, and the linear system to be solved will also increase in number of equations. The computational effort also increases, and in an almost exponential way, what requires special algorithms, such as multigrid, to avoid excessive CPU time.

Figure 3.1 exemplifies the task of the numerical method, which is to transform a differential equation, defined in a domain D, into a system of algebraic equations. To this end, the derivatives in the differential equation must be replaced by the discrete values of the function. Transforming the derivatives into terms that contain the function means integrating the differential equation, and the various ways of doing this is what characterizes the type of the numerical method. Our concern in this text will be only with the finite volume method, but others can be employed, such as finite differences and finite elements, for example, in spite of the former is rarely used.

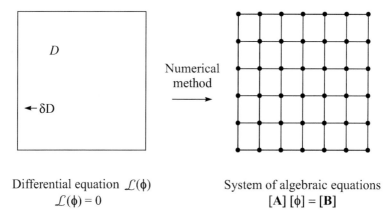

Fig. 3.1 From continuous to discrete

3.3 Why Finite Volume Methods is a Good Choice

When a physical problem needs to be numerically approximated, and there is a mathematical representation of it, it is compulsory that the physics be respected in any level of discretization. In other words, this means that the conservation of mass, for example, must be satisfied independently of the size of the grid employed. To satisfy this rule the only alternative is to have the approximated equations (discrete equations) obtained through balances at the finite volumes. It is known that any numerical method, if consistent, will produce the same results if a very fine grid is used. The question is that this kind of solution is not attained due to the large memory and computer time required. Thus, we should be happy with a discrete level solution which contain discretization errors, but retains the conservation principles which, by no means, can be sacrificed.

Besides the procedure of creating the approximated equations through balances, it is possible to obtain them by integrating the differential equations over the control volume in space and time if they are in conservative form. In a conservative form the fluxes of the property are within the sign of the derivative and, in the first integration, the fluxes appear at the boundaries of the control volumes, what is equivalent, therefore, of performing balances. Let's use the mass conservation equation to exemplify, which can be written in conservative form as

$$\frac{\partial}{\partial t}(\rho) + \frac{\partial}{\partial x}(\rho u) + \frac{\partial}{\partial y}(\rho v) + \frac{\partial}{\partial z}(\rho w) = 0, \quad (3.1)$$

in which it is seen that the fluxes (mass in this case) are inside the derivatives. This equation, expanding the derivatives, can be written as

$$\frac{\partial}{\partial t}(\rho) + \rho\frac{\partial}{\partial x}(u) + \rho\frac{\partial}{\partial y}(v) + \rho\frac{\partial}{\partial z}(w) \\ + u\frac{\partial}{\partial x}(\rho) + v\frac{\partial}{\partial y}(\rho) + w\frac{\partial}{\partial z}(\rho) = o \quad (3.2)$$

Equations (3.1) and (3.2) are, obviously, the same equations and will give the same results if an analytical solution would be done using either one. Numerically, however, they have a marked difference, since when integrating Eq. (3.2) in a numerical procedure, what appears at the boundaries of the control volumes are not the fluxes exactly, therefore, not obeying conservation. This learning tells that we should try, always, in any numerical method, use a conservative form of the differential equations to be integrated. It is expected that the numerical method does not destroy this characteristic. Not obeying discrete conservation, besides violating the physics, may create convergence difficulties.

To illustrate the connection between the approximate equations used in the finite volume method and the differential equations in the conservative form, consider the two-dimensional control volume shown in Fig. 3.2. Our interest at this point is to derive the differential equation representing the conservation of mass.

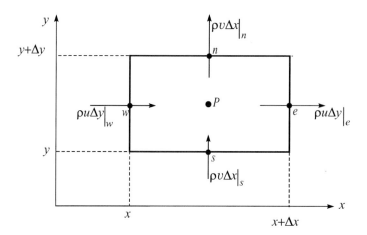

Fig. 3.2 Mass conservation balance

The mass balance, as result of the Reynolds Transport Theorem, applied to the elementary volume shown in Fig. 3.2 is given by,

$$\dot{m}_w - \dot{m}_e + \dot{m}_s - \dot{m}_n = \left.\frac{\Delta m}{\Delta t}\right|_{CV}, \qquad (3.3)$$

in which the lower-case letters at the control volume boundaries were identified in the previous chapter. Recall, from the undergraduate engineering lectures and textbooks, that the application of a conservation principle to a control volume, as done by Eq. (3.3), is the first step when deriving the corresponding partial differential equation. Translating the terms in Eq. (3.3) using the velocity field, the result is,

$$\rho u \Delta y|_w - \rho u \Delta y|_e + \rho v \Delta x|_s - \rho v \Delta x|_n = \left.\frac{\Delta}{\Delta t}(\rho \Delta x \Delta y)\right|_{CV}, \qquad (3.4)$$

in which it is implicit a unit of thickness to comply with the units of each term with the unit of mass flow in Eq. (3.3). Dividing Eq. (3.4) by the product $\Delta x \Delta y$, it gives

$$\frac{\rho u|_w - \rho u|_e}{\Delta x} + \frac{\rho v|_s - \rho v|_n}{\Delta y} = \frac{\Delta \rho}{\Delta t}, \qquad (3.5)$$

which, after applying the limit for the spatial and temporal coordinates, and using the definition of the derivative, the differential equation for the mass conservation equation in conservative form is obtained,

$$\frac{\partial \rho}{\partial t} + \frac{\partial}{\partial x}(\rho u) + \frac{\partial}{\partial y}(\rho v) = 0 \qquad (3.6)$$

3.3 Why Finite Volume Methods is a Good Choice

It is seen, therefore, that the conservation equation for a finite volume is an intermediate step to obtain the conservation equation at the infinitesimal level, that is, the differential equations. Therefore, the finite volume equations are the discrete counterpart of the differential equation. It can be also realized that the differential equation is a balance in a control volume with the size of a point.

As, in practice, it is impossible to solve de balance equations for a control volume with the size of a point (continuous differential equation), it is physically consistent to solve its discrete counterpart, the finite volume equations. Refining the grid is equivalent to the limiting approach to the analytical solution in a physically consistent way.

Now, let us go the way back to obtain the finite volume equations starting from the differential equation written in conservative form. Performing the integration over the control volume and in time, as indicated in Eq. (3.7),

$$\int_{t}^{t+\Delta t}\int_{w}^{e}\int_{s}^{n}\left(\frac{\partial\rho}{\partial t}+\frac{\partial}{\partial x}(\rho u)+\frac{\partial}{\partial y}(\rho v)\right)dxdydt = 0, \quad (3.7)$$

and recalling that the velocities at the control volume boundaries are calculated at the mid-point of the face, what represents an average of the velocity on that face, one obtains

$$\left(\frac{\rho^{t+\Delta t}-\rho^{t}}{\Delta t}\right)\Delta x\Delta y + \left[\rho u|_{e}-\rho u|_{w}\right]\Delta y + \left[\rho v|_{n}-\rho v|_{s}\right]\Delta x = 0, \quad (3.8)$$

which is, exactly, Eq. 3.4, obtained through a mass balance in a finite volume (discrete level). This procedure can be applied to any other entity, like momentum or energy. Figure 3.3 explain the link between mathematical limiting and grid refining when the equations are kept in conservative form. This is the beauty of finite volume methods, easy to understand, easy to obtain the approximated equations which respects the physics at discrete levels, and robustness, all reflexes of its physical background.

By performing the integration for all control volumes, we obtain an algebraic equation for each control volume and, thus, the system of algebraic equations to be solved. The preference to obtain the approximate equations by integrating the differential equation comes from the fact that not all balances are easy to derive as was the conservation of mass. A momentum balance, for example, requires the identification and summation of all stresses acting on the control volume. For irregular control volumes, this is not an easy task, so, it is preferable to start with the differential equation in conservative form.

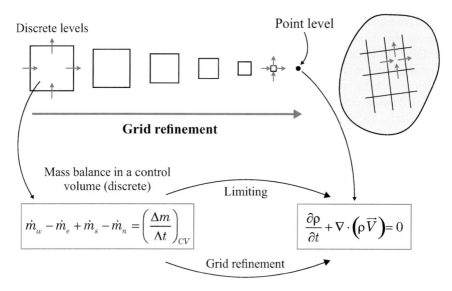

Fig. 3.3 Why finite volume methods

3.4 Few Words About the Conservative Property

It was stressed the importance of having the conservation principles obeyed at discrete level in a numerical simulation. Let's see now how the finite volume method can be seen in the framework of the weighted residual approach when applied to the control volume. The statement of a weighted residual is a way of seeing all numerical methods under the same umbrella and helps to understand when the conservation principles are broken.

When solving a partial differential equation numerically, the solution is approximated. Let $\tilde{\phi}$ be this approximation. Inserting this approximation into the partial differential equation a residual is obtained, as

$$\pounds(\tilde{\phi}) = R(\tilde{\phi}), \tag{3.9}$$

in which £ is the differential operator, given by, for example,

$$\pounds(\phi) = \frac{\partial}{\partial t}(\rho\phi) + \frac{\partial}{\partial x_j}(\rho u_j \phi) + \frac{\partial}{\partial x_j}\left(\Gamma^\phi \frac{\partial \phi}{\partial x_j}\right) + S^\phi \tag{3.10}$$

The weighted residual approach states that the average (integral in the domain) of the residue weighted by an arbitrary function w is forced to zero, that is,

$$\int_V w \left[\frac{\partial}{\partial t}(\rho\tilde{\phi}) + \frac{\partial}{\partial x_j}(\rho u_j \tilde{\phi}) + \frac{\partial}{\partial x_j}\left(\Gamma^\phi \frac{\partial \tilde{\phi}}{\partial x_j}\right) + S^\phi\right] dV = 0 \tag{3.11}$$

3.4 Few Words About the Conservative Property

According to what was learned from the previous section, the integration of the partial differential equation in conservative form is equivalent as to perform balances in the control volume. Paying attention to Eq. (3.11) we can see that if $w = 1$ it reproduces the finite volume approach. In the scope of the weighted residual approach when $w = 1$ the method is called sub-domain. Therefore, any weighting function different from 1 destroy the conservation, since the integration no longer reproduces the balances and, therefore, it is not a conservative scheme.

When w is different from 1 it gives rise of a series of method in the broad class called finite element methods. The most popular and widely used in the solution of solid mechanics problems is the Galerkin method, in which the weighting function (w) is the same as the interpolation functions (ψ_i) used to expand the solution inside the element. The solution in the finite element method can be represented by

$$\tilde{\phi} = \sum_{i=1}^{n} \psi_i \phi_i, \quad (3.12)$$

in which n is the number of nodes in the element. The functions ψ interpolate the values of the unknown variable ϕ inside the element based on the values located at its vertices. These interpolation functions are also called shape functions. Introducing Eq. (3.12) into Eq. (3.11), choosing the weighting function and integrating in the domain, it is obtained the local matrix for the element, which is assembled element-by-element to construct the global matrix. It is used to lower the order of the differential equation using integral by parts and getting the weak form.

Discussions in the literature and internet blogs try to compare finite element, finite differences and finite volume methods. All methods, with their specific features, are suitable for solving fluid mechanics problems, and many of the arguments come from a lack of knowledge about the methods, specially about finite volumes. It is not rare in the literature to find misunderstandings between finite differences and finite volumes. They are completely different methods, and only in very simplified cases, they give rise to the same linear system. It is true that finite difference developers migrated to finite volumes along time and, perhaps, linked to the fact that finite differences were always applied for orthogonal coordinate systems, mainly Cartesian, it is said that finite volume method has difficulties in dealing with unstructured grids.

This may have been true five decades ago, when finite difference and finite volume analysts were deeply involved in treating the nonlinearities and the difficult couplings of the Navier–Stokes equation, period in which they left behind the development of algorithms for complex geometries.

Unstructured grids are nowadays a standard practice in cell-center as well as in cell-vertex finite volume methods. In the class of cell-vertex methods, the EbFVM-Element-based Finite Volume Method has as fundamental geometrical entities the element (exactly as in finite element) and the control volume (entity not existing in finite element).

The order of interpolation is also free to choose in both methodologies, since it depends solely on the order of approximation of the interpolation function. In

EbFVM the interpolation functions are applied in the element, exactly as done in finite element and, therefore, any order of interpolation can be used. First order, second order or higher order elements can be applied, irrespectively, in EbFVM as well in any finite element method. Even for cell-center finite volumes, high order interpolation can be used, however, with a more cumbersome implementation of the computer code, not as clean as in finite element or EbFVM.

But, as already stressed, the important difference of the methods lies on the local discrete conservation. It is our feeling that this is of utmost importance, since the physics must be obeyed when numerically solving a physical problem. Perhaps, this is the reason why all widespread commercial CFD codes used nowadays in engineering simulation employs finite volume methods. Finite elements, one could say, was born using unstructured grids and, since conservation is not a big issue when there is no mass flow, it is a perfect tool for solving structural problems, even if the force balances are not strictly respected. When conservation is required, difficulties appears. Of course, in finite element methods it is possible to enforce local conservation, but at expenses of more complex algorithms, more mathematically elaborate elements and more CPU time for the solution.

As final words, it is worth to mention that it is frequently reported in the literature that finite volume is a particular case of finite element. In our understanding it is not, it is a particular case of the application of the weighted residue statement with weighting function equal to 1. Of course, it is free to call all methods that use the weighted residual statement applied to an element, as being a finite element method, since the element is always involved. But it is interesting to note that Galerkin, least squares, co-location and others, are in the same hierarchy as the finite volume method, or the sub-domain method, all under the umbrella of the weighted residual approach. In fact, the denomination finite element comes from the fact that the weighted residual statement is applied to finite elements, and not to the whole domain [1]. The interest reader in finite element is referred to the corresponding extensive literature, in which excellent books are encountered.

3.5 Cell-Center and Cell-Vertex Methods

In a finite volume technique, the construction of the control volume is made using the elements furnished by the grid generator. What we get from the grid generator are the elements defined by the coordinates of their vertices, in its simplest form. Depending on how dedicated the generator is, connectivity, geometric data, node numbering and element numbering can be furnished to the numerical method by the grid generator. Using the elements, two versions of the finite volume method are possible, cell-center and cell-vertex, with the previous one known as the conventional finite volume by the numerical community.

Considering a 3D mesh generator, the minimum information provided by the generator is the coordinates of the 8 vertices of the element for a hexahedron, 4 for a tetrahedron and so on. If the element is used as control volume, it is called

3.5 Cell-Center and Cell-Vertex Methods

a cell-center method, what means that the unknown variable is at the center of the element, and, of course, at the center of the control volume, as in any method. Cell, in this denomination, is the element, what would suggest that a better denomination would be an element-center method. On the other hand, if the unknowns are stored at the vertices of the element, the method is called cell-vertex. In a cell vertex method, as before, one needs to decide on the construction of the control volume based on the elements. One possible construction will be seen when EbFVM is described. In 2D we have triangles and rectangles, and in 3D tetrahedra, hexagons, prisms and pyramids, as elements frequently used, being possible to use other type of elements.

To exemplify the cell-center method, consider Fig. 3.4, in which the element is taken as the control volume. For the cell-vertex case, in Fig. 3.5 the control volume is constructed using part of each element which shares the same node (unknown node P). For the cell-center there are four fluxes to be calculated to perform the balances of the properties, while for the cell-vertex for each element taking part of the control volume, two fluxes need to be calculated and will enter the conservation equation. In the example of Fig. 3.5, six elements take part in the control volume, therefore, 12 fluxes will enter the control volume balance. Details will be given when the EbFVM will be discussed, in Chap. 13.

Between the cell-center and cell-vertex methods, the key difference is in the application of boundary conditions, which will be discussed at the end of this chapter. By now, it is enough to recognize that in the cell-vertex method there are unknowns lying on the boundary, while in the cell-center there are not, what makes the application of boundary conditions different in the two methods.

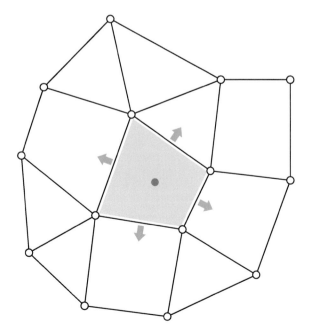

Fig. 3.4 Cell-center method. Element as control volume

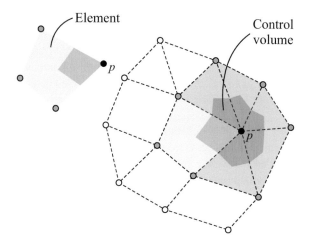

Fig. 3.5 Cell-vertex method

Next, let's begin the construction of our knowledge in applying a finite volume method using the 1D transient heat conduction as the working equation. With this equation we will learn the approximation of the transient and diffusive terms. In Chap. 5 we will tackle the treatment of the advective term, which is the responsible for the main difficulties in fluid flow problems. The developments will start using the cell-center method, since it is better known by the scientific community. When boundary conditions are treated few comments will be done related to the cell-vertex.

3.6 One Dimensional Transient Heat Diffusion

The 1D transient heat diffusion equation with source term will be used by several reasons: being 1D, it is easy to deal with, it contains both time and space coordinates, it is a problem normally known by the beginners in heat transfer and fluid flow, and we will have the chance to deal with the linearization of possible source terms. The equation is

$$\frac{\partial}{\partial t}(\rho c T) = \frac{\partial}{\partial x}\left(k \frac{\partial T}{\partial x}\right) + S^T, \qquad (3.13)$$

which can also represent others physical transient with diffusion problems. The lower P index on the specific heat at constant pressure was removed to use this symbol as location on the grid when the equation will be integrated. The mesh employed is shown in Fig. 3.6, in which, just for the sake appearance, the mesh is shown with some dimension in the other direction.

Note that the mesh has full volumes throughout the domain, characterizing the cell-center method. The other choice would be to have half-volumes at the boundaries,

3.6 One Dimensional Transient Heat Diffusion

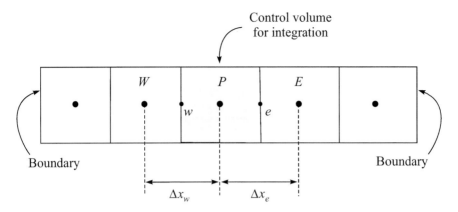

Fig. 3.6 Cell-center discretization. Element as control volume

with the nodal points (unknowns) on the boundaries, which would be the cell-vertex approach.

Equation (3.13) is in conservative form, thus, its integration will result in approximate equations that represents conservation balances at control volume level. Following the procedure inherent to the finite volume method, the integration of Eq. (3.13), indicated in Eq. (3.14),

$$\int_t^{t+\Delta t}\int_w^e \frac{\partial}{\partial t}(\rho c T)\, dx dt = \int_t^{t+\Delta t}\int_w^e \frac{\partial}{\partial x}\left(k\frac{\partial T}{\partial x}\right) dx dt + \int_t^{t+\Delta t}\int_w^e S\, dx dt, \quad (3.14)$$

in the left-hand side in time and the right-hand side in space, reads,

$$\int_w^e (\rho c T - \rho^o c^o T^o)\, dx = \int_t^{t+\Delta t}\left(k\frac{\partial T}{\partial x}\bigg|_e - k\frac{\partial T}{\partial x}\bigg|_w\right) dt + \int_t^{t+\Delta t}(S_p T_p + S_c)\Delta x\, dt$$

(3.15)

The absence of upper index on the variables means they are being evaluated at the time level in which one wants to have the solution, that is, at $t + \Delta t$, while the superscript "o", means evaluated at the "old" time level, level in which all variables are known. The source term was linearized, which is the highest order this term can be expressed, since it will take part in a linear system of algebraic equations. S_P is evaluated at point P and may be function of T if the source term is not linear, situation in which it should be update during iterations. During the integration procedure, it should be exercised to include in the matrix coefficients as much as possible the relevant information of the source term and not left it entirely in the independent vector, in a lazy attitude [2]. Convergence is affected if the independent term in the linear system is to "heavy", leaving the system more explicit, as will be seen in a coming section. The interpolation function in time uses two time levels, thus, it is a first order approximation. Of course, one could apply a second order using three-time levels, if desired.

The reader is invited to dedicate special attention to what follows, since it will contain relevant details of the finite volume procedure. The integration of the left hand-side term in space, from "w" to "e" uses the temperature at the central point P, what assumes a linear variation inside the control volume. This integration reads,

$$M_P T_P - M_P^o T_P^o = \frac{1}{c_P} \left[\int_t^{t+\Delta t} \left(k \frac{\partial T}{\partial x} \bigg|_e - k \frac{\partial T}{\partial x} \bigg|_w \right) dt \right] + \int_t^{t+\Delta t} \left(S_p T_p + S_c \right) \Delta x \, dt,$$

(3.16)

in which M_P is the mass contained in the control volume and c_P stands for the specific heat at constant pressure evaluate at point P. Recall that the volume in this 1D problem is $(\Delta x.A)$ and the cross-sectional area $A = 1$.

The integration in time of the fluxes at the faces of the control volume would requires a function stating the variation of these fluxes from time level t to $t + \Delta t$, such that the integration can be performed. Since we are using two time levels the interpolation function is linear and one needs to decide in which position in the time interval will the fluxes be evaluated. It is a decision that must be made based on the physics of the problem, as commented shortly. This choice gives rise to the well-known formulations in time, namely, explicit, implicit and fully implicit. The superscript θ identifies the point in the time interval in which the fluxes are evaluated.

Therefore, removing the last integral of Eq. (3.16), it is obtained

$$M_P T_P - M_P^o T_P^o = \frac{1}{c_P} \left[\left(k \frac{\partial T}{\partial x} \bigg|_e^\theta - k \frac{\partial T}{\partial x} \bigg|_w^\theta \right) \right] \Delta t + \left(S_p T_p^\theta + S_c \right) \Delta x \Delta t, \quad (3.17)$$

in which when $\theta = 0$ gives rise to the explicit formulation, when $0 < \theta < 1$, the implicit formulation, and when $\theta = 1$, to the fully implicit formulation, to be presented shortly.

It is time now of defining the spatial interpolation, since we need to get rid of the temperature derivative in space. This means to specify how temperature changes inside the control volume, and it is worth to mention that this is the second integration required and, in this case, it uses central difference scheme, which assumes a linear variation in temperature. This is one situation in which finite differences are used as an auxiliary technique. Let's keep in mind that the interpolation function for diffusive problems (second order derivative terms) poses no instabilities difficulties, since linear interpolation always suffices. Thus, using central differencing for evaluation of the first derivative at the interfaces of the control volume, the interpolation functions are

3.7 Explicit, Implicit and Fully Implicit Formulations

$$\left.\frac{\partial T}{\partial x}\right|_e^\theta = \frac{T_E^\theta - T_P^\theta}{\Delta x_e}$$
$$\left.\frac{\partial T}{\partial x}\right|_w^\theta = \frac{T_P^\theta - T_W^\theta}{\Delta x_w} \tag{3.18}$$

Substituting Eqs. (3.18) into Eq. (3.17), it is obtained

$$\frac{M_P T_P - M_P^o T_P^o}{\Delta t} = \frac{1}{c_p}\left[\left.\frac{k}{\Delta x}\right|_e (T_E^\theta - T_P^\theta)\right]$$
$$- \frac{1}{c_p}\left[\left.\frac{k}{\Delta x}\right|_w (T_P^\theta - T_W^\theta)\right] + S_p T_P^\theta \Delta x + S_c \Delta x, \tag{3.19}$$

which, after some algebraic arrangement, it becomes

$$\frac{M_P T_P}{\Delta t} = \frac{1}{c_P}\left(\frac{k}{\Delta x}\right)_e T_E^\theta + \frac{1}{c_P}\left(\frac{k}{\Delta x}\right)_w T_W^\theta + \frac{1}{c_P}\left[-\left(\frac{k}{\Delta x}\right)_e - \left(\frac{k}{\Delta x}\right)_w\right] T_P^\theta$$
$$+ \frac{M_P^o T_P^o}{\Delta t} + S_p T_P^\theta \Delta x + S_c \Delta x \tag{3.20}$$

Using only two time levels, the determination of the temperature, T_P^θ, inside the interval Δt follows a linear equation, given by $T^\theta = \theta T + (1 - \theta)T^o$. For the derivative, a first order approximation is possible, as

$$\frac{\partial T}{\partial t} = \frac{T_P^{n+1} - T_P^n}{\Delta t} = \frac{T_P - T_P^o}{\Delta t}, \tag{3.21}$$

following the convention for the superscripts. This approximation was already used in Eq. (3.15).

Higher order interpolation may be employed, allowing to follow the transient with larger time steps and keeping fidelity, having in mind that the true transient requires a grid resolution study in time. Figure 3.7 depicts the three-time levels which can be used to have a second order interpolation in time. There is a general equation using three-time levels with different time steps sizes. For the equal step sizes, the approximation is given in the insert of Fig. 3.7.

3.7 Explicit, Implicit and Fully Implicit Formulations

3.7.1 Explicit Formulation

As Eq. (3.20) reveals, when $\theta = 0$, that it is possible to explicit the unknown temperature as function of all other temperatures known from the previous time

Fig. 3.7 Time levels for a second order approximation

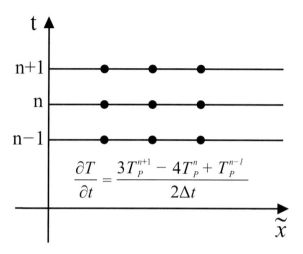

level. We will then have an equation for each control volume and these equations are uncoupled, so, the solution marches in time solving each equation, one-by-one, in each time level. There is no need of solving a linear system, since the equations are uncoupled. When all the equations were solved, it is jumped to a new time level and the process is repeated. Equation (3.20), applying the explicit formulation, takes the form

$$A_P T_P = A_e T_E^o + A_w T_W^o + (A_P^o - A_e - A_w + S_P \Delta x) T_P^o + S_c \Delta x, \quad (3.22)$$

in which the coefficients are

$$A_P = \frac{M_P}{\Delta t}; \quad A_P^o = \frac{M_P^o}{\Delta t}; \quad A_e = \frac{1}{c_P} \frac{k}{\Delta x}\bigg|_e; \quad A_w = \frac{1}{c_P} \frac{k}{\Delta x}\bigg|_w \quad (3.23)$$

It is well known that for 1D problems the condition for stability is to have the coefficient of T_P^o positive. If ρ, c_P and k are constant in the domain, $S_P = 0$, with a uniform mesh, the condition for the positivity is,

$$\frac{\alpha \Delta t}{\Delta x^2} \leq \frac{1}{2}, \quad (3.24)$$

in which α is the thermal diffusivity. To deep the understanding of the implications in violating the positivity of T_P^o, consider the solution of the 1D problem in the grid shown in Fig. 3.8, in which there are four elements and five grid nodes, characterizing a cell-vertex configuration. This was used just for simplicity in applying the Dirichlet boundary conditions.[1] The source term is made equal to zero and $\Delta x_e = \Delta x_w$. The

---

[1] In fact, the difference in cell-vertex and cell-center are mainly in the application of the boundary conditions, since T_1 and T_5 are known temperatures.

3.7 Explicit, Implicit and Fully Implicit Formulations

initial condition is $T = 0$ for the whole domain, when, suddenly, T_1 is raised to 100 and T_5 is kept in 0. We are not carrying the units of temperature, since it is immaterial for our purpose. The physics of this problem teach us that heat will start to penetrate in the domain by diffusion, heating up the whole domain until the steady state is reached. The steady state solution for this problem is, obviously, a linear profile, as depicted in Fig. 3.9, in which intermediate transient profiles are qualitatively shown.

Considering the explicit formulation Eq. (3.20) takes the form

$$T_P = (1 - 2r)T_P^o + rT_E^o + rT_W^o \tag{3.25}$$

Assuming $r = \alpha \Delta t / \Delta x^2 = 0.5$, for simplicity, but still satisfying the criterion given by Eq. (3.24), it results,

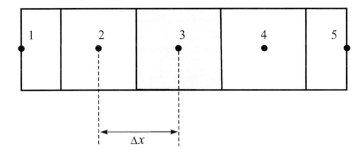

Fig. 3.8 Cell-vertex discretization

Fig. 3.9 Transient temperature profiles

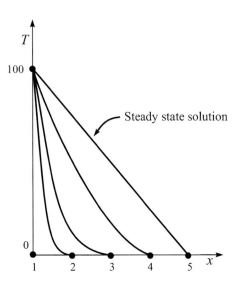

Table 3.1 Explicit solution. One dimensional problem

Time level	0	1	2	3	4	Steady state
Point 1	100	100	100	100	100	100
Point 2	0	50	50	62.5	62.5	75
Point 3	0	0	25	25	37.5	50
Point 4	0	0	0	12.5	12.5	25
Point 5	0	0	0	0	0	0

$$T_P = \frac{1}{2}\left(T_E^o + T_W^o\right) \tag{3.26}$$

Equation (3.26) is the general equation which applies for the three unknowns, as

$$T_2 = \frac{1}{2}\left(T_3^o + T_1^o\right)$$
$$T_3 = \frac{1}{2}\left(T_4^o + T_2^o\right)$$
$$T_4 = \frac{1}{2}\left(T_5^o + T_3^o\right) \tag{3.27}$$

Note that these equations are not coupled, since the right-hand side depend only on temperatures from the previous time level, all known. The procedure follows by marching in time with the possibility of changing the time step as the transient progresses. Table 3.1 presents the solution for four time levels, and it is observed that the heat flow advances one volume in each calculation. The numerical values of the temperature for the steady state solution obtained with the explicit formulation for this problem are 100, 75, 50, 25 and 0. This is the exact solution, and this happens because the interpolation function used is the exact solution of the steady state problem.

In this case it is said that the interpolation function is exact. The spatial temperature distribution during the transient is not correct, since its behavior is exponential in space and time, and the interpolation function is kept linear. To capture the transient properly, meshes refined in space and in time are required.

Inspecting Eq. (3.22), we conclude that stability was possible by keeping the coefficient of T_P^o positive. For simple problems, as the one under analysis, this can be proved using von Neumann analysis [3, 4]. An important practice, thus, in explicit schemes, is to keep this coefficient always positive. Accuracy, what requires very small time steps and spatial grids, is mandatory in numerical solutions. In explicit formulations the time step is limited by stability, and since this limit is quite restrictive, accuracy, related to the refinement of the time step, is almost always satisfied. The reader is invited to repeat this exercise with $r > 0.5$ and see what happens with the solution.

There is a general feeling that explicit formulation are not recommended due to its time step limitations, however, it depends, of course, on the physics under analysis. In fact, it is a formulation widely used when very fast transient should be captured, like

3.7 Explicit, Implicit and Fully Implicit Formulations

in car crashing, for example. In these problems the time step required is extremely small, and accuracy, therefore, is in general satisfied. There are disciplines, using finite element methods specialized in car crashing and other fast transient phenomena, named "explicit transient methods", dealing with such a kind of physics, what shows the importance of this formulation.

3.7.2 Fully Implicit Formulation

Opposed to the explicit, which calculates all fluxes on the previous time level, the fully implicit formulation, does that on the time level in which the solution is sought. Consequently, the approximate equation for each control volume becomes coupled with the neighboring ones on that time level. The general equation, Eq. (3.20) takes, then, the following form,

$$A_P T_P = A_e T_E + A_w T_W + A_P^o T_P^o + S_c \Delta x, \qquad (3.28)$$

in which

$$A_P = \frac{M_P}{\Delta t} + A_e + A_w - S_P \Delta x, \qquad (3.29)$$

with A_e and A_w keeping the same expressions. Notice now that, in the fully implicit formulation, it is desirable that the S_P term be negative to help the magnitude of the diagonal of the linear system. Linear systems with dominant diagonal are amenable to solution even by non-powerful solvers, such as line-by-line or even point-by-point iterative solvers. Since in general the formulation used is the fully implicit, one always should tries to linearize the source term with negative S_P, despite the fact that in complex mathematical model it is not always possible to do that, and some artificial negativity need be imposed [2]. Since the last two terms on the right-hand side of Eq. (3.28) are known, they can be lumped together, resulting in the linear system that must be solved to obtain the temperature distribution in the slab for that time level,

$$A_P T_P = A_e T_E + A_w T_W + B_P \qquad (3.30)$$

When solving a true transient iteratively, it should be kept in mind that the variable at the previous time level $\left(T_P^o\right)$ taking part in B_P can't be changed during the iterations, and should be modified only when the solution is considered converged (attending a specified tolerance) and it is advanced to a new time level. It seems a trivial comment, but it is a common mistake done by beginners.

On the other hand, when the interest is the steady solution only, we can follow the transient in a distorted manner, since makes no sense to solve the linear system with a tight tolerance in each time level if the solution is not need at those time levels.

What is done is just iterate the linear system a couple of times and jumping to a new time level changing the T_P^o, or others variable being solved. The time step should be as larger as possible, enough for maintaining convergence of the process trying to reach the steady state solution with a minimum CPU time. Unfortunately, there are no mathematical rules on how many iterations or size of the time steps should be used. What counts, in fact, is the experience of the user in simulating fluid flows. It is also obvious that a direct method for solving the linear system shouldn't be used, since there is no interest in accurate solutions along the distorted transient and a direct method would give us the exact solution of the linear system in each time level.

The question which arises is: why going through a distorted transient if it is possible to make $\Delta t = \infty$ and solve just one linear system and get the steady state solution? The reason is that for the solution of complex problems which involves a system of partial differential equations, convergence would not be achieved. Even using a fully implicit formulation for every equation of the system as, normally the full system is solved sequentially, some explicitness is always embodied in the solution procedure, what precludes the use of large time steps. Using $\Delta t = \infty$ may work for linear and simple problems. Doing that in Eq. (3.28) and Eq. (3.29), they result,

$$A_P T_P = A_e T_E + A_w T_W + S_c \Delta x \qquad (3.31)$$

with

$$A_P = A_e + A_w - S_P \Delta x, \qquad (3.32)$$

in which one sees that the term T_P^o disappeared, since the steady state solution doesn't depend on temperatures at the previous time levels, but only on boundary conditions. This is logical, because the time no longer takes part in the equation. Keep in mind that the steady state solution doesn't depend on how was the transient, any initial (estimated) condition, following any distorted transient will always reach the same steady state solution.

Returning to the one-dimensional diffusion problem, there is no longer the possibility of the negative coefficient for T_P^o in the fully implicit formulation. This formulation gives rise to a system of equations, since the equations are now coupled together. In Eq. (3.28), the temperatures T_E and T_W are being calculated at the same time level as T_P, which characterizes the coupling between them. For this simple equation under consideration, such a formulation is unconditionally stable, and the time interval is limited by accuracy. Note that this formulation is called fully implicit, because the values of the temperatures entering the diffusive flux calculation are made equal to the values at the end of the time interval.

Considering again $r = 1/2$ the resulting system of equations for this formulation is given by

3.7 Explicit, Implicit and Fully Implicit Formulations

$$T_2 = \frac{1}{4}\left((T_3 + T_1) + 2T_2^o\right)$$

$$T_3 = \frac{1}{4}\left((T_4 + T_2) + 2T_3^o\right),$$

$$T_4 = \frac{1}{4}\left((T_5 + T_3) + 2T_4^o\right) \tag{3.33}$$

which is the 3×3 linear system that must be solved. Observe that Eq. (3.28) could be written as,

$$A_P T_P = A_e T_E + A_w T_W + B_P, \tag{3.34}$$

with P equal to 2, 3 and 4, for the linear system expressed by Eq. (3.33) and reproduced below, with a slightly different nomenclature,

$$A_P^2 T_2 = A_e^2 T_3 + B_2$$
$$A_P^3 T_3 = A_e^3 T_4 + A_w^3 T_2 + B_3$$
$$A_P^4 T_4 = A_w^4 T_3 + B_4 \tag{3.35}$$

in which the boundary conditions are, of course, included in the B_2 and B_4 terms. In Eq. (3.35) a more rigorous notation for the coefficients was adopted, which will not be followed, but deserves to be explained while we are still at the beginning of the text. First, recognize that each of the equations forming the system (3.35) was obtained by applying the numerical method to a control volume of the mesh. Therefore, all coefficients that appear in an equation are coefficients belonging to that volume for which the equation was written, e.g., the coefficient A_P^3 is the central coefficient of volume 3, while A_e^3 and A_w^3 are coefficients connecting the volume 3 with their neighbors volumes. The superscript indicating the control volume will be omitted from now on, for simplicity, since there is a little chance of misunderstanding, since all coefficients in one equation are referred to that control volume.

The solution of the linear system given by Eq. (3.34) or Eq. (3.35), when solved, furnish the temperatures T_2, T_3, and T_4. It should also be recognized that the system given by Eq. (3.33) can be written in matrix form as

$$[A][T] = [B] \tag{3.36}$$

or, for the problem in consideration,

$$\begin{pmatrix} 1,00 & -0,25 & 0,00 \\ -0,25 & 1,00 & -0,25 \\ 0,00 & -0,25 & 1,00 \end{pmatrix} \begin{pmatrix} T_2 \\ T_3 \\ T_4 \end{pmatrix} = \begin{pmatrix} B_2 \\ B_3 \\ B_4 \end{pmatrix} \tag{3.37}$$

Note that the zeros that appear in the matrix of coefficients do not explicitly exist in Eqs. (3.33), because the form of writing these equations involves only the temperatures that are connected to the cell in question, and it was chosen to have just one connection in each side of the control volume. Considering T_4, since it has no connection with T_2, the connecting coefficient is, of course, zero. The same is true for T_2 with respect to T_4, what originates the zeros coefficients in the system (3.37). Now imagine that the grid would have not only 3 control volumes, but say, one hundred unknowns. It is easy to see that the coefficient matrix would have 100 rows by 100 columns. In each row, there would be only 3 nonzero coefficients, with the remaining 97 being zero on that row.

Anticipating the events of a later section of this chapter, it is worth remembering that when iterative methods are used to solve the linear system, we work only with the non-zeros of the matrix, while in direct solutions, such as Gauss elimination, for example, all the entries of the matrix take part in the operations. Since, in general, the matrices obtained when applying numerical methods are very sparse, it is advisable to use iterative methods to avoid operations with zeros. For the simple case shown here, with three equations and three unknowns, it is easy to solve the linear system directly by substitution. When the number of points increases considerably, efficient methods of solving linear systems must be employed. Note that the system (3.37) must be solved for each time level, because the problem under consideration is transient. If the steady state solution is of interest, it is enough to make $r = \infty$ (infinite advancing in time) and solve the resulting linear system. Again, for this problem, the numerical solution is equal to the analytical one for the reasons already stated. Recall that the alternative, of using $\Delta t = \infty$ may hardly work for complex systems of equation, as explained.

3.7.3 Implicit Formulation

In the implicit formulation the temperatures involved in the fluxes are neither evaluated at the beginning or at the end of the time step, but weighting them according to the interpolation function, repeated herein,

$$T_P^\theta = \theta T_P + (1 - \theta) T_P^o \qquad (3.38)$$

with the most famous algorithm being of the Crank-Nicolson, in which $\theta = 0.5$. It is pertinent note that it is enough θ be non-zero for the equations to be coupled, characterizing the implicitness between them. The implicit formulation travels from the explicit to the fully implicit. When θ approaches zero, the explicitness of the system increases, with less influence of the matrix coefficients, being the opposite when it approaches unity. It is common in the literature to call the formulation with $\theta = 1$ simply an implicit formulation and not fully implicit, as called here. The reason for this is that the majority of methods use $\theta = 1$ for stability reasons.

Figure 3.10 illustrates, for the three types of formulations, the connections between point P with its neighbors at the time level of the calculation and at the previous time. The figure shows that when there are connections between point P and the

3.7 Explicit, Implicit and Fully Implicit Formulations

neighbors points at the level in which the solution is sought, there will be always the need of solving a linear system. Its left to the reader as exercise to obtain the coefficients, like Eq. (3.33), for the implicit case. It will be seen that the coefficients are a sum of two parts weighted by θ.

Along the presentation of the formulations in time we made comments about the real and distorted transients. Maintaining the time coordinate in the equations is of utmost importance, since, even if only the steady state solution is of interest, the time is used as a physically consistent relaxation parameter. In fact, makes no sense to remove the time from the equations. Figure 3.11depicts the four alternatives for marching in time pointing out the important characteristics.

Fig. 3.10 Explicit, implicit and fully implicit formulations

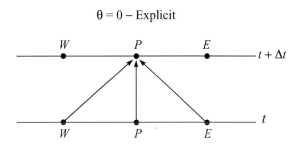

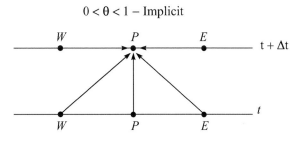

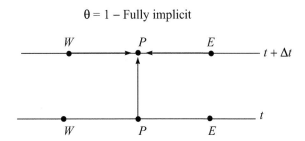

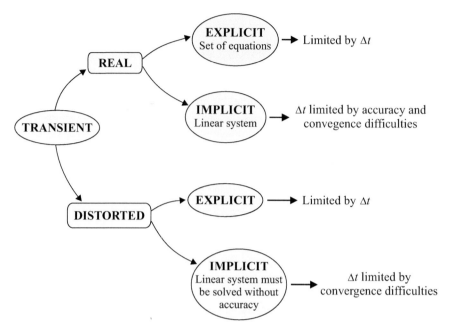

Fig. 3.11 Alternatives for advancing in time

3.8 Linearization of the Source Term

It may happen that significant terms which does not fit as advective or diffusive are included in the source term. The easiest move would be to keep them as a constant and updating it each iteration, without an analysis about the possibility of linearizing it. Linearization reinforces the diagonal of the linear system and may prevent divergence, avoiding that the B term dominates the iterative procedure. As an example, a typical flow in which important effects are included in the source term is high-rotation flow. In such flows, the centripetal force, dependent on the azimuthal velocity in the radial equation of motion, and the Coriolis force dependent on the radial velocity in the azimuthal equation of motion, are part of the source term. These source terms are particularly difficult to handle because the variable that appears in the source term of one equation is the principal variable of the other equation. The source term in this case must be treated with extreme care, because the centrifugal and Coriolis effects are determining factors in the phenomenon [5]. Source terms of this nature are more complex than the case we are considering herein, where the variable that appears in the source term is the variable of the equation itself.

The first rule to be followed is to take the source term as implicit as possible. However, often, depending on the importance of the source term, linearization alone is not enough, and it may be necessary to update it more frequently than the rest of the coefficients. Our goal is to obtain a linearization of the type

3.8 Linearization of the Source Term

$$S = S_P T_P + S_c, \tag{3.39}$$

in which it would appropriate to have the coefficient S_P negative. Notice from Eq. (3.39) that S_P is the slope of the straight line in a given point in a $S x T$ plot. Considering the function $S x T$, there are two possibilities of behavior of S with T. The first one is when the tangent is naturally negative, characteristic presented by most physical problems. In this case, the recommended method of linearization [2] is to expand the source term in Taylor series, as

$$S = S^* + \left. \frac{dS}{dT} \right|_P^* (T_P - T_P^*), \tag{3.40}$$

and finding S_P and S_C. The second, is when the $S x T$ behavior has positive slope. In this case, it is necessary to artificially create a linearization with negative S_P. It is not difficult to understand that this process is only possible through an increase, also artificial, in S_C, which will imply in an increase in the independent term of the linear system. The following examples, taken from [2], clarify. Let the source terms be given by

$$S = 5 - 4T \tag{3.41}$$

$$S = 5 - 4T^2 \tag{3.42}$$

The first one, by visual inspection or through the series expansion gives, $S_P = -4$ and $S_C = 5$, while for the second expression, using the series expansion gives $S_P = -8T_P^*$ and $S_C = 5 + 4(T_P^*)^2$. To exemplify the creation of an artificial negative S_P, consider

$$S = 3 + 7T \tag{3.43}$$

One possible linearization is $S_P = -2$ and $S_C = 3 + 9T_P^*$. There are others, of course, but all of them may impair convergence, and the trade-off in these cases is the improvement of the diagonal of the matrix (increase the negativity of S_P) against the increase in S_C, which may slow down convergence. It is not easy, if not impossible, to determine a good parameter in this situation because of the numerous factors that act concurrently in the process. It is always possible, of course, to make $S_P = 0$. This practice, however, does not utilize the possibilities of improving the numerical scheme via implicit treatment of all or part of the source term. For real problems, with several partial differential equations and different source terms, it is difficult to foresee the gains. The important is to be aware of the influence the source term does on the linear system.

3.9 Boundary Conditions

3.9.1 Balances for the Boundary Volumes

The natural way of applying boundary conditions in a finite volume method is to perform balances of the quantity in the volumes at the boundary, exactly as done for the internal control volumes. incorporating the boundary conditions in the balance. Therefore, continuing with our 1D transient problem, consider Fig. 3.12, in which a boundary control volume is shown for a cell-center construction. The flux appearing in the figure can be given as boundary condition or can be calculated if a Dirichlet condition is prescribed. Using a fully implicit formulation, the discretized equations reads,

$$\frac{M_P T_P - M_P^o T_P^o}{\Delta t} = \frac{q_f''}{c_p} - \frac{1}{c_p}\left(\frac{k}{\Delta x}\right)_e (T_P - T_E) \qquad (3.44)$$

in which the flux, expressed by the derivatives in our general equation, Eq. (3.17), at the left boundary (w), was substituted by q_f''. We can now work on the three boundary conditions for the diffusion problem.

1. Prescribed Boundary Condition (Dirichlet)

    In this case the flux can be calculated by

$$q_f'' = k_f \frac{T_f - T_P}{\Delta x_f} \qquad (3.45)$$

    in which T_f is the prescribed temperature at the boundary. The flux is introduced in Eq. (3.44) and the final equation obtained.

2. Prescribed Heat Flux (Neumann)

    In this case just plug into Eq. (3.44) the known value of the flux.

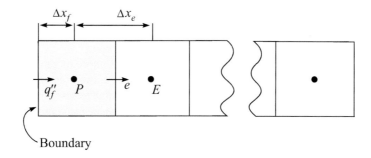

Fig. 3.12 Boundary control volume. Cell-center

3.9 Boundary Conditions

3. Convection Boundary Condition (Robin)

This condition tells us that the heat flux by convection enter the domain by diffusion, as

$$q''_f = h(T_\infty - T_f) = k_f \frac{T_f - T_P}{\Delta x_f} \tag{3.46}$$

in which h is the convection heat transfer coefficient and T_∞ the ambient temperature. Isolating T_f and substituting in any of the two equations, the flux is

$$q''_f = \frac{h}{1 + \frac{h \Delta x_f}{k_f}} (T_\infty - T_P), \tag{3.47}$$

which should be introduced in Eq. (3.44) to obtain the final equation in the form,

$$A_P T_P = A_e T_E + B_P, \tag{3.48}$$

3.9.2 Using Fictitious Volumes

It is also possible to use fictitious volumes (or points) to help the application of the boundary conditions. In this approach all volumes of the domain "feel" as an interior control volume, and the procedure for obtaining the approximate equations is, therefore, the same as for the real interior control volumes. The equations for the fictitious points are constructed via boundary conditions.

Figure 3.13 shows in dashed lines the fictitious control volume, arbitrarily created for applying the boundary conditions. The practice of using fictitious volumes is attractive and easy to apply. The disadvantage is the creation of new unknowns, increasing the size of the linear system, a situation that gets worse as the size of the problem increases. In a 1D problem with 1,000 unknowns, we will have only two fictitious volumes, representing, therefore, 0.2% of the total number of volumes. In a 2D situation, also with 1,000 unknowns, in a mesh of 33 × 33, approximately, we will have 132 fictitious volumes, representing 13.2% of the total. In three dimensions, for 1000 unknowns the mesh will be 10 × 10 × 10, with 600 fictitious volumes, a significant increase in the number of equations of the linear system. For a simple practical situation of a 3D mesh with 30 × 30 × 30 volumes, we would have an increase of about 20% in the number of equations.

With the creation of the fictitious volumes, we must create the equations for these volumes as a function of the existing boundary conditions. It is natural to write these equations in the same form as the equations for the internal volumes, such as

$$A_P T_P = A_e T_E + B_P, \tag{3.49}$$

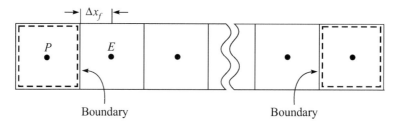

Fig. 3.13 Fictitious boundary volume

in which, of course, the W grid point doesn't exist. The coefficients of the above equation depend upon the type of boundary condition. For prescribe temperature one has,

$$T_f = \frac{T_P + T_E}{2}; \quad A_P = 1; \quad A_e = -1; \quad B_P = 2T_f \tag{3.50}$$

For a prescribed heat flux, the boundary condition and coefficients are,

$$q_f'' = -k_f \frac{T_E - T_P}{2\Delta x_f}; \quad A_P = 1; \quad A_e = 1; \quad B_P = \frac{2q_f'' \Delta x_f}{k_f}, \tag{3.51}$$

and for the domain exchanging heat with the surrounding environment by convection, the boundary conditions and coefficients are,

$$h\left(T_\infty - \frac{T_P + T_E}{2}\right) = k_f \frac{(T_P - T_E)}{2\Delta x_f}$$

$$A_P = \frac{h}{2} + \frac{k_f}{2\Delta x_f};$$

$$A_e = \frac{k_f}{2\Delta x_f} - \frac{h}{2};$$

$$B_P = hT_\infty \tag{3.52}$$

It is worth to mention that the approximated equations for the fictitious points, which are new unknowns, irrespectively if the problem is 1D, 2D or 3D, are found using only two grid nodes, the fictitious and the neighbor internal control volume. The additional unknowns are not, necessarily, added to the linear system and can be solved separately in an iterative way. Firstly, based on estimate values for all points, internal and fictitious, the interior unknowns are solved and, secondly, in a sweep, the values of all fictitious points are determined, one-by-one, and the process goes on until a specified tolerance is achieved. For nonorthogonal grids at the boundaries this single connection of a fictitious point and the interior is not possible, and the recommended approach is to perform balances at the boundary control volumes, as seen in Sect. 3.9.1.

3.9.3 About Boundary Conditions in Cell-Vertex

We have seen the application of boundary conditions for cell-center methods, which means not having unknowns over the boundary and no half or a quarter of volumes inside the domain. The application of boundary conditions for cell-vertex construction deserves few words. Consider the 1D grid for the 1D diffusion problem under discussion, shown in Fig. 3.14 in which the element as well as the control volume are depicted. In this cell-vertex construction of the control volume, the unknowns lie on the boundary creating a half-control volume.

Concerning the application of a Dirichlet boundary condition, since the unknown lies on the boundary, it is usual to remove this equation from the linear system. In this case, the apparent advantage of not being necessary to create an equation for the boundary volume, since T_f is known, translates into the non-observance of the conservation, since, for the boundary half-volume, a balance was not done to obtain the approximated equation.

Figure 3.15 depicts the boundary with half control volumes in gray, in a 2D region, that will not satisfy the conservation if all boundary conditions are of the Dirichlet type. For Neumann boundary conditions the flux at the boundary is given, and prescribing fluxes is the natural boundary condition for the finite volume method.

The conservation at the boundaries for Dirichlet boundary conditions in cell-vertex can be circumvented when the fluxes are calculated for the elements and then assembled to obtain the equation for the control volume, as in EbFVM. A conservation equation is created for the boundary volumes using the prescribed boundary condition. We will be back to this topic when cell-vertex method for unstructured grids will be presented.

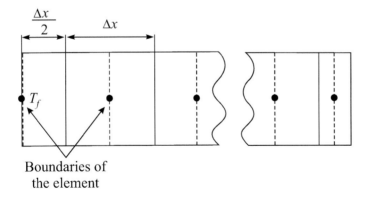

Fig. 3.14 Dirichlet boundary condition. Cell-vertex

Fig. 3.15 Non-conservative region in 2D

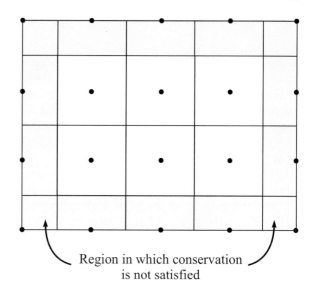

Region in which conservation is not satisfied

3.10 Discretization of the 3D Diffusion Equation

There will be nothing new in this section, just the presentation, for the sake of completeness, of the 3D diffusion equation, given by

$$\frac{\partial}{\partial t}(\rho T) = \frac{\partial}{\partial x}\left(\frac{k}{c_p}\frac{\partial T}{\partial x}\right) + \frac{\partial}{\partial y}\left(\frac{k}{c_p}\frac{\partial T}{\partial y}\right) + \frac{\partial}{\partial z}\left(\frac{k}{c_p}\frac{\partial T}{\partial z}\right) + S \qquad (3.53)$$

The same discretization that was done for the diffusion term in the x coordinate will now be done for the other two axis, y and z. Figure 3.16 illustrate the 3D control volume with the six connections to the neighboring ones, in which the cardinal points, as before, are used added with the Front and Back control volumes.

The integration using a fully implicit formulation, doing exactly what was done for the 1D problem, gives

$$A_P T_P = A_e T_E + A_w T_W + A_n T_N + A_s T_S + A_f T_F + A_b T_B + B_P \qquad (3.54)$$

with the following coefficients,

$$A_e = \frac{1}{c_P}\left.\frac{k}{\Delta x}\right|_e \Delta y \Delta z \qquad (3.55)$$

$$A_w = \frac{1}{c_P}\left.\frac{k}{\Delta x}\right|_w \Delta y \Delta z \qquad (3.56)$$

3.10 Discretization of the 3D Diffusion Equation

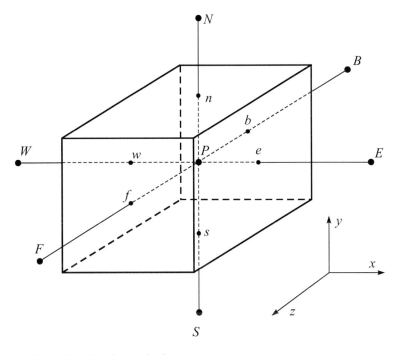

Fig. 3.16 Three-dimensional control volume

$$A_n = \frac{1}{c_p} \frac{k}{\Delta y}\bigg|_n \Delta x \Delta z \tag{3.57}$$

$$A_s = \frac{1}{c_p} \frac{k}{\Delta y}\bigg|_s \Delta x \Delta z \tag{3.58}$$

$$A_f = \frac{1}{c_p} \frac{k}{\Delta z}\bigg|_f \Delta x \Delta y \tag{3.59}$$

$$A_b = \frac{1}{c_p} \frac{k}{\Delta z}\bigg|_b \Delta x \Delta y \tag{3.60}$$

$$B_P = A_P^o T_P^o + S_c \Delta x \Delta y \Delta z \tag{3.61}$$

$$A_P^o = \frac{M_P^o}{\Delta t} = \frac{\rho^o}{\Delta t} \Delta x \Delta y \Delta z \tag{3.62}$$

$$A_P = A_e + A_w + A_n + A_s + A_f + A_b + \frac{M_P}{\Delta t} - S_P \Delta x \Delta y \Delta z \tag{3.63}$$

To build an adequate matrix, which can accept even weak iterative solvers, taking care of the diagonal is recommended. Besides the diagonal dominance, care should be exercised to have all the connecting coefficients positive. In finite volume techniques, in most of the algorithms, this is accomplished automatically. The secret is to create numerical schemes which follows the physics and not merely artificial numerics. One can appreciate in Eq. (3.63) other two relevant points, being one of them the required negativeness of S_P or, if not possible, to keep it small compared with the other terms. The second one is confirming the importance of keeping the coordinate time in the equations, even for steady state calculations. Using small time steps increases the value of $M_P/\Delta t$, augmenting the diagonal terms of the matrix represented by A_P, helping the solution of the linear system. The important fact is that this is a controlled parameter by the numerical analyst. In Chap. 4 this matter will be considered again.

3.11 Structure of the Matrix of Coefficients

The structure of the matrix of coefficients resulting from the numerical approximation has important influences in the choice of the method for solving the linear system.

First, remember that, whether the problem is one, two, or three-dimensional, the result is always a linear system. The coefficient matrix changes its structure according to the dimension of the problem, the ordering of the elementary volumes and the interpolation function. In the formulations shown, with the volumes numbered in order along the coordinates, the matrix is tridiagonal for one-dimensional problems, pentadiagonal for two-dimensional, and heptadiagonal for three-dimensional situations.

To understand the reason for this structure, it is enough to recall that in the discretization of the diffusion equation, the connection of point P with its neighbors appeared when numerically approximating the fluxes at the interfaces. Since we used a central-difference approximation, only the adjacent volumes participated in this approximation. As the ordering of the volumes is in sequence, if the problem is 1D, only two neighbors take part, resulting in an equation with only three terms. Therefore, the structure of the matrix is tridiagonal, with the diagonals closed to each other, because of the ordering. In the same 1D problem, if the approximation of the fluxes would involve two points in front and two points behind, we would have a pentadiagonal matrix, with the all five diagonals close to each other. In a 2D problem with sequential ordering and using one points in each side of the 2D volume, we would again have a pentadiagonal matrix, but now with three diagonals together and another two apart. Figure 3.17 shows these structures for the one and two-dimensional situations.

It is possible to use more neighboring points to establish the link of the central volume P with its neighbors. There are methods that approximate the fluxes using polynomials involving more points. If this is done, the number of non-zeros in that row changes. The limit would be to use an approximation in which the volume P would be connected to all other volumes in the domain, which would give a full matrix of coefficients. High order schemes are prone to oscillation easier than schemes involving just few points, raising the compromise among stability and size of the grid.

3.11 Structure of the Matrix of Coefficients

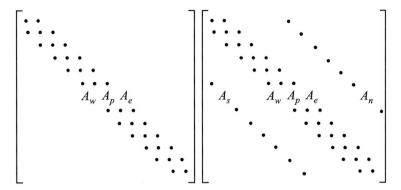

Fig. 3.17 Matrix structure. 1D and 2D

Since in an engineering problem we need to refine the grid to capture the physics, perhaps, it would be advantageous to have simple and stable schemes, reducing the truncation errors via refinement instead of with higher order approximations.

Another important detail about the matrix is its sparsity. For the 1D problem, with only two neighbors used in the interpolation, considering that the mesh has 100 volumes, we have a matrix with 10,000 entries and, in each row of just 3 non-zeros and 97 zeros, that is, the matrix is highly sparse. The sparsity index of the matrix influences in the choice of the method used to solve the linear system. If direct solution methods are employed, they require the handling of the non-zeros, greatly influencing the computation time. Quantifying the sparsity of the matrix can be done by the sparsity index, defined by

$$\left[1 - \left(\frac{number\ of\ non-zero\ entries}{total\ number\ of\ entries}\right)\right]$$

Table 3.2 presents the sparsity index for three different meshes, showing that for a 40 × 40 mesh, which is a coarse one in practical applications, the sparsity index is close to 1. The same table shows the storage required when the matrix is full and when only the non-zeros are considered. Logically, a full matrix has zero sparsity index.

Table 3.2 Sparsity index of a matrix

Grid	10 × 10	20 × 20	40 × 40
Total number of entries	1.10e4	16.10e4	256.10e4
Non-zeros	500	2000	8000
% of non-zeros	5	1.25	0.312
Sparsity index	0.95	0.9875	0.9968
Required memory (Kb)	40/2	640/8	10,240/2

The structures discussed so far is valid for numerical approximations using structured meshes, that is, meshes whose volumes always have the same number of neighbors. In unstructured discretization we can have different numbers of neighbors for each volume, giving matrices that are not tri, penta or heptadiagonal, but rather have a variable band, the so-called skyline matrices. Skyline matrices have special storage algorithms, SKS (skyline storage) in order to save memory, not storing the whole matrix. If the matrix is symmetric, as in finite element method using the weak form of the differential equation, just the upper diagonal is stored. The solution methods for linear systems with matrices of this nature are more elaborate. The use of unstructured discretization will be discussed in Chap. 13.

3.12 Handling Non-linearities

Numerical approximation of a linear partial differential equation gives rise to a linear system of equations whose matrix has constant coefficients. When the problem is nonlinear, again the approximation gives rise to a linear system of equations, but this time the matrix contains variable-dependent coefficients and must therefore be updated over iterations.

In the case of the heat conduction problems discussed in this chapter, common nonlinearities are the dependence of k with T and a possible nonlinearity in the source term. In heat convection problems, which involve the solution of the Navier–Stokes equations, important nonlinearities appear. In all cases the equation is linearized, transferring the nonlinearity to the coefficient matrix, to be updated in the next iteration. It is also possible to use a Newton-like method creating the Jacobian matrix and solving for the variation of the unknowns. The method continues to be iterative, since the Jacobian matrix must be updated if the problem is non-linear. If the problem is linear, this class of methods finds the solution in one outer iteration.

Note that even when considering only one differential equation, and solving the linear system by a direct method, the nonlinearity introduces an additional iterative level in the process, which is precisely the matrix update. When the solution method of the linear system is iterative, we may misunderstand the iterations due to the nonlinearities with those of the solution method of the linear system. These two loops of iteration should be not misinterpreted. They may be done in the same or separated iterative loops, depending on the problem. The conception of efficient iterative loops assumes importance when nonlinear and coupled systems of equations are solved, as is the case of CFD calculations. The physics will dictate the emphasis to give in each iterative loop, if on the coupling or on the nonlinearities.

The analyst's experience and familiarity with the physical problem will allow him to decide on the number of times to update the matrix compared to the iterations of the solution method.

3.13 Relevant Issues When Discretizing the Equations

The option made, in this text, to obtain the approximate equations was the finite volume method, i.e., the central issue is the conservation of the property at the discrete level. We also saw that the structure of the resulting matrix must respect certain criteria, like diagonal dominance and positivity of the coefficients, so that iterative methods can be applied. In the following, these rules [2] will be presented and discussed so that the reader can keep them in mind when developing his numerical method.

3.13.1 Positivity of Coefficients

The positivity of the coefficients is of fundamental importance for obtaining physically consistent solutions. To aid in this reasoning, consider a two-dimensional problem whose temperatures of neighboring volumes of P are greater than the temperature of volume P. Now imagine that the connection coefficients of P with its neighbors are negative and A_P is positive. The correct physics of the problem requires the increase of T_P. By the expression

$$A_P T_P = A_e T_E + A_w T_W + A_n T_N + A_s T_S + B_P, \tag{3.64}$$

which is the approximate equation for point P, there is no such guarantee, if the coefficients are negative. The positivity of the coefficients also greatly helps the overall performance of the method, because the central coefficient, being the sum of the connection coefficients with the neighbors, will have diagonal dominance, and any iterative method can be employed. It is also important to emphasize that the existence of negative coefficients does not always indicate that the solution will be incorrect or physically inconsistent. It is possible to have numerical approximations with negative coefficients converging to correct solutions, provided that the approximation is consistent, as already defined. In these cases, the penalty comes from the need for more robust methods for solving the linear system. The possibility of the solution divergence is, therefore, also strongly related to the use of methods not robust enough for certain matrices.

3.13.2 Fluxes Continuity at Interfaces

Flux continuity at the interfaces says that the flow of any property (advective or diffusive), leaving a given control volume, should be calculated in the same way as the flow entering the neighboring control volume. If this does not happen, we will have generation/disappearance of the property at the interface, obviously changing

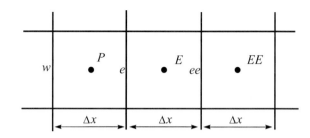

Fig. 3.18 Flux continuity at the interfaces

the value of the function locally. When interpolation functions are employed that do not use the same polynomial fitting points for the flow leaving and entering the neighboring control volume, non-conservation at the boundary may appears.

Using the equations in nonconservative form also entails problems for conservation at the elementary volume level. The advective terms, present on the left-hand side of Eq. (2.64), can be written in the following forms: conservative

$$\frac{\partial}{\partial x}(\rho u \phi) + \frac{\partial}{\partial y}(\rho v \phi) + \frac{\partial}{\partial z}(\rho w \phi) \qquad (3.65)$$

and non-conservative

$$\rho u \frac{\partial \phi}{\partial x} + \rho v \frac{\partial \phi}{\partial y} + \rho w \frac{\partial \phi}{\partial z} \qquad (3.66)$$

Consider the first term of the conservative form being computed at the control volumes shown in Fig. 3.18. Integrating this term in the volume centered at P with dimensions Δx, it is obtained

$$\int \frac{\partial}{\partial x}(\rho u \phi)\, dxdy = \int [(\rho u \phi)_e - (\rho u \phi)_w]\, dy = \dot{m}_e \phi_e - \dot{m}_w \phi_w \qquad (3.67)$$

Applying the same procedure for the control volume centered in E with the same dimension, it is obtained

$$\int \frac{\partial}{\partial x}(\rho u \phi)\, dxdy = \int [(\rho u \phi)_{ee} - (\rho u \phi)_e]\, dy = \dot{m}_{ee} \phi_{ee} - \dot{m}_e \phi_e \qquad (3.68)$$

Finally, let's consider a control volume centered in "e" with dimension $2\Delta x$. The result is,

$$\int \frac{\partial}{\partial x}(\rho u \phi)\, dxdy = \int [(\rho u \phi)_{ee} - (\rho u \phi)_w]\, dy = \dot{m}_{ee} \phi_{ee} - \dot{m}_w \phi_w \qquad (3.69)$$

Summing up Eqs. (3.67) and (3.68) results in Eq. (3.69), demonstrating that at the interfaces of control volumes P and E there is no generation or disappearance of

3.13 Relevant Issues When Discretizing the Equations

the property ϕ. That is, the flux leaving P enters E, and we say that the scheme is conservative.

Now, let's integrate the term

$$\rho u \frac{\partial \phi}{\partial x} \qquad (3.70)$$

For the control volume centered in P, the integration of the flux reads,

$$\int \rho u \frac{\partial \phi}{\partial x} dxdy = \int \rho_P u_P (\phi_e - \phi_w) \, dy = \rho_P u_P \phi_e \Delta y - \rho_P u_P \phi_w \Delta y \qquad (3.71)$$

For the control volume centered in E,

$$\int \rho u \frac{\partial \phi}{\partial x} dxdy = \int \rho_E u_E (\phi_{ee} - \phi_e) \, dy = \rho_E u_E \phi_{ee} \Delta y - \rho_E u_E \phi_e \Delta y \qquad (3.72)$$

And, finally, for the centered in "e", the integral is,

$$\int \rho u \frac{\partial \phi}{\partial x} dxdy = \int \rho_e u_e (\phi_{ee} - \phi_w) \, dy = \rho_e u_e \phi_{ee} \Delta y - \rho_e u_e \phi_w \Delta y \qquad (3.73)$$

Equation (3.73) clearly shows that the conservation was broken. There are, thus, creation and disappearance of the property at the interfaces, which may destroy completely the solution, since any generation or disappearance of a physical quantity changes the local profile of the variable. We should, therefore, always to develop schemes which are conservative at discrete level. This issue was already discussed, but this section demonstrates what happens when conservation is not obeyed. Of course, to have this condition imposed on the numerical scheme the concept of control volume must be used.

For the diffusion terms the same comments apply. We should always have these terms in the form,

$$\frac{\partial}{\partial x}\left(\Gamma \frac{\partial \phi}{\partial x}\right) \qquad (3.74)$$

and not in the form,

$$\Gamma \frac{\partial^2 \phi}{\partial x^2} + \frac{\partial \phi}{\partial x} \frac{\partial \Gamma}{\partial x} \qquad (3.75)$$

The integration of Eqs. (3.74) and (3.75) will show that conservation is not respected. The physical properties should be evaluated at the same local in which the fluxes are evaluated, that is, at the interfaces.

3.13.3 Linearization of Source Term with S_P negative

The importance of having negative S_P in the linearization of the source term has been discussed previously in considerable depth. It suffices here to recall that negative S_P increases the value of A_P, giving the matrix a stronger diagonal dominance, which is extremely beneficial for convergence. On the other hand, the need for a negative S_P is in accordance with physical processes, since these processes are in general limited, which would not be the case with a positive S_P. The expression of the A_P coefficient below shows that the negative S_P helps in the diagonal dominance,

$$A_P = \sum A_{nb} - S_P \Delta V \tag{3.76}$$

The possibility of creating numerical schemes where the central coefficient is equal, at least, to the sum of the influence coefficients of the neighboring volumes is a factor that also contributes to satisfy Scarborough's criterion. The equations obtained, so far, with finite volume discretization satisfy this criterion.

3.13.4 Truncation Errors

The numerical solution of a differential equation, being discrete, has approximation errors that distance it from the exact solution. These errors, called truncation errors (TE), can be determined by using the Taylor series expansion of the function around a point to obtain the numerical expressions of the derivatives of the differential operator. This is exactly the finite difference method, which, when replacing the differential operator by finite differences, puts clear the truncation errors involved. As an example, take the one-dimensional transient diffusion equation, given by

$$\frac{\partial T}{\partial t} = \alpha \frac{\partial^2 T}{\partial x^2} \tag{3.77}$$

The truncation error of a numerical approximation depends logically on the order of approximation chosen for the derivatives of the differential operator. For this, let us use Fig. 3.19, with the following Taylor series expansions of the temperature around point P

$$T_E = T_P + \left.\frac{\partial T}{\partial x}\right|_P \Delta x + \left.\frac{\partial^2 T}{\partial x^2}\right|_P \frac{\Delta x^2}{2!} + \left.\frac{\partial^3 T}{\partial x^3}\right|_P \frac{\Delta x^3}{3!} + \cdots + \tag{3.78}$$

$$T_W = T_P - \left.\frac{\partial T}{\partial x}\right|_P \Delta x + \left.\frac{\partial^2 T}{\partial x^2}\right|_P \frac{\Delta x^2}{2!} - \left.\frac{\partial^3 T}{\partial x^3}\right|_P \frac{\Delta x^3}{3!} + \cdots - \tag{3.79}$$

3.13 Relevant Issues When Discretizing the Equations

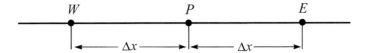

Fig. 3.19 Truncation error analysis

From these equations we can find the numerical approximations of the partial derivatives. Using Eqs. (3.78) and (3.79), we find, respectively,

$$\left.\frac{\partial T}{\partial x}\right|_P = \frac{T_E - T_P}{\Delta x} + O(\Delta x) \tag{3.80}$$

$$\left.\frac{\partial T}{\partial x}\right|_P = \frac{T_P - T_W}{\Delta x} + O(\Delta x), \tag{3.81}$$

which are the forward and backward numerical approximations of the first-order derivative. Note that the truncation errors are of the order of Δx. Adding Eq. (3.78) with Eq. (3.79), we obtain,

$$\left.\frac{\partial^2 T}{\partial x^2}\right|_P = \frac{T_E + T_W - 2T_P}{\Delta x^2} + O(\Delta x)^2, \tag{3.82}$$

which is the numerical approximation for the second-order derivative using central differencing. In this case, the truncation error is of the order of Δx^2.

Working with the Taylor series expansions of the function, it is possible, using more terms of the series, to represent derivatives of any order and to determine the truncation errors of any numerical approximation. Of course, the higher the order of the derivative, and according to the desired order of truncation error, the more points are needed around P. The approximations given by Eqs. (3.80)–(3.82) suffices for our example. The interested reader can consult [3], where a complete table of numerical approximations of derivatives are available. Using Eq. (3.80), to obtain a similar equation for the first order expansion in time,

$$\left.\frac{\partial T}{\partial t}\right|_P = \frac{T_P - T_P^o}{\Delta t} + O(\Delta t), \tag{3.83}$$

and using a 2nd order approximation of the second order derivative in space, Eq. (3.77) results,

$$\frac{\partial T}{\partial t} - \alpha \frac{\partial^2 T}{\partial x^2} = \frac{T_P - T_P^o}{\Delta t} - \alpha \frac{T_E + T_W - 2T_P}{\Delta x^2} + TE, \tag{3.84}$$

which can be written, relating the continuous and discrete operators, as

$$\pounds(\phi) = L_N(\phi) + TE, \tag{3.85}$$

in which TE are the truncations errors in each coordinate direction, given by, in this example,

$$TE = O(\Delta t, \Delta x^2) \tag{3.86}$$

Based on this brief presentation of the discrete approximation of a continuous operator we could see that any order of approximation can be used, with any variable grid dimensions, and the resulting error can be determined using Taylor series. Just for a recap, recall that when integrating the heat diffusion equation using finite volumes, the second integration, after the conservation was satisfied, was done using a central differencing scheme (Eq. (3.18)). It was not mentioned the truncation error introduced in our interpolation at that time, and we know now it is of second order (Δx^2), if the grid is uniform. The first three exercises of this chapter practice a little bit with finite differences.

3.13.5 Consistency, Stability and Convergence

In general, practical problems of interest in engineering and physics give rise to complex systems of equations about whose mathematical behavior little is known. When the problem governed by a single equation and linear, there are mathematical tools that can prove whether a given numerical approximation is stable and convergent. When working with systems of nonlinear equations, usually solved sequentially, where delicate couplings are present, it is very difficult to prove mathematically that a numerical approximation is stable and convergent. It would be a wonderful gift to users of numerical methods if mathematical analysts could provide the conditions (mesh and time interval size for types of schemes) for the numerical approximations of coupled, nonlinear problems to be stable and convergent.

Not having these parameters is why the task of performing numerical simulations, besides requiring a deep knowledge of the physics, requires experience to find the parameters that lead the iterative process to convergence. One of the fundamental requirements of a numerical approximation is that it reproduces the differential equation when the spatial and temporal mesh sizes tend to zero. That is, the truncation errors must tend to zero when the mesh tends to an infinite number of points. A numerical approximation that possesses this characteristic is said to be **consistent**. In short, the discretized equations should tend to the differential equations when the space and time discretization tend to zero.

Apparently, this is an obvious question, but there are approximations in which the truncation errors increase with mesh refinement [3]. Fortunately, any numerical model developed from the equations in conservative form using finite volumes is consistent.

Another required characteristic is that the numerical solution obtained be the exact solution of the discretized equations. This means that the solution obtained solving the linear systems is the exact solution of these linear systems. It sounds strange, but it is possible to solve a linear system and not to get the exact solution of this system. This is called **stability**. Here, several factors come into play, such as machine round-off errors, which keep growing and may destabilize the solution; difficulties in treating couplings between variables, causing some variables to evolve faster than others, causing instabilities etc. The issue of stability is the most serious problem in obtaining a numerical solution, precisely because of the lack of knowledge of the mathematical characteristics of the approximations.

Consistency and stability are necessary and sufficient conditions for **convergence**. The numerical solution is convergent when it is stable and tends towards the solution of the differential equations when the grid in space and time is refined.

3.14 Conclusions

Throughout this chapter we have seen that the construction of a finite volume method follows a series of steps over which the numerical analyst has control. All operations were under control. This intimacy with the method has benefits, because the analysis of the numerical results can be done in light of the balances. Work at the level of discrete control volumes helps in detecting errors in the solution. For example, it is usual among finite volume practitioners, when the error persists to choose some control volumes in the solution domain and check, for example, the mass conservation. This can be done using very coarse meshes, it is inexpensive and extremely helpful. All these strategies in finding code mistakes or code misconception requires qualified users with physical and numerical background. These are key requirements to perform a useful simulation. If the control volume doesn't exist, this error finding procedure, of course, cannot be done.

It is worth noting that, although we have dealt in this chapter with a one-dimensional conduction problem, while practicing the integration using finite volumes, there was outlined a large quantity of details whose concepts apply for any numerical model developed using finite volumes. For example, the concepts of explicit, implicit and fully implicit formulations apply for any equation or system of equations. Cell-center and cell-vertex methods, structure of the coefficient matrix, positivity of coefficients, non-linearities, truncation errors etc. were topics described which apply in general for finite volume methods.

The coming chapter deals with the solution of the linear systems obtained with the discretization, when, then, we will have all ingredients for developing a computer code to solve diffusion problems. Let's keep in mind that the discretization of the diffusive term can be done using central differencing. Details of the discretization of advective terms is left for Chap. 5.

3.15 Exercises

3.1 Show that the approximation of $\partial f/\partial x$ at point P with truncation errors of the order of (Δx^2), for the configuration shown in Fig. 3.20, is given by,

$$\left.\frac{\partial f}{\partial x}\right|_P = \frac{\alpha^2 f_E - \beta^2 f_W + (\beta^2 - \alpha^2) f_P}{(\alpha\beta^2 + \alpha^2\beta)\Delta x} + O(\Delta x^2) \quad (3.87)$$

3.2 Using the same points of Fig. 3.20, with $\alpha = \beta = 1$, and knowing the values of f in P and E, and the $(\partial f/\partial x)_W$, find the value of f in W with an approximation of second order (Δx^2).

3.3 Obtain the following relations considering Δx and Δy equally spaced,

$$\left.\frac{\partial^2 f}{\partial x^2}\right|_i = \frac{-f_{i+3} + 4f_{i+2} - 5f_{i+1} + 2f_i}{\Delta x^2} + O(\Delta x^2), \quad (3.88)$$

$$\left.\frac{\partial f}{\partial x}\right|_i = \frac{-f_{i+2} + 8f_{i+1} - 8f_{i-1} + f_{i-2}}{12\Delta x} + O(\Delta x^4), \quad (3.89)$$

$$\left.\frac{\partial^2 f}{\partial x \partial y}\right|_{ij} = \frac{f_{i+1,j+1} - f_{i+1,j-1} - f_{i-1,j+1} + f_{i-1,j-1}}{4\Delta x \Delta y} + O(\Delta x^2, \Delta y^2) \quad (3.90)$$

3.4 Obtain the numerical approximation using finite differences for Eq. (3.77) for any θ. Refer to other literature [3, 4] to study the von Neumann stability criterion, and show that for any θ the criterion is:

$$\frac{\alpha \Delta t}{\Delta x^2} \leq \frac{1}{2 - 4\theta}, \quad (3.91)$$

for $\theta \leq 0.5$ and

$$\frac{\alpha \Delta t}{\Delta x^2} \leq \infty \quad (3.92)$$

for $0.5 \leq \theta \leq 1$

3.5 Using a polynomial relation to approximate a function through points of Fig. 3.20, and considering $\alpha = \beta = 1$, obtain $\partial^2 f/\partial x^2$ evaluated at the point

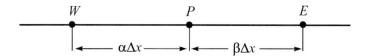

Fig. 3.20 Non-uniform mesh

3.15 Exercises

E with the respective order of approximation. Compare your results with those obtained through a Taylor series.

3.6 Solve numerically Eq. (3.77) using an explicit formulation and a grid with unknowns over the boundary. The initial temperature is $T(x) = 0$ and. suddenly, both boundaries go to $T = 1$. Considering $\alpha = 1$ and $\Delta x = 1$, use $\Delta t = 0.25$, 0.50 and 0.75 and advance the solution in time for 6 time steps and comment your findings. Solve the problem for steady state.

3.7 Obtain an expression for $\partial^2 f/\partial x^2$ with order of approximation of Δx^4. Using the same expression for the second derivative of f with respect to y, write the approximate equation for a point P for the two-dimensional Laplace's equation and sketch the structure of the matrix of coefficients.

3.8 Solve analytically and numerically, using finite volumes and finite differences, the one-dimensional heat conduction problem with heat generation shown in Fig. 3.21, given by

$$\frac{d^2T}{dx^2} + \frac{q'''}{k} = 0 \qquad (3.93)$$

and compare the solutions. For the finite difference method, use the points marked by filled circles, while for the finite volume method the temperature is stored in the open circles. You will see that the solution using finite differences is exact, i.e. without truncation errors, so it is independent of the mesh size, whereas for finite volumes the numerical solution depends on the mesh. Explain the reason and make a proposition that allows to get the solution by finite volumes also exact.

Obs.1 In the case of the Laplaces's equation, the application of finite differences or finite volumes with central difference interpolation function, results in identical approximate equations when the volumes are internal.

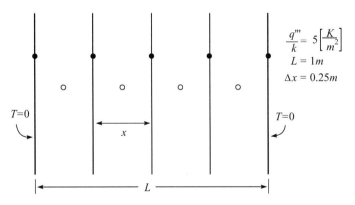

Fig. 3.21 Grid for Problem 3.8

Obs.2. Obtaining a numerical solution identical to the exact one is easy only in one-dimensional problems. Therefore, it is always convenient to have control volumes, in which balances are performed that cover the entire computational domain.

3.9 Starting from Eq. (3.93) adding the transient term, obtain the approximate equations for the explicit and fully implicit formulations. In the explicit formulation, there are three usual ways to distort the transient. One is to advance the solution with the maximum Δt allowed for each cell; another is to make use of the newly calculated values of the variables in the same iteration; and the third one is to overrelax these values according to

$$T^{n+1} = \omega T^* + (1 - \omega) T^n \qquad (3.94)$$

in which T* represents the value obtained from solving the linear system by any solver. Show that if the linear system obtained with the fully implicit formulation and infinite Δt is solved point by point, the three forms of distortion of the explicit transient are exactly the Jacobi, Gauss–Seidel, and SOR methods, respectively. This exercise shows that iterative point-by-point methods for solving linear system take the same time as an explicit formulation, since the operations are similar. Hint: See the Jacobi, Gauss–Seidel and SOR methods in Chap. 4.

3.10 Starting from Eq. (3.53) in two dimensions, obtain the approximate equations using explicit formulation. Show that the maximum possible time advance for the solution, according to von Neumann's criterion, is given by

$$\Delta t \leq 0.5/\alpha \left(\frac{1}{\Delta x^2} + \frac{1}{\Delta y^2} \right) \qquad (3.95)$$

3.11 For anisotropic materials, the equation for the heat conduction in two dimensions reads

$$\frac{\partial}{\partial t}(\rho c_p T) = \frac{\partial}{\partial x}\left(k_{11}\frac{\partial T}{\partial x} + k_{12}\frac{\partial T}{\partial y}\right) + \frac{\partial}{\partial y}\left(k_{21}\frac{\partial T}{\partial x} + k_{22}\frac{\partial T}{\partial y}\right) + q''' \qquad (3.96)$$

in which k_{ij} is the thermal conductivity tensor. Number the elementary volumes from left to right and from bottom to top; integrate the above equation using finite volumes with the fully implicit formulation and obtain the coefficients of the discretized equation. Show what the structure of the coefficient matrix looks like and calculate the sparsity index for a mesh of 20×20 volumes.

References

1. Reddy JN (2006) An introduction to finite element methods, 3rd edn. Mac Graw Hill
2. Patankar SV (1980) Numerical heat transfer and fluid flow. Hemisphere Publishing Corporation
3. Anderson DA, Tannehill JC, Pletcher R (1984) Computational fluid mechanics and heat transfer. Hemisphere Publishing Corporation
4. Roache PJ (1976) Computational fluid dynamics. Hermosa
5. Maliska CR, Silva AFC, Andrade D (1992) A strong coupling procedure for the segregated solutions of rotating flows. In: Wood HG (ed) Separation phenomena in liquids and gases. Engineering Academic Outreach Publication

Chapter 4
Solution of the Linear System

4.1 Introduction

After the presentation of the finite volume method for approximating partial differential equations, pointing out several aspects of the formulation, this chapter is dedicated for discussing few methods for solving the system of linear algebraic equations arising from the application of the numerical method. Considering the applications in computational fluid mechanics, it is worth to mention that the computation time required to solve a given problem is concentrated around to 60/70% in the solution of the linear system. Therefore, it is necessary that the user, or whoever develops an application for this purpose, invests in the quality of the method for solving the linear system.

The methods can be classified into direct and iterative. Direct methods are all those that work with the full matrix and require, in one way or another, processes equivalent to the inversion of the matrix. Being direct, these methods do not require, logically, an initial estimate of the variables to obtain the solution, and it is the exact one.

As seen in Sect. 3.11, the matrices obtained by applying numerical methods are very sparse, and because they are of large size, the operations performed in the inversion process work mainly with the null elements of the matrix. The computational effort is therefore very large, and this class of methods is not normally used in computational fluid mechanics applications. Moreover, because almost always the differential equations are nonlinear, the coefficient matrix of the algebraic linear system must be updated throughout the process, and therefore it makes no sense to directly (i.e., exactly) solve a linear system whose coefficients are not correct. Also, when the distorted transient is followed, the linear system does not need to be solved exactly at each time level.

Therefore, for CFD applications, the preferred methods are the iterative ones, and direct methods will not be discussed. In this section two basic goals are foreseen: to present simple iterative methods, so that the reader's activities in developing their academic codes can be carried out, and to advance a very efficient multigrid accelerator to work with iterative methods. The development of solvers for linear system is in the branch of linear algebra, a very important research topic, since the applied mathematicians, physicists and engineers are always looking for efficient solvers. When applications to large problems are needed, methods based on the multigrid concept should be used [1–3] By no means this chapter is intended to serve as a collection of several iterative solvers to be chosen by the numerical analyst. There are excellent libraries available online in which a myriad of solvers can be found [4].

4.2 Iterative Methods

Iterative methods are those that solve the system of equations considering that the equations are uncoupled and using the values from the last iteration to go on with the solution. Therefore, they require an initial estimate to start and proceed until a prescribed tolerance is achieved. They are generally classified as point-by-point, line-by-line, or plane-by-plane. It is logical that a point-by-point iterative method will be a direct method if the problem has just one control volume. Similarly, the line-by-line method is a direct method when the problem is one-dimensional, and a plane-by-plane is direct for a two-dimensional problem.

Following, some iterative methods are presented, remembering that iterative methods are extremely slow in their rate of convergence, especially when faced with large system of equations. They should be always supported by an accelerator, like multigrid.

4.2.1 Jacobi

The most known point-by-point solver is the Jacobi's method, which solves iteratively the linear system visiting equation-by-equation using the known variables from the previous iteration level. Re-writing Eq. (3.54) in the form

$$A_P T_P^{k+1} = \sum A_{nb} T_{NB}^k + B_P \qquad (4.1)$$

the following iterative procedure applies:

Give an initial estimate of the variables to start;
Perform iterations in k;
Calculate T_P^{k+1} using Eq. (4.1);
Check convergence;
Return and **repeat**, if tolerance is not met.

4.2 Iterative Methods

The Jacobi's method is of very slow convergence and requires a diagonal dominant matrix to attain convergence.

4.2.2 Gauss-Seidel

For the Gauss-Seidel Method, Eq. (3.54) is written as

$$A_P T_P^{k+1} = \sum A_{nb} T_{NB}^{k+1} + A_e T_E^k + A_n T_N^k + A_f T_F^k + B_P \qquad (4.2)$$

This method is essentially the same as the previous one, except that it makes use, during the same iterative cycle, of values of variables already calculated in that cycle. This speeds up convergence relative to the Jacobi method, but retains all the difficulties of a point-by-point iterative method. The procedure is

Give an initial estimate of the variables to start;
Iterate in k;
Calculate T_P^{k+1} by Eq. (4.2) in which a sweep was considered from back to front, west to east and south to north, being possible to consider as known, in the same sweep, the temperatures T_W, T_S and T_B, in the 3D case.
Check convergence;
Return and **repeat**, if the criterion is not met.

The same comments about the structure of the matrix made for the previous method applies here.

4.2.3 SOR-Successive Over Relaxation

The SOR method tries to further accelerate the convergence process of the previous methods. This is done by applying an overrelaxation to the values obtained with the Gauss-Seidel method. The iterative cycle has the following structure,

Give an initial estimate of the variables to start;
Iterate in k;
Calculate T_P^{k+1} by

$$A_P T_P^{k+1} = \sum A_{nb} T_{NB}^{k+1} + A_e T_E^k + A_n T_N^k + A_f T_F^k + B_P \qquad (4.3)$$

Apply over-relaxation using

$$T_P^{k+1} = w T_P^{k+1}\big|_{GS} + (1-w) T_P^k \qquad (4.4)$$

Check convergence;
Return and **repeat**, if the criterion is not met.

In Eq. (4.4) $T_P^{k+1}\big|_{GS}$ represents the value calculated with the Gauss-Seidel method, Eq. (4.3), and w, the relaxation parameter. The relaxation parameter serves to faster advance the solution when the process is slow, or "hold" the variable, when it is advancing too far and can cause divergence. Recommended values of w to advance the solution faster are between 1.5 and 1.7, although this value is dependent on the mesh size. Values smaller than 1.0 under-relax the solution. For the three methods described above, the end of iterations can be established by checking the value of the residual given by,

$$R^{k+1} = \sqrt{\sum \left(B_P + \sum A_{nb} T_{NB}^{k+1} - A_P T_P^{k+1}\right)^2} \tag{4.5}$$

attending a specified tolerance.

4.2.4 Alternating Direction Implicit Methods

Alternating direction implicit, or line-by-line methods, solve the linear system sweeping the domain line-by-line, solving implicitly, or directly, on that line, and repeating the procedure in alternating directions. This means that if a 1D problem is being solved, the solution is direct, since in 1D problems one has just one line of control volumes. For 2D and 3D problems it becomes iterative, because it is needed to sweep, line-by-line all domain, keeping in the independent vector the influence of the neighbors control volumes which are off-line. All ADI (Alternating Direction Implicit) algorithms for solving 2D and 3D problems take advantage of solving the tridiagonal problem directly, or implicitly [5]. The Thomas's algorithm, also known as the TDMA (Tri Diagonal Matrix Algorithm) [2, 6, 7] is largely used for this task. Of course, the discretization should obtain a tridiagonal, penta or hepta matrices such that the recursive relations, as will be seen, can be applied. It is worth quoting that implicit solution is also used in the literature as synonymous of direct solution.

Consider Fig. 4.1 in which a 2D problem showing the line in which a TDMA algorithm is applied.

The equation to be solved is

$$A_P T_P = A_e T_E + A_w T_W + A_n T_N + A_s T_S + B_P \tag{4.6}$$

Since only one line will be solved the values of the variables in the north and south lines will be accommodated in the independent term. Equation (4.6), then, is written as,

$$A_m T_m + B_m T_{m+1} + C_m T_{m-1} = D_m \tag{4.7}$$

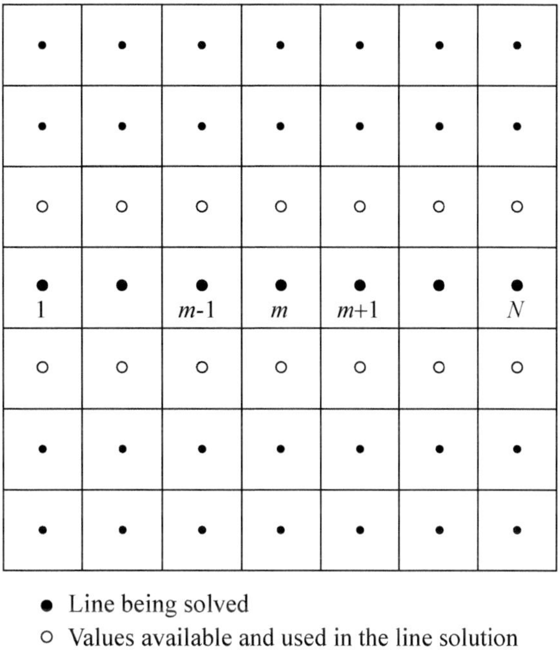

4.2 Iterative Methods

Fig. 4.1 Line in which TDMA is applied

- Line being solved
- Values available and used in the line solution

The interest is to find a recursive relation in the form,

$$T_m = P_m T_{m+1} + Q_m, \quad (4.8)$$

which permits, with the use of the boundary conditions, to sweep the line in one direction, from west to east, for example, calculating the P and Q parameters, and coming back from east to west calculating the variable, which is the desired solution.

The TDMA algorithm is extremely simple to implement and efficient. Lowering one index in Eq. (4.8), one obtains

$$T_{m-1} = P_{m-1} T_m + Q_{m-1} \quad (4.9)$$

Substituting Eq. (4.9) into Eq. (4.7) and comparing with Eq. (4.8) the expressions for calculating the coefficients P and Q, are

$$P_m = \frac{-B_m}{A_m + C_m P_{m-1}} \quad (4.10)$$

$$Q_m = \frac{D_m - C_m Q_{m-1}}{A_m + C_m P_{m-1}} \quad (4.11)$$

Equations (4.10) and (4.11) are recursive relation which permits, after knowing P_1 and Q_1 find all values of P and Q. Comparing the coefficients of Eqs. (4.6) and

(4.7), it gives,

$$A_m = A_P; \quad B_m = -A_e; \quad C_m = -A_w \quad (4.12)$$

$$D_m = A_n T_N + A_s T_S + B_P \quad (4.13)$$

Note that the line-by-line method is iterative in a 2D problem recognizing that the coefficient D_m must be updated, because T_N and T_S are unknowns and an estimate was used in that iteration. Alternating the direction, the coefficient become

$$A_m = A_P; \quad B_m = -A_n; \quad C_m = -A_s \quad (4.14)$$

$$D_m = A_w T_W + A_e T_E + B_P \quad (4.15)$$

The values of P_1 and Q_1 are easy to infer inspecting Eqs. (4.10) and (4.11). Considering that the indexes grow according Fig. 4.1, the approximate equation for the boundary volume (volume 1) can't depend on values of variables in the left, since there is no left neighbor. Thus C_1 must be zero, resulting in

$$P_1 = -\frac{B_1}{A_1}; \quad Q_1 = \frac{D_1}{A_1} \quad (4.16)$$

For the other boundary volume (volume N), the approximate equation can't depend on volumes at right. Therefore, B_N must be zero by Eq. (4.10), what results, from Eq. (4.8)

$$T_N = Q_N \quad (4.17)$$

For 3D problems it is enough to add in the D_m coefficient the contribution of the third direction. Of course, we now have one more direction to sweep. As mentioned, TDMA is an efficient solver and widely used in numerical method, provided that the number of control volumes are not too big, in which case it would need to be associated with a multigrid accelerator. The TDMA algorithm can be summarized as follow,

Estimate the variable for starting.
Calculate P_1 and Q_1 using Eq. (4.16);
Calculate all P_m and Q_m with m from 2 to N using Eqs. (4.10) and (4.11);
Make $T_N = Q_N$;
Calculate the variables for points $N - 1$ through 1 using Eq. (4.9);
Check the convergence. If the criterion is not satisfied, repeat or alternate the direction.

In the TDMA method it is also important to observe the dominant boundary conditions (Dirichlet) to perform the process, in that direction, since the boundary

4.2 Iterative Methods

Fig. 4.2 TDMA for a bank of lines

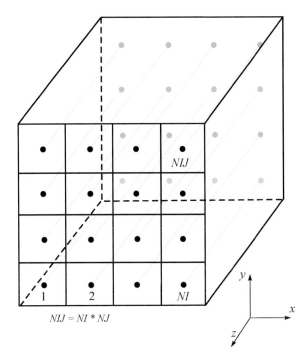

condition will more rapidly influence the interior of the domain. In three-dimensional problems it is possible to apply TDMA to a bank of lines, as shown in Fig. 4.2. The all 1D problem along z characterized in the (x, y) plane, can be swept together in the implicit direction (z), instead of solving one line at a time [8]. The bank of lines alternates in the z, x and y directions. It is also possible to choose the direction in which the mesh is smallest and iterate more in that direction, taking advantage of the recursive relationships along the smallest number of meshes. In short, one should choose the row bank that has the shortest possible rows. Computer implementing alternatives can be devised to sweep by bank of lines instead of sweeping line-by-line.

4.2.5 Incomplete LU Decomposition

Before talking about the incomplete LU decomposition, let's review the exact LU decomposition, which recovers the matrix [**A**], and allows a direct solution. The decomposition is an operation in which a [**L**](lower triangular) and a [**U**] (upper triangular) matrices are determined, such that,

$$[\mathbf{L}][\mathbf{U}]=[\mathbf{A}] \tag{4.18}$$

Being possible to find this decomposition, the linear system can be solved in two steps. Still using T as variable, the original linear system can be written as,

$$[\mathbf{A}][\mathbf{T}] = [\mathbf{L}][\mathbf{U}][\mathbf{T}] = \mathbf{B}, \tag{4.19}$$

in which two problems can be defined,

$$[\mathbf{U}][\mathbf{T}] = [\mathbf{D}] \tag{4.20}$$

and

$$[\mathbf{L}][\mathbf{D}] = [\mathbf{B}] \tag{4.21}$$

As both systems are diagonal (upper and low), through Eq. (4.21) by a backward substitution $[D]$ is found. With $[D]$ as independent vector in Eq. (4.20) the unknown $[T]$ is obtained by a forward substitution. This is an exact solution found through an exact LU decomposition, what implies in a direct method of solution. However, to find the exact LU decomposition of a matrix is not an easy task, since the operations are equivalent of inverting the matrix [**A**] and, unless some special algorithms are used, the operation work with all elements of the matrix, which is practically filled with zeros. Gauss elimination, by the way, is a kind of LU decomposition which creates an upper diagonal matrix after several operations in the [**A**] matrix.

In computational fluid dynamics, in which system of partial differential equations with nonlinearities are common, direct method doesn't play and important rule, since it is frequent in the solution procedure not to solve the linear system exactly since it will be changed many times in the iterations cycles due to the nonlinearities. It would be a waste of computer time to do so.

Based on these comments and because studying methods for solving linear systems is not in the scope of our text, direct methods will not be described herein. The point solvers and TDMA methods are good enough for those interested in building their own computer code for classes exercises or learning activities.

Back to our topic of iterative methods, an alternative is to have and incomplete LU decomposition, thus avoiding expending to much effort in finding an exact decomposition. The incomplete LU decomposition finds a **L** and a **U** which doesn't reproduces the [**A**] matrix, but a similar one, as

$$[\mathbf{L}][\mathbf{U}] = [\mathbf{A}] + [\mathbf{A}'], \tag{4.22}$$

which is, in fact, an approximation of the exact decomposition. The secret, of course, is to easily find the matrices $[L]$ and $[U]$ such that the error in the decomposition, the matrix $[\mathbf{A}']$, does not impair, or even preclude the convergence of the iterative process. The rationale of this iterative procedure is

$$[\mathbf{A} + \mathbf{A}'][\mathbf{T}]^{k+1} = [\mathbf{A} + \mathbf{A}'][\mathbf{T}]^k - \{[\mathbf{A}][\mathbf{T}]^k - [\mathbf{B}]\} \tag{4.23}$$

4.2 Iterative Methods

in which k indicates the iterative level. When the converged solution is obtained, the second term on the right-hand side of Eq. (4.23) goes to zero, resulting in $T^{k+1} = T^k$ Note that the matrix $[A']$ can be anyone, as it has no influence on the solution of the system, as expected, but it is determinant in the speed of convergence of the iterative method. Defining a correction for the variable T as,

$$[\delta]^{k+1} = [\mathbf{T}]^{k+1} - [\mathbf{T}]^k, \quad (4.24)$$

and a residue, or the error in the solution, by

$$[\mathscr{R}]^k = [\mathbf{A}][\mathbf{T}]^k - [\mathbf{B}] \quad (4.25)$$

Equation (4.23) results

$$[\mathbf{A} + \mathbf{A}'][\delta]^{k+1} = -[\mathscr{R}]^k \quad (4.26)$$

Using the decomposition $[\mathbf{L}][\mathbf{U}]$ of $[\mathbf{A} + \mathbf{A}']$, the following system should be solved,

$$[\mathbf{L}][\mathbf{U}][\delta]^{k+1} = -[\mathscr{R}]^k \quad (4.27)$$

The solution can be obtained by two processes of successive substitutions, since $[\mathbf{L}]$ and $[\mathbf{U}]$ are lower and upper diagonal, respectively. Defining a new vector $[\mathbf{V}]$, it can be found by,

$$[\mathbf{L}][\mathbf{V}] = -[\mathscr{R}]^k \quad (4.28)$$

Knowing $[\mathbf{V}]$ it can be solved for $[\delta]^{k+1}$ using

$$[\mathbf{U}][\delta]^{k+1} = [\mathbf{V}], \quad (4.29)$$

and, using, Eq. (4.24) to find the value for T in the next iterative level. The iterative process continues until the residual, calculated via Eq. (4.25), is smaller than the specified tolerance. These methods are quite powerful, but because they are strongly implicit the algorithm is no parallelizable due to its recursive structure. One of the pioneering methods in this category is the SIP—Strong Implicit Procedure [9], with many variants, being one of them the MSIP-Modified Strongly Implicit [10].

There are many direct methods that work only with the non-zeros of the matrix available in the literature. These are the so-called sparsity techniques, in which a direct solution of the linear system is sought in reasonable computational times and with storage compatible with iterative methods. The use of sparsity techniques to solve linear systems should also be considered by users of numerical methodologies [11, 12].

4.2.6 A Note on Convergence of Iterative Methods

The point-by-point iterative methods are techniques that can be classified as "weak" from the point of view of convergence of the solution. They are slow in transmitting information from the boundary condition to inside the domain. It is easy to realize that when sweeping the domain point by point, the information "travels" inside the domain at the rate of one mesh at each iterative cycle. This is similar, therefore, in having a time marching explicit solution with an equivalent Δt. In other words, iterative solution point-by-point is equivalent of explicit transient and, therefore, are restrictive.

Recalling that during an iterative process the error embedded in a solution can be decomposed into a spectrum of frequency modes [1, 2], iterative methods can only efficiently decrease the errors whose wavelengths are equivalent to the mesh size. Errors with other wavelengths, are hardly reduced with that mesh, and this is the reason why it is needed multigrid methods to accelerate convergence. Iterative methods, after they have reduced the errors with wavelengths equivalent to the size of the grid they stall. Several filters are necessary, and the filter with the size of the grids, to filter up all errors. To observe this behavior, the reader is suggested to resolve Exercise 4.7. In these methods, it is also important to choose the sweeping direction properly, starting from Dirichlet boundary conditions, i.e., strong boundary conditions.

The main requirement for convergence of a point-by-point solver is the diagonal dominance of the coefficient matrix, as discussed when discretizing the 1D equation. As example, consider the following problem [7], which can be understood as the linear system obtained from a discretization with two nodal points,

$$T_1 = 0.4T_2 + 0.2 \quad (4.30)$$

$$T_2 = T_1 + 1, \quad (4.31)$$

in which, for the first equation, $A_P > \sum A_{nb}$ and for the second equation, $A_P = \sum A_{nb}$. Applying the Gauss–Seidel iterative method, the result will be $T_1 = 1$ and $T_2 = 2$, which is the exact solution obtained by a direct method. In the system of equations given by Eqs. (4.30) and (4.31), the first equation was chosen as the evolution equation for the variable T_1 and the second equation for T_2. If the order is now reversed, making the second equation the evolutionary equation for T_1 and the first equation for T_2, the following system emerges,

$$T_1 = T_2 - 1 \quad (4.32)$$

$$T_2 = 2.5T_1 - 0.5 \quad (4.33)$$

4.2 Iterative Methods

Applying the same iterative procedure, the solution diverges, since the Scarborough criterion [7, 13] requires the following sufficient condition for convergence of the Gauss-Seidel iterative solver,

$$\frac{\sum |A_{nb}|}{|A_P|} \begin{cases} \leq 1 & \text{for all equations} \\ < 1 & \text{at least for one of the equations} \end{cases}$$

In finite volume methods, due to its physical background it generates matrices with diagonal dominance, as could be seen in Eq. (3.63) taking care of the negativity of the S_P term and of the interpolation function for high Reynolds number. Of course, a positive S_P can always be avoided, by leaving the whole source term in the independent term of the linear system. This may have a price to pay in convergence. More details about iterative methods will be given in the coming section, when multigrid methods are discussed.

4.2.7 Multigrid Method

In the previous section it was commented that point-by-point iterative methods do not have good convergence rates. There are two main reasons for this behavior. First, the anisotropy of the coefficients[1] can create matrices ill-conditioned, with the large ratio of the largest to the smallest coefficient. This can be appreciate obtaining the approximation of the Laplacian of ϕ. This equation is,

$$\frac{\partial^2 \phi}{\partial x^2} + \frac{\partial^2 \phi}{\partial y^2} = 0 \qquad (4.34)$$

Using finite differences[2] the discretized equation, considering a sweep from left to right and south to north, in a Cartesian grid, according Fig. 4.3, is

$$2\phi_P^{k+1}\left[\left(\frac{\Delta x}{\Delta y}\right)^2 + 1\right] = \phi_E^k + \phi_W^{k+1} + \left(\frac{\Delta x}{\Delta y}\right)^2 \phi_N^k + \left(\frac{\Delta x}{\Delta y}\right)^2 \phi_S^{k+1} \qquad (4.35)$$

Inspecting Eq. (4.35), it can be seen that the ratio of the coefficients connecting the variable in the y direction, to those connecting the variable in the x direction is of the order of $(\Delta x/\Delta y)^2$. If this ratio is large,[3] the south and north coefficients will be larger than the west and east coefficients, therefore, the approximate equation exhibits

---

[1] We are borrowing from physics the word anisotropy for identifying a matrix with coefficients with a large variation in magnitude.
[2] There is no worry about conservation now.
[3] For having a clear drawing, Fig. 4.3 does not exaggerate on the aspect ratio.

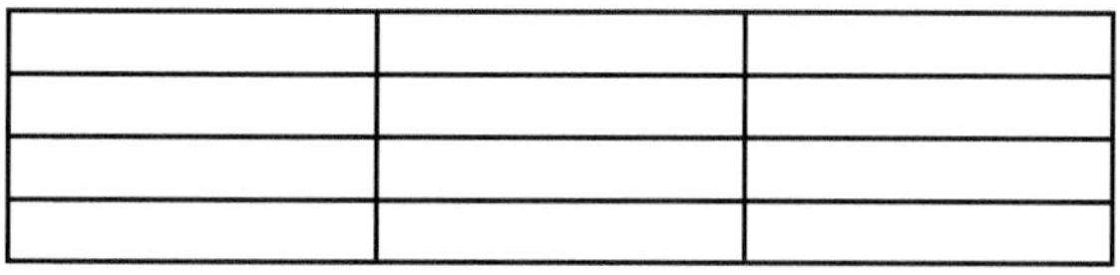

Fig. 4.3 Grid with large aspect ratio

anisotropy in its coefficients [2]. The consequence is that the solution advances faster in the direction of the larger coefficients and slower in the other, causing the whole process to slow down. Therefore, a balanced or isotropic matrix is desired. A problem with different physical properties in each direction also produces anisotropic coefficients, even with a mesh with aspect ratio close to unity, since a coefficient is always a combination of geometry and physics. If the physics is anisotropic, it doesn't help to have the grid regular, because, at the end, the coefficients will be anisotropic.

The second reason, briefly described in the previous section, is that iterative methods are efficient only in reducing errors with wavelengths of the order of the mesh size. In the solution of a real problem one uses the finer possible grid for capturing the spatial variations, what means that only errors with small wavelength will be eliminated. A numerical solution, however, during the iterative process, contains errors of all wavelength, what suggests it is needed several grids to damp out all errors. Both effects, anisotropy of the coefficients and the ability of reducing errors with specified wavelength only with grids of comparable size, tell us that different grids (multi-grids) should be used in conjunction with iterative methods.

The key question is: how to solve the linear system using several grids? There are several multigrid methods in the literature to face this task. The crucial operation in these methods is the agglomeration of the grids, such that multi-grids are created. In the geometric methods, the name is saying that, the agglomeration of the volumes is done based on the geometry of the mesh. In the algebraic ones the agglomeration is done considering the magnitude of the coefficients, trying to create isotropic matrices. This encompasses both effects of slow convergence, since geometry and physical properties appear in the coefficients.

The basic idea of accelerating convergence by multigrid methods is to recognize that iterative methods can only eliminate errors with wavelengths of the order of the mesh size. Thus, if meshes ranging from very fine to very coarse are used, the errors of all frequencies (all wavelengths) will be eliminated, thus speeding up the convergence process. The procedure is conceptually simple, it is enough to create new meshes using an agglomeration process which can be geometric or algebraic.

Consider Fig. 4.4 in which an agglomeration is realized without considering the physics to be solved, that is, based only in the geometry of the grid, which works well for isotropic physics. Since the physics to be solved in fluid mechanics problem are always anisotropic, the agglomeration must be done based on the magnitude of the coefficients.

Our goal in this section is to provide the reader with an algebraic multigrid method to be used in conjunction with iterative methods. To start, consider a simple problem to verify that the solution progress only when the errors wavelength is of the equivalent

4.2 Iterative Methods

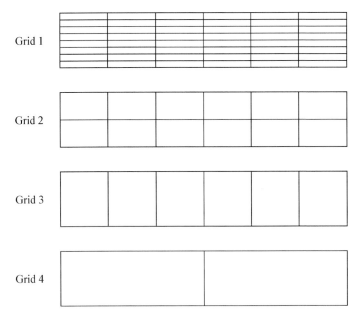

Fig. 4.4 Geometric agglomeration

size of the grid. Be the following 1D problem [14],

$$\frac{d^2\phi}{dx^2} = 0, \text{ with } \phi = 0 \text{ in } x = 0 \text{ and } x = 1, \tag{4.36}$$

whose solution is $\phi = 0$. Let's solve this equation numerically, iteratively, and doing so we need and initial estimate of the variable, which will be chosen containing errors of two wavelength, one of the size of the domain, while the other of the size of the grid following a sinusoidal shape. The exact solution lies on the x-axis, therefore the errors to be reduced during the iterations should bring the initial estimate to fit the analytical solution, which is $\phi = 0$. Figure 4.5 shows the initial estimate with two error components, one of small wavelength (high frequency) created, by purpose with the size of the grid on the top of a large wavelength error with the size of the domain. The other curves appearing on the plot are the distribution of ϕ after several iterations. It is clearly seen that the error with wavelength of the side of the mesh rapidly disappeared, while the other error with a large wavelength will not be damped out with this mesh.

According to the wavelength of the remaining error, a very coarse grid would be the ideal one. Indeed, one point in the middle of the domain would bring the solution to the exact one in just one iteration. This exemplifies that one needs grids of several sizes to eliminate errors of all wavelengths to accelerate the convergence of iterative methods.

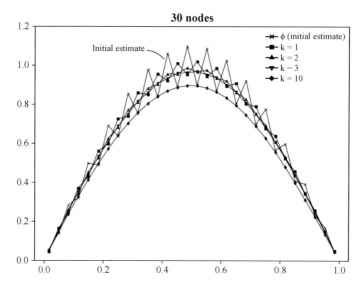

Fig. 4.5 Errors with different wavelengths

The reader can consult the works in [2, 14–16], to mention a few. The method to be presented herein is the ACM-Addictive Correction Multigrid [2, 14] whose principle, besides the essence of the multigrid approach, maintain the conservative property in the agglomerated volumes created from the fine mesh. In the initial mesh, which is the fine one, conservation is satisfied by the finite volume technique of obtaining the approximate equations.

4.2.7.1 ACM Multigrid for a Scalar Variable

Consider the following discretized equation

$$A_P^i \phi_P^i = \sum_{nb} A_{nb}^i + B_P^i, \qquad (4.37)$$

which is the linear system to be solved in the i grid. It was introduced this upper index i just to refer this linear system to the i grid. A_{nb}, as before are all the coefficients of volume P that connects it to its neighboring volumes nb. The idea is to obtain an approximate solution on the fine mesh in a few iterations to eliminate high frequency errors.[4]

One should not iterate too much on this mesh, because, after the errors on that grid have been eliminated, the other errors, with a lower frequency, decrease very

---

[4] High and low frequencies are not absolute. We are calling high frequency when related to the finest grid employed, usually the grid on which one is seeking for the solution.

4.2 Iterative Methods

Fig. 4.6 Iterative stall during iterations

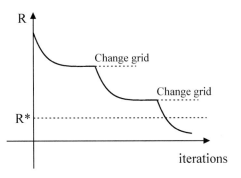

slowly, and the method is said to stall, as represented in Fig. 4.6. When an iterative method stalls, what it is needed to do, in fact, is to modify the field of variable and keep iterating. But the key question is how to modify, based on what? The multigrid methods have the answer, which is finding a new solution on a coarse grid which corrects the solution of the previous grid allowing to progress in the iterations. The way the new linear system for the coarser grid is found is on what the several multigrid methods differ. The ACM creates the new linear system based on conservation principles in the volumes of the coarse grid.

In the ACM procedure, an agglomeration is realized based on the magnitude of the coefficients to obtain a coarser grid, identified by grid I. How to obtain the linear system on the new agglomerated grid is now presented.

Let ϕ_P^I be a correction made over ϕ_P^i, as

$$\tilde{\phi}_P^i = \phi_P^i + \phi_P^I \tag{4.38}$$

in which ϕ_P^I is found on the grid I. Recall that after obtaining ϕ_P^I and using Eq. (4.38), the iterations could continue on grid i considering $\tilde{\phi}_P^i$ the new estimated variables for the iterative process. This could be done, but to reduce all errors components one needs go on coarsening the grid in several levels, working always with two grids, i and I. Therefore, the task of the method is finding ϕ_P^I for the next coarse grid. Introducing $\tilde{\phi}_P^i$ into the linear system given by Eq. (4.37) it produces a residue given by

$$\tilde{r}_P^i = B_P^i - A_P^i (\phi_P^i + \phi_P^I) + \sum_{nb \in i} A_{nb}^i (\phi_{NB}^i + \phi_{NB}^I), \tag{4.39}$$

which, after expanded, reads

$$\tilde{r}_P^i = B_P^i - A_P^i \phi_P^i + \sum_{nb \in i} A_{nb}^i \phi_{NB}^i - A_P^i \phi_P^I + \sum_{nb \in i} A_{nb}^i \phi_{NB}^I \tag{4.40}$$

Recognizing that the first three terms in the right-hand side of Eq. (4.40) is the residue of the ϕ_P^i,

$$\tilde{r}_P^i = r_P^i - A_P^i \phi_P^I + \sum_{nb \in i} A_{nb}^i \phi_{NB}^I \qquad (4.41)$$

To have a conservative scheme also in the coarse grid, the residue of the grid I, which contains several control volumes of the previous grid i, is forced to zero, as

$$\sum_{i \in I} \tilde{r}_p^i = 0 \qquad (4.42)$$

Summing up Eq. (4.40) for all i inside I,

$$0 = \sum_{i \in I} r_P^i - \sum_{i \in I} A_P^i \phi_P^I + \sum_{i \in I} \sum_{nb \in I} A_{nb}^i \phi_{NB}^I, \qquad (4.43)$$

which is the linear system to be solved in grid I, given by

$$A_P^I \phi_P^I = \sum A_{nb}^I \phi_{NB}^I + B_P^I, \qquad (4.44)$$

with the coefficients given by,

$$A_{nb}^I = \sum_{i \in \delta I} A_{nb}^i, \quad \delta I \to \text{boundary of } I \qquad (4.45)$$

$$A_P^I = \sum_{i \in I} A_P^i - \sum_{i \in \Omega I} A_{nb}^i, \quad \Omega I \to \text{interior of } I \qquad (4.46)$$

$$B_P^I = \sum_{i \in I} r_p^i \qquad (4.47)$$

There are several key features of the multigrid technique just demonstrated, but two of them, very impacting, are the conservative approach used to derive the linear system for the coarse grid, and the calculations of the coefficients in the coarse grid, which uses the coefficients of the previous grid. This means that the coefficients are calculated only once, for the fine grid when the process starts. Other features will be commented at the end of this section.

To demonstrate the use of the methodology let us use the grids shown in Figs. 4.7 and 4.8, in which the (i) grid and the agglomerated grid (I) are shown. For our goal, it does not matter how the agglomeration is done. According to the numbering on those figures, the control volumes 21, 27 and 33 in the fine grid are agglomerated into the volume 9 in the coarse grid. Therefore, all coefficients are known on the fine grid, since this is supposed to be the mesh on which the solution of the problem is sought. Using Eqs. (4.45–4.47), the coefficients on the grid I can be calculated, as

$$A_P^9 = A_P^{21} + A_P^{27} + A_P^{33} - A_n^{21} - A_n^{27} - A_s^{33} - A_s^{27} \qquad (4.48)$$

4.2 Iterative Methods

$$A_e^9 = A_e^{21} + A_e^{27} + A_e^{33} \tag{4.49}$$

$$A_w^9 = A_w^{21} + A_w^{27} + A_w^{33} \tag{4.50}$$

$$A_n^9 = A_n^{33} \tag{4.51}$$

$$A_s^9 = A_s^{21} \tag{4.52}$$

$$B_P^9 = r_{21} + r_{27} + r_{33} \tag{4.53}$$

The discretized equation for the volume 9 on the coarse grid can, therefore, be written as,

$$A_P^9 \phi_9 = A_e^9 \phi_{10} + A_s^9 \phi_3 + A_w^9 \phi_8 + A_n^9 \phi_{15} + B_P^9 \tag{4.54}$$

Figure 4.9a tries to exemplify visually the coefficients, represented by wavy arrows, which connects a control volume P in grid i with its neighbors control volumes. In Fig. 4.9b it is shown the coefficients connecting the control volume P (number 9 in this example) with its neighbors. It is possible to visualize in Eq. (4.49)

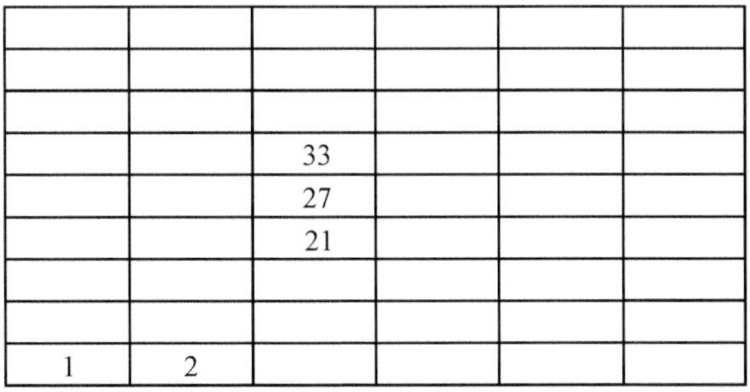

Fig. 4.7 Grid i for the multigrid method

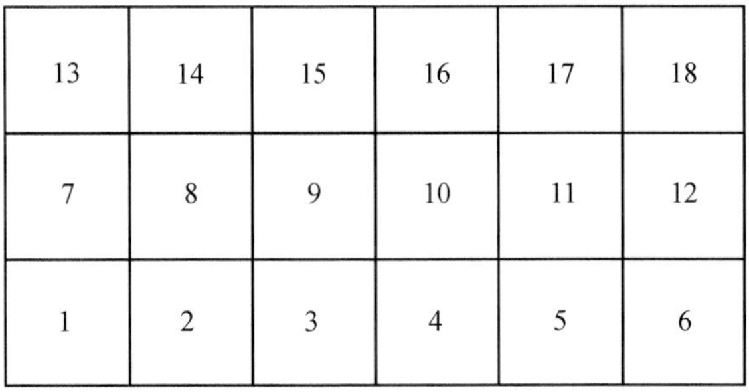

		33				
		27				
		21				
	1	2				

Fig. 4.8 Grid I. Agglomerated grid of Fig. 4.7

13	14	15	16	17	18
7	8	9	10	11	12
1	2	3	4	5	6

Fig. 4.9 Connections (coefficients) in the i and I grids

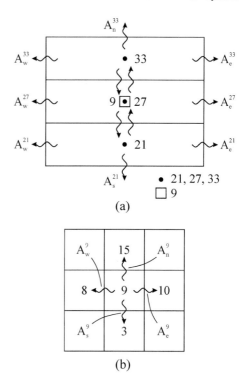

which the A_e^9 is composed of the three arrows (A_e^{21}, A_e^{27}, A_e^{33}) forming the east coefficient which connects the volume 9 with the volume 10 in the coarse grid. The same is applied to the other three boundaries of volume 9. Note that the coefficients which are removed from the A_P^9 in Eq. (4.48) are coefficients of the i volumes internal to grid I (4 of them in this example).

Similar equations are obtained for all the 18 control volumes of grid I, and the linear system can be solved. If one wants to go further with new agglomeration the process is repeated until a very coarse grid is reached which may, even, allow a direct solution of the linear system. In the way back from the coarsest mesh, the solution is corrected in each grid until reaching the original grid, on which iterations are performed until convergence is achieved. If it stalls again, the process can be repeated.

In the given example, each ϕ_P^i needs to be corrected with the solution obtained in the I grid, as

$$\begin{aligned}
\tilde{\phi}_{21}^i &= \phi_{21}^i + \phi_9^I \\
\tilde{\phi}_{27}^i &= \phi_{27}^i + \phi_9^I, \\
\tilde{\phi}_{33}^i &= \phi_{33}^i + \phi_9^I
\end{aligned} \qquad (4.55)$$

with the same correction applied to all control volumes. The iteration on the fine grid could continue with this new estimate $\tilde{\phi}$. As mentioned, the correction couldn't be done in this level and go forward with a new agglomeration. Note that it is always used only two grids in each level.

During the application of the multigrid accelerator it is necessary to establish a convergence criterion to guide the number of iterations and number of levels to be used, what is not an easy task, since many variables are involved. What is done is to monitor the normalized residue of the equations, being the normalized mass residue the usual choice.

4.2.7.2 ACM Multigrid for a Vector Variable

The multigrid can also be applied to linear system originated from simultaneous solution, that is, system in which the unknown is not a scalar, as considered up to now, but a vector. For example, let the vector $[u, v, p]$ be the unknowns of a 2D fluid flow problem whose simultaneous solution is sought. The multigrid method just advanced can be applied to this vector the same way it was applied to the scalar ϕ. In a vector form, Eq. (4.37) can be written as,

$$\begin{bmatrix} A^{uu} & A^{uv} & A^{up} \\ A^{vu} & A^{vv} & A^{vp} \\ A^{pu} & A^{pv} & A^{pp} \end{bmatrix}_P^i \begin{bmatrix} u \\ v \\ p \end{bmatrix}_P^i = \begin{bmatrix} B^u \\ B^v \\ B^p \end{bmatrix}_P^i, \qquad (4.56)$$

while the corrected variable, by

$$\begin{bmatrix} \tilde{u} \\ \tilde{v} \\ \tilde{p} \end{bmatrix}_P^i = \begin{bmatrix} u \\ v \\ p \end{bmatrix}_P^i + \begin{bmatrix} u \\ v \\ p \end{bmatrix}_P^I \qquad (4.57)$$

Following the same steps as done for the scalar ϕ, one obtains

$$\begin{bmatrix} \tilde{r}^u \\ \tilde{r}^v \\ \tilde{r}^p \end{bmatrix}_P^i = \begin{bmatrix} r^u \\ r^v \\ r^p \end{bmatrix}_P^i - \begin{bmatrix} A^{uu} & A^{uv} & A^{up} \\ A^{vu} & A^{vv} & A^{vp} \\ A^{pu} & A^{pv} & A^{pp} \end{bmatrix}_P^i \begin{bmatrix} u \\ v \\ p \end{bmatrix}_P^I$$

$$+ \sum \begin{bmatrix} A^{uu} & A^{uv} & A^{up} \\ A^{vu} & A^{vv} & A^{vp} \\ A^{pu} & A^{pv} & A^{pp} \end{bmatrix}_{nb}^i \begin{bmatrix} u \\ v \\ p \end{bmatrix}_{NB}^I \qquad (4.58)$$

Forcing the residue of the i volumes inside the I grid to be zero, and after some algebraic manipulation, same as done for the scalar variable, the final system of equation results,

$$\begin{bmatrix} A^{uu} & A^{uv} & A^{up} \\ A^{vu} & A^{vv} & A^{vp} \\ A^{pu} & A^{pv} & A^{pp} \end{bmatrix}_P^I \begin{bmatrix} u \\ v \\ p \end{bmatrix}_P^I = \sum \begin{bmatrix} A^{uu} & A^{uv} & A^{up} \\ A^{vu} & A^{vv} & A^{vp} \\ A^{pu} & A^{pv} & A^{pp} \end{bmatrix}_{nb}^I \begin{bmatrix} u \\ v \\ p \end{bmatrix}_{NB}^I + \begin{bmatrix} B^u \\ B^v \\ B^p \end{bmatrix}_P^I$$
(4.59)

which should be compared with its scalar counterpart, Eq. (4.44). Similar equations as done for the scalar are used to obtain A_{nb}^I, A_P^I and B_P^I.

As discussed, the solution can proceed using several levels of grids. If just one or more coarse grids correcting the variables in the way back to the original grid are used, it is called a V cycle, as in Fig. 4.10a, b. If the problem is solved for one coarse grid, then back to the original one and then solve again for other levels of coarse grid, it called a W cycle, as in Fig. 4.10c, d. It is called a F cycle (flexible) when the procedure goes back and forth on several grids, as shown in Fig. 4.10e, f.

To appreciate the types of grids generated using the algebraic and geometric multigrid, Fig. 4.11 shows the grid obtained solving an anisotropic heat conduction problem in a flat plate [17] with both approaches, algebraic ACM and geometric agglomeration. The base solver is Gauss-Seidel, and the computing time was close to ten times less using the algebraic multigrid compared to geometric. Following, it is shown several results, Figs. 4.12, 4.13, 4.14 and 4.15, obtained when solving a fluid flow in a curved duct using multigrid [14].

Figure 4.12, shows the cost by grid nodes using a line solver, the gradient conjugate and the multigrid accelerator, revealing an impacting reduction in the computer costs. Another very important results are the independence of the convergence rate with the number of unknowns in the problem, as can be seen in Fig. 4.13 using 25 k, 100 k and 400 k nodes.

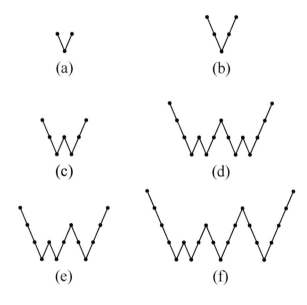

Fig. 4.10 Possible cycles in the multigrid method

4.2 Iterative Methods

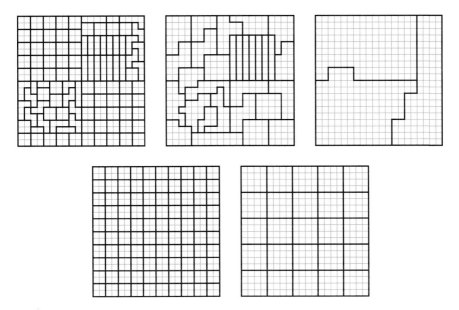

Fig. 4.11 Algebraic (top) and geometric (bottom) agglomerations

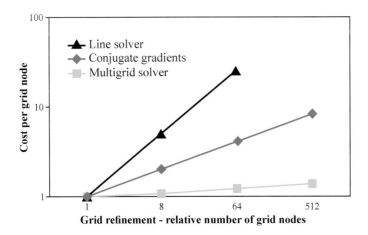

Fig. 4.12 Cost per grid node

It is not too much to enforce that the independence of convergence rate with the size of the problem is of utmost importance in the CPU efforts. Figure 4.14 brings the comparison of the algebraic and the geometric multigrid focusing on the aspect ratio of the mesh dimensions, since it is a parameter which impacts the coefficients anisotropy. We can appreciate that the convergence rate of the algebraic method is superior than the geometric, which, even shows a stalling behavior when the mesh has large aspect ratio.

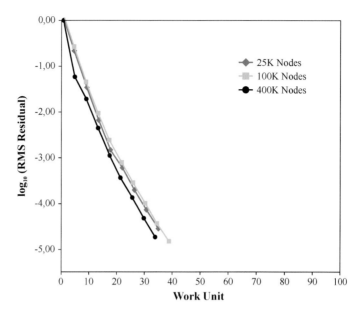

Fig. 4.13 Convergence rate. Number of unknowns

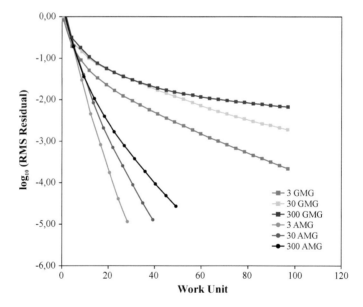

Fig. 4.14 AMGxGMG. Convergence rate for different aspect ratios

Fig. 4.15 Gauss-Seidel versus ACM multigrid

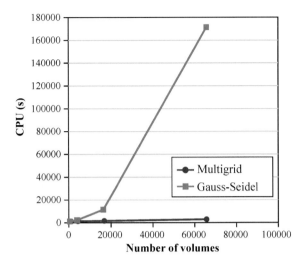

And, finally, it is brought a result in Fig. 4.15 comparing a Gauss-Seidel method working alone and with the multigrid accelerator. It is amazing the enormous difference in CPU time. Those results clearly demonstrate that in numerical simulation of fluid flows it is almost impossible to be competitive without a multigrid to accelerate the solution of the linear systems.

4.3 Conclusions

This chapter provided a brief discussion about the characteristics of iterative solvers in connection with the solution of large sparse matrix encountered in computational fluid dynamics Very simple point-by-point solver were presented, which can help students to solve simple exercises and to test their computer algorithm. One important line-by-line method, known as TDMA, easy to implement and efficient, which solves a tridiagonal matrix was also presented. Finally, as the most important part of the chapter, it was discussed why iterative methods fail, explaining the connection of the errors in the solution with the size of the grid, and presented the ACM multigrid method, a very robust and efficient tool for solving large linear systems of equations.

4.4 Exercises

4.1 Solve, using finite volumes the transient heat conduction in a fin with the top insulated and the base at temperature T_b, according to Fig. 4.16, with the following data:

Fig. 4.16 Figure for Problem 4.1

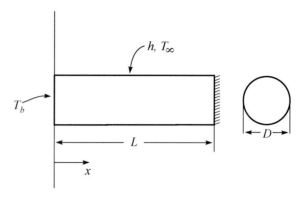

$T_b = 373$ K, $T_\infty = 293$ K, $k = 10$ W/mK, $D = 0.01$ m, $h = 5$ W/m²K and $\alpha = 10^{-6}$ m²/s.

The fin is at ambient temperature at $t = 0$ and suddenly its base temperature is raised to T_b. Don't forget to do a grid resolution study and compare the numerical solution with the analytical one available in any textbook of basic heat transfer. For comparison make your results dimensionless.

4.2 Solve the 2D steady state heat conduction Problem Given by

$$\frac{\partial^2 T}{\partial x^2} + \frac{\partial^2 T}{\partial y^2} = 0 \qquad (4.60)$$

with boundary conditions as given in Fig. 4.17. Compare with the analytical solution given by

$$T(x, y) = \frac{\operatorname{senh}\left(\frac{\pi y}{a}\right)}{\operatorname{senh}\left(\frac{\pi b}{a}\right)} \operatorname{sen}\left(\frac{\pi x}{a}\right) \qquad (4.61)$$

Fig. 4.17 Figure for Problem 4.2

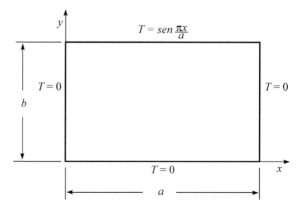

4.4 Exercises

4.3 Solve numerically the one-dimensional transient heat conduction problem described by the following equation

$$\frac{\partial T}{\partial t} = \alpha \frac{\partial^2 T}{\partial x^2} \qquad (4.62)$$

When the initial temperature distribution is

$$T(x,0) = T_o \operatorname{sen}\left(\frac{\pi x}{L}\right), \qquad (4.63)$$

the analytical solution is

$$T(x,t) = T_o e^{-\alpha \lambda_1^2 t} \operatorname{sen}(\lambda_1 x), \text{ with } \lambda_1 = \pi/L \qquad (4.64)$$

and when it is T_i, constant, the analytical solution is

$$T(x,t) = \frac{4}{\pi} T_i \sum_{n=1,3,5,7}^{\infty} \frac{1}{n} e^{-\alpha \lambda_n^2 t} \operatorname{sen}(\lambda_n x), \text{ with } \lambda_n = n\pi/L \qquad (4.65)$$

The boundary conditions are $T(0,t) = T(L,t) = 0$.

Derive the analytical expression to obtain the heat flux at the boundaries and compare with the heat flux numerically calculated. This is an interesting exercise because de refinement in time can be exercised, because one wants to follow the real transient for both situations. The refinement in space close to the base of the fin helps in calculating the heat flux more precisely.

4.4 When the flow between parallel plates becomes fully developed, the resulting equation is mathematically similar to the heat conduction problem with uniform heat generation,

$$\frac{1}{\mu}\frac{dp}{dx} = \frac{d^2 u}{dy^2}, \qquad (4.66)$$

in which dP/dx, the pressure gradient is constant, and μ is the absolute viscosity of the fluid. Notice that the left side of the equation plays the role of a heat generation in a heat conduction problem. Solve this problem numerically and compare the solution with the exact one. All comments made for Exercise 3.8 apply to this problem.

4.5 A flat plate of thickness $L = 3$ m has on its left face a heat flux entering the plate of 10 W/m², while on the right face a heat flux leaving the plate equal to 20 W/m². There is a uniform heat generation equal to 7 W/m³ inside the plate. The thermal conductivity of the plate is equal to 1 W/mK.

(a) Determine the temperature distribution, using the methods of Jacobi, Gauss-Seidel, S.O.R. and TDMA methods. For this situation, the problem has no steady state solution. Explain why.
(b) Now, make the heat flux leaving the wall equal to 31 W/m². Why is it not possible to find the solution using TDMA? Set the value of the temperature of the last volume on the right to 10 K and solve again. Comment.
(c) Solve by Jacobi, Gauss-Seidel and S.O.R. methods, fixing and not fixing the value of the volume on the right. Analyze the behavior of the methods regarding the number of iterations required.

4.6 For Problem 4.1, with 16 control volumes, and for a two-dimensional conduction problem, also with 16 volumes and numbered according to Fig. 4.18a, always using central differences, sketch the structure of the matrix of coefficients for the two cases. Imagine now that in obtaining the approximate equations for the fin problem, two volumes on the right and two volumes on the left are used in the equation for P, as shown in Fig. 4.18b. What does the structure of the matrix look like now? What is the similarity of this matrix with the two-dimensional problem?

4.7 To recognize that errors in a numerical solution using iterative methods decay faster only when the grid has comparable size of the error wavelength, solve, using Gauss-Seidel, the steady state one-dimensional heat conduction problem given by $d^2T/dx^2 = 0$, with boundary conditions $T = 0$, at $x = 0$ and $x = L$ with an initial estimated condition of the type,

$$T(x) = \text{sen}(\pi x)(1 + 0.1(-1)^n), \tag{4.67}$$

which represents the composition of a high and a low frequency errors, being n ranging in the interval $[0, N]$, with N being the number of nodes employed. The higher frequency error is, by purpose, created according to the number N of volumes. With the iterative process marching from left to right and a mesh

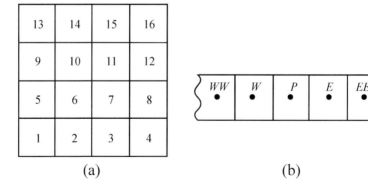

Fig. 4.18 Figure for Problem 4.6

of 30 volumes, $N = 30$, for example, plot the initial distribution according to Eq. (4.67) and observe that after a few iterations (3 or 4) the high-frequency errors will disappear and the low-frequency errors will decrease very slowly. Refine the mesh further and see that it gets even harder to decrease the low frequency errors. Use a coarse mesh, $N = 3$, and see how quickly the low frequency errors, now, decrease.

References

1. Brandt A (1977) Multi-level adaptative solutions to boundary value problems. Math Comput 31:333–390
2. Hutchinson BR, Raithby GD (1986) A multigrid method based on the additive correction strategy. Numer Heat Transfer 9:511–537
3. Ferziger JH, Peric M (1999) Computational methods for fluid dynamics, 2nd edn. Springer
4. Balay S, Gropp W, McInnes LC, Smith BF (1998) PETSc, the portable, extensible toolkit for scientific computation. Argonne National Laboratory 2(17)
5. Peaceman DW, Rachford HH Jr (1955) The numerical solution of parabolic and elliptic differential equations. J Soc Ind Appl Math 3:28–41
6. Richtmyer RD, Morton KW (1967) Difference methods for initial-value problems. Wiley, New York
7. Patankar SV (1980) Numerical heat transfer and fluid flow. Hemisphere Publishing Co.
8. Silva AFC, Marchi CH, Livramento MA and Azevedo JLF (1991) On the effects of vectorization for efficient computation of three-dimensional segregated finite volume solutions. In: XI Brazilian congress of mechanical engineering, pp 109–112, São Paulo, Brasil
9. Stone HL (1968) Iterative solution of implicit approximation of multidimensional partial differential equations. SIAM J Numer Anal 5:530–558
10. Schneider GE, Zedan M (1981) A modified strongly implicit procedure for numerical solution of field problems. Numer Heat Transfer 4:1–19
11. Davis TA (2006) Direct methods for sparse linear systems. SIAM_Society for Industrial and Applied Mathematics, Philadelphia, PA
12. George AJ, Liu JW (1981) Computer solutions of large sparse positive definite systems. Prentice-Hall, Englewoods Cliffs
13. Scarborough JB (1958) Numerical mathematical analysis. Johns Hopkins Press
14. CFX-TascFlow Theory Documentation (2000) AEA Technology, Canada
15. Philips RE, Schmidt FW (1984) Multigrid techniques for numerical solution of diffusion equations. Numer Heat Transfer 7:251–268
16. Settari A, Aziz K (1973) A generalization of the additive correction methods for the iterative solution of matrix equations. SIAM J Numer Anal 10:506–521
17. Keller SC (2007) Multigrid method with addictive correction for the simultaneous solution of the Navier-Stokes equation in unstructured grids. Ph.D. Thesis, Mech. Eng. Dept. UFSC (in Portuguese). https://repositorio.ufsc.br/handle/123456789/89788

Chapter 5
Advection and Diffusion—Interpolation Functions

5.1 Introduction

The previous chapters introduced the reader to the basics of the finite volume method, pointing out the strong characteristics of the method and providing the key tools for exercising the development of basic computational codes. The method was elaborated working with the transient, the diffusive and the source terms. No advection was involved.

The most in-depth discussion was done with the transient term, what gives rise to the explicit, implicit, and fully implicit formulations. We were not concerned to emphasize the use of interpolation functions, although these have been used in the approximation of the transient term and the diffusive fluxes at the interfaces of the control volumes. It has been stressed that for the evaluation of diffusive fluxes (of any property), the interpolation functions can be of central differences, since it do not pose any stability problem for the numerical method.

Although dealing with problems involving advection, it is not of concern, at this point, to know how the velocity field which carries the mass flow was obtained. It should be interpreted that this field is known and available. In other words, one is interested on how a property ϕ is transported by a known mass flow. Later, it will devote attention to methods that teach us how to numerically calculate the velocity and pressure fields.

The coming section is devoted to present the general conservation equation pointing out the necessary interpolation functions for the advective and diffusive terms. Following, it is addressed the relationship that the interpolation function has with the physics of these terms for succeeding in the task. Along the chapter, types of interpolation functions will be analyzed, in the beginning, in order to build the knowledge piece-by-piece, for a unidimensional problem and, with the knowledge gained, extrapolate to two and three dimensions. A reasonable deep discussion about numerical diffusion and numerical oscillation, a controversial topic in the literature, will be also on the board. The cell-center method continues being used in this chapter.

5.2 The General Equation

Before starting to deal with the interpolation functions for the advective terms, it is didactic integrate the general conservation for ϕ, , see all terms and identify the ones that deserves attention after the integration of the conservation equations. Initially, let's consider Eq. (2.64) in its vector form,

$$\frac{\partial}{\partial t}(\rho\phi) + \nabla.(\rho\mathbf{V}\phi) = \nabla.(\Gamma^\phi\nabla\phi) + S^\phi, \tag{5.1}$$

in which are identified, in order, the transient, advective, diffusive and source terms. The advective and diffusive fluxes are the vectors of the divergence operator, characterizing the conservative form of the equation. Following the procedure of a finite volume technique, this equation is integrated in space and in time, as

$$\int_t \int_V \left\{ \frac{\partial}{\partial t}(\rho\phi) + \nabla.(\rho\mathbf{V}\phi) = \nabla.(\Gamma^\phi\nabla\phi) + S^\phi \right\} dV dt, \tag{5.2}$$

Applying the divergence theorem to transform the volume integrals into surface integrals, linearizing the source term and substituting the surface integrals by its numerical counterpart, one obtains

$$\frac{M_P\phi_P - M_P^o\phi_P^o}{\Delta t} = -\sum_{ip}(\rho(\mathbf{V}.\mathbf{n})\phi\Delta s)_{ip} + \sum_{ip}\left(\Gamma^\phi\nabla\phi.\mathbf{n}\Delta s\right)_{ip} + (S_P\phi_P + S_C)\Delta V \tag{5.3}$$

Equation (5.3) can be applied in a control volume of arbitrary shape, as shown in Fig. 5.1, in which are identified the integration points (ip), lying at the middle of the surfaces, locals where the advective and diffusive fluxes are calculated.

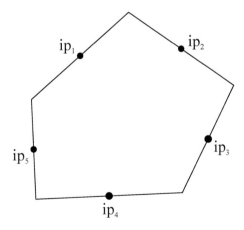

Fig. 5.1 General control volume

It is interesting to compare Eq. (5.3) with Eq. (3.17), which was the discrete equation of the 1D heat conduction problem used to introduce the finite volume technique. In that case, ϕ was T and, as it was 1D, the integration points (ip) were e and w, being important to recognize that the components of $(\nabla T)_{ip}$ required by Eq. (5.3) were $(\partial T/\partial x)_e$ and $(\partial T/\partial x)_w$, which were discretized using central differencing scheme by Eqs. (3.18) and (3.19). No advection was considered in Eq. (3.17).

Hence, in any finite volume discretization of fluid flow equations, ϕ_{ip} and $(\nabla \phi)_{ip}$ must be determined and, as they are required at the interfaces of the control volume, an interpolation function must be applied to describe the behavior of the function among grid points. The gradient, $(\nabla \phi)_{ip}$, poses no difficulties for its determination, being our efforts now devoted in finding ϕ_{ip}.

5.3 The Difficulty of the Advective-Dominant Problem

To begin the discussion of the difficulties of treating advective-dominant problems, let us consider the one-dimensional advection/diffusion of the entity ϕ without the transient and source terms, given by

$$\frac{\partial}{\partial x}(\rho u \phi) = \frac{\partial}{\partial x}\left(\Gamma^\phi \frac{\partial \phi}{\partial x}\right) \quad (5.4)$$

in which ϕ represents a transported property by unit of mass and Γ^ϕ the corresponding diffusion coefficient. The 1D grid, familiar to us, is shown in Fig. 5.2. The integration of Eq. (5.4) gives,

$$\rho u \phi|_e - \rho u \phi|_w = \Gamma^\phi \frac{\partial \phi}{\partial x}\bigg|_e - \Gamma^\phi \frac{\partial \phi}{\partial x}\bigg|_w \quad (5.5)$$

in which e and w are, again, the integration points for this 1D problem. The calculation of the fluxes in Eq. (5.5) must be done as a function of the values at the nodes using an interpolation function which has the role of connecting the nodal points, allowing to interpolate ϕ at the integration points. The attempt is always to propose an interpolation function with the smallest possible error, that is, representing the local physics, and, at the same time, does not involve too many nodal points to avoid complex matrix structure. The ideal interpolation function would be the exact solution of the problem to be solved. This interpolation function, called exact, of course, is not available. If it was, there was no reason for solving the problem numerically. What is usually done is to use as interpolation functions the exact solution of some simplifications of the real problem or find ways of inserting in the interpolation functions terms which represents important parts of the physics.

The natural tendency is to choose central differencing scheme (CDS) as interpolation function for all terms of the equation, taking advantage of its 2nd order accuracy and the involvement of only two nodal points in the calculation of each flux.

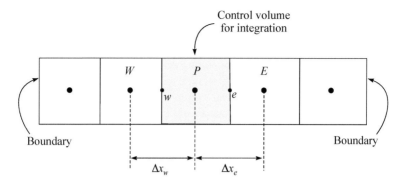

Fig. 5.2 Control volume for integration

Considering constant physical properties, employing the fully implicit formulation and approximating the fluxes using central differences (CDS), it gives, considering a uniform mesh,

$$\rho u \frac{(\phi_E + \phi_P)}{2} - \rho u \frac{(\phi_P + \phi_W)}{2} = \Gamma^\phi \frac{(\phi_E - \phi_P)}{\Delta x} - \Gamma^\phi \frac{(\phi_P - \phi_W)}{\Delta x}, \quad (5.6)$$

which, after rearranging terms, take the form,

$$A_P \phi_P = A_e \phi_E + A_w \phi_W, \quad (5.7)$$

with the coefficients given by,

$$A_P = \frac{2\Gamma^\phi}{\Delta x^2} \quad (5.8)$$

$$A_e = -\frac{\rho u}{2\Delta x} + \frac{\Gamma^\phi}{\Delta x^2} \quad (5.9)$$

$$A_w = \frac{\rho u}{2\Delta x} + \frac{\Gamma^\phi}{\Delta x^2} \quad (5.10)$$

Considering the velocity u positive, the following relation must be satisfied for having the A_e coefficient positive,

$$\frac{\rho u \Delta x}{\Gamma^\phi} \leq 2 \quad (5.11)$$

in which the parameter is recognized as the Reynolds number of the cell. Note that if it was the energy equation with temperature as variable, it would be the Peclet number, since Γ^ϕ would be k/c_P in Eq. (5.11).

5.3 The Difficulty of the Advective-Dominant Problem

When the velocity u increases, the mesh must be reduced proportionally, if the coefficient A_e is to be kept positive. Keeping the coefficients positive is a desired characteristic for any numerical method. Therefore, the use of central differences in the approximation of the advective terms almost always creates negative coefficients, because it is difficult in real problems to refine the mesh to keep $Re_{\Delta x} \leq 2$ for all meshes of the domain.

The presence of negative coefficients immediately brings two difficulties. The first is associated with the nature of the iterative method used to solve the linear system. If the method is not robust, such as point solvers, for example, the solution may diverge. The second is linked to the order of approximation of the interpolation function. High-order approximations, such as central differences, in the advective terms, when these are dominant, generate instabilities, producing solutions that present numerical oscillations in regions of large gradients. The characteristic of these oscillations is shown in Fig. 5.3a, in which a pulse of ϕ should be reproduced numerically, but under and overshoots are present.

Note that the solution showing the numerical oscillations may be a converged solution. The impossibility of dissipating the oscillations is a characteristic of high-order schemes, including the central-difference approximation.

It is relevant to point out that the existence of negative coefficients does not mean that it is impossible to obtain the solution. The use of robust methods to solve the linear system, and the way to advance the calculation procedure, allow the solution to be obtained, even with negative coefficients.

The way to avoid this negative coefficient is to use another approximation for the advective term. A one-sided, first-order approximation, known as upwind differencing scheme (UDS), for example, solves the problem. For positive u, the upwind approximation, that is, $u_e = u_P$ and $u_w = u_W$, results in the following numerical

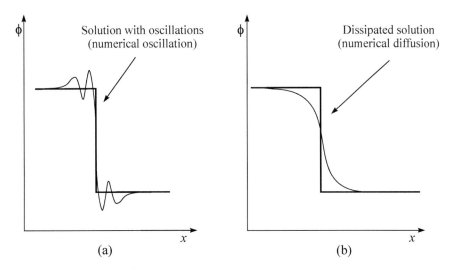

Fig. 5.3 Numerical oscillation (**a**) and numerical diffusion (**b**)

approximation for Eq. (5.4),

$$\rho u \phi_P - \rho u \phi_W = \Gamma^\phi \frac{(\phi_E - \phi_P)}{\Delta x} - \Gamma^\phi \frac{(\phi_P - \phi_W)}{\Delta x}, \tag{5.12}$$

with the following coefficients

$$A_P = \frac{2\Gamma^\phi}{\Delta x^2} + \frac{\rho u}{\Delta x} \tag{5.13}$$

$$A_e = \frac{\Gamma^\phi}{\Delta x^2} \tag{5.14}$$

$$A_w = \frac{\rho u}{\Delta x} + \frac{\Gamma^\phi}{\Delta x^2} \tag{5.15}$$

Inspecting the coefficients, one sees they are all positive. For a negative u velocity the integration of Eq. (5.4) gives,

$$\rho u \phi_E - \rho u \phi_P = \Gamma^\phi \left(\frac{\phi_E - \phi_P}{\Delta x} \right) - \Gamma^\phi \left(\frac{\phi_P - \phi_W}{\Delta x} \right), \tag{5.16}$$

with the following coefficients,

$$A_P = \frac{2\Gamma^\phi}{\Delta x^2} - \frac{\rho u}{\Delta x} \tag{5.17}$$

$$A_e = -\frac{\rho u}{\Delta x} + \frac{\Gamma^\phi}{\Delta x^2} \tag{5.18}$$

$$A_w = \frac{\Gamma^\phi}{\Delta x^2}, \tag{5.19}$$

with all coefficient positives, since u is negative. The numerical solution of the pulse of ϕ using upwind is sketched in Fig. 5.3b, disappearing the oscillations, but a new discretization error, called numerical diffusion, takes place. The upwind scheme is stable, but the price to be paid is the smearing of the large gradients in the domain. These two behaviors obey a physical consistency that will be discussed in a next section. For now, it is important to keep in mind the following three facts that will be discussed in more depth later in this chapter:

1. The use of central differences (CDS), and other high-order schemes, in dominant advection problems in which the grid is not refined enough, generally generates unrealistic solutions with oscillations, because they are non-dissipative schemes, amplifying the errors;
2. UDS schemes produce physically coherent solutions, but have the property of smoothing high gradients, because they are dissipative.

3. CDS and UDS and other interpolations schemes that will be seen are approximations of the real function, hence, they introduce discretization errors. Numerical diffusion and numerical oscillations are, therefore, discretization errors.

When using these schemes care should be exercised because if the physical problem contains oscillations, using upwind they may be damped out. Using central differencing it may show up oscillations which do not exist in the real physics. Following, the physics behind an interpolation function is briefly discussed with the presentation of several 1D interpolation functions usually employed in numerical simulations.

5.4 Interpolation Functions for ϕ

In the previous section it was exercised two types of interpolation, CDS and UDS. In finite volume methods an interpolation function is needed to calculate the fluxes at the boundaries of the control volume which lies among two grid points. Figure 5.4 sketches a function which represents some physical problem, and two possible discretization, a coarse and a refined one. In the coarse grid it is evident that the interpolation among two grid points using CDS (dashed line) will contain large errors. On the other hand, if CDS is used in the fine grid, the interpolation produces better results. This means that there would be no worries about interpolation functions if the grid were fine enough such that CDS could be used without restrictions, satisfying the criterion given by Eq. (5.11). Perhaps this is what the future deserves for the interpolation functions in CFD simulations as the computer power increases.

But this is not the real picture in the simulation world nowadays, in which fine enough grids cannot be thoroughly used, and it is needed to live with coarse meshes and finding ways of minimizing its effects. The interpolation function is the agent which can help on that. Back to Fig. 5.4, if the mesh is coarse, the physics among the grid points should enter the interpolation function such that the interpolated value

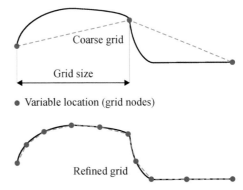

Fig. 5.4 Interpolation functions: coarse and fine grids

can be better calculated. This is why, as will be seen shortly, the differential equation to be solved is used to create the interpolation function of the problem.

5.4.1 The Physics Behind the Interpolation Functions

The adequacy of the CDS scheme for purely diffusive problems and the UDS scheme for dominant advection problems has a strong physical motivation. To interpret it, let us again consider Eq. (5.4) in which these two terms are present. As discussed in Chap. 2, the terms of a differential equation always represent the physics of the phenomenon, and for each term we can associate a mathematical interpretation. The diffusive term is elliptic, with second order derivatives and, therefore, requiring two boundary conditions at the extremes of the axis. This requirement means that a disturbance provoked in a point of the domain is transmitted (diffused) in all directions by this term. Therefore, the right-hand side of Eq. (5.4) transmits a disturbance equally in both directions of the x-axis. The advective term on the left-hand side of the equation, is of a first order derivative, is parabolic and, consequently, requires just one boundary condition. In fluid flows, the disturbance travels from upstream to downstream, and transmits disturbances only in the direction of the velocity. The effects of these terms logically influence the variable profile and must be considered when creating an interpolation function.

To help in this analysis, the idealized problem, shown in Fig. 5.5, is useful. Consider a one-dimensional flow with constant velocity u with temperature at $x = 0$ and $x = L$ equal to 1 and 0, respectively. The problem is idealized, and therefore, should have no questioning how temperatures equal to 1 and 0 are physically maintained; they are boundary conditions.

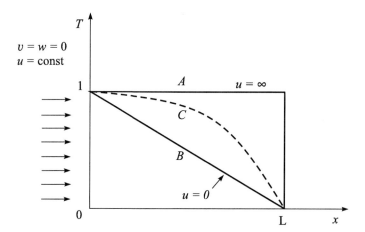

Fig. 5.5 Advection/diffusion problem

5.4 Interpolation Functions for ϕ

In Eq. (5.4) consider ϕ equals to T^1 for this 1D advection/diffusion problem.

Consider, initially, that the velocity u is equal to zero. In this situation the fluid is at rest, and the problem is the same as the heat conduction in a solid plate with temperatures 1 and 0 at the faces. In this case only the diffusive effects remain, and the solution of the equation is a straight line, since the prescribed temperatures at $x = 0$ and $x = 1$ will have the same influence in establishing the profile within the domain, due to the elliptical characteristics of diffusion. The exact solution, which is a straight line, coincides with our physical feelings.

The other limit happens when the velocity is very large, positive, tending to infinity. Our physical intuition tells us that a temperature equal to 1 will be established over the entire domain, since the downstream boundary condition, $T = 0$, will not influence the solution, because the diffusive effects (being small, or zero) cannot transmit information of the boundary conditions in the direction opposite to the flow. The advective effects, very strong in this direction, do not allow, and force over the entire domain the temperature equal to 1.

Between these two limits there are infinite solutions where the balance between the diffusive and advective effects exists, as shown by the dashed line in Fig. 5.5.

This problem teaches us that the use of central differences (CDS) is consistent for the diffusive terms, while the use of the one-sided approximation (UDS) is physically consistent for the advective term. It seems, therefore, coherent that the interpolation functions in advective/diffusive problems take this physical characteristic into account. Thus, for an advection-dominant problem, with large positive velocity, the appropriate interpolation function is curve A, for dominant diffusion, curve B, and for intermediate situations a curve, dashed line, that has as parameter the Peclet number, which is nothing more than the relationship between the advective and diffusive fluxes [1]. The upstream value on the UDS approximation, which has total influence in stablishing the solution is called the donor cell, since it donates the information. In 1D problems the donor cell is, clearly, the upwind cell, but in 2D and 3D it is not easily known, and to discover from where comes the information in the flow field is of utmost importance for creating interpolation functions which produces positive coefficients. This will be seen in Chap. 13.

5.4.2 One Dimensional Interpolation Functions

5.4.2.1 Exponential Scheme

As suggested in [2], the analytical solution of Eq. (5.4), serves as a family of 1D interpolation functions, with CDS and UDS being the limiting cases. Setting the position $x = 0$ and $x = L$, in Fig. 5.5, as the nodal points P and E, respectively, shown in Fig. 5.6, the boundary conditions for Eq. (5.4) are given by

---

[1] Temperature is used here just for having a familiar physical variable for analysis of diffusion and advection.

Fig. 5.6 Exponential interpolation functions

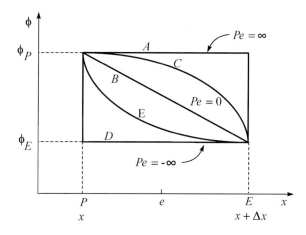

$$\phi = \phi_P \quad \text{in} \quad x = x_P$$
$$\phi = \phi_E \quad \text{in} \quad x = x_E \tag{5.20}$$

Defining the following dimensionless variables,

$$\xi = \frac{x - x_P}{\Delta x} \left(\rho \frac{u \Delta x}{\Gamma^\phi} \right)$$
$$\phi^* = \frac{\phi - \phi_P}{\phi_E - \phi_P} \tag{5.21}$$

the differential equation takes the form

$$\frac{\partial \phi^*}{\partial \xi} = \frac{\partial^2 \phi^*}{\partial \xi^2}, \tag{5.22}$$

with the following boundary conditions

$$\phi^* = 0 \quad \text{in} \quad \xi = 0$$
$$\phi^* = 1 \quad \text{in} \quad \xi = Pe \tag{5.23}$$

in which the Peclet number based on Δx is

$$Pe = \frac{\rho u \Delta x}{\Gamma^\phi} \tag{5.24}$$

and the analytical solution of Eq. (5.22) is,

$$\phi^* = \frac{e^\xi - 1}{e^{Pe} - 1} \tag{5.25}$$

5.4 Interpolation Functions for ϕ

As seen, Fig. 5.6 shows the two limiting cases for u positive, curves A (pure advection) and B (pure diffusion), and curve C for a problem in which both, advection and diffusion take place. Curve D is the pure advection situation with negative velocity, and curve E for the situation when both effects are present and negative velocity. It can be advanced that in real fluid flow problems the cell Reynolds (or Peclet) reaches values high enough, even in problems with moderate velocities, such that the solution tends to be close to the limiting upwind case. In these cases, it is easier to implement just upwind and not to introduce more complex relations which will not help in minimizing numerical diffusion.

The interpolation function given by Eq. (5.25) allows to find the value of the function and its derivative (gradient) in the integration points lying on the surface of the control volume. For the exponential scheme, the value of ϕ_e^* and $(\partial \phi^*/\partial \xi)_e$, for example, are given by

$$\phi_e^* = \frac{e^{\frac{Pe}{2}} - 1}{e^{Pe} - 1} \tag{5.26}$$

$$\left.\frac{\partial \phi^*}{\partial \xi}\right|_e = \frac{e^{\frac{Pe}{2}}}{e^{Pe} - 1} \tag{5.27}$$

If the same one-dimensional advection/diffusion problem that gave rise to the interpolation function is solved numerically using this interpolation function, the numerical solution obtained will be exact, regardless of the number of meshes employed. A similar observation was made in Chap. 3, in which the numerical solution of the one-dimensional steady state heat conduction in a flat plate with central differencing, provides the exact solution, regardless of mesh size. This happened because the interpolation function used (CDS) is the exact solution of the problem. Equation (5.21) must be used to obtain the dimensional expressions of ϕ. The same procedure should be done to obtain the values of ϕ and its derivatives at w, for subsequent substitution in Eq. (5.5). Note that the Pe number is calculated with Δx_e, for its evaluation at face e, and with Δx_w, for face w, that is, it is a local Peclet number.

The drawback of the exponential scheme is the computational time to evaluate the exponentials. Since the interpolation function depends on the velocity (Peclet), it will be necessary to compute exponentials for all interfaces of the control volumes. A variant of this method, with simplifications in the calculations, creating expressions that try to follow the exact solution by ranges of Peclet numbers, called Power-Law, is described in [3]. For completeness, the limiting cases of the exponential scheme are now reported.

5.4.2.2 Central Differencing Scheme

The CDS scheme uses a linear interpolation. Considering an equally spaced discretization, the values of ϕ at the integration points are,

$$\phi_e = \frac{\phi_E + \phi_P}{2}; \quad \phi_w = \frac{\phi_W + \phi_P}{2} \qquad (5.28)$$

As consequence, the derivatives are,

$$\left.\frac{\partial \phi}{\partial x}\right|_e = \frac{\phi_E - \phi_P}{\Delta x_e}; \quad \left.\frac{\partial \phi}{\partial x}\right|_w = \frac{\phi_P - \phi_W}{\Delta x_w} \qquad (5.29)$$

Substituting Eqs. (5.28) and (5.29) into Eq. (5.5) one obtains the coefficients given by Eqs. (5.8–5.10), which will be always negative regardless the sign of the velocity when Pe, or Re for the momentum equation, are greater than 2. All the implications of this scheme have been discussed. In obtaining Eqs. (5.8–5.10) it was used equally spaced meshes only for convenience and not as a necessity, since the expression can be weighted by the grid dimensions.

5.4.2.3 Upwind Scheme

To avoid the appearance of negative coefficients and numerical oscillations the upwind scheme (UDS) is used. Now, the interpolation functions have the following expressions, using Fig. 5.6,

$$\phi_w = \phi_W; \quad \phi_e = \phi_P; \quad u > 0 \qquad (5.30)$$

$$\phi_w = \phi_P; \quad \phi_e = \phi_E; \quad u < 0 \qquad (5.31)$$

with exactly the same coefficients as given by Eqs. (5.13–5.15) and Eqs. (5.17–5.19) for positive and negative u, respectively. Note that the diffusive term continued to be approximated by central differences. When the flow is advective-dominant it would be coherent to remove from the equation the diffusive term, but this is not required, the diffusive term will play no role in the solution and can be approximated by CDS, as if it was a pure diffusive problem.

5.4.2.4 WUDS—Weighted Upstream Differencing Scheme

In this scheme, the exact interpolation function is associated with two coefficients, α and β, depending on the Peclet number, and serve as weighting factors for advection and diffusion [4]. The values of ϕ and its derivative at the interfaces are written, taking again the east face example, are

$$\phi_e = \left(\frac{1}{2} + \alpha_e\right)\phi_P + \left(\frac{1}{2} - \alpha_e\right)\phi_E \qquad (5.32)$$

5.4 Interpolation Functions for ϕ

$$\Gamma_e^\phi \left.\frac{\partial \phi}{\partial x}\right|_e = \beta_e \Gamma_e^\phi \left(\frac{\phi_E - \phi_P}{\Delta x_e}\right) \qquad (5.33)$$

Inspecting the previous equations, one sees that for $\alpha = 0$ and $\beta = 1$ the CDS scheme is recovered, while for $\alpha = 0.5$ and for $\alpha = -0.5$ with $\beta = 0$ for both, the upwind scheme for positive and negative velocities, respectively, is recovered. The values of these coefficients are determined using Eq. (5.25) and Eqs. (5.32) and (5.33) rewritten for ϕ^*, and applying the boundary conditions given by Eq. (5.23). For Eq. (5.32), one finds

$$\phi_e^* = \left(\frac{1}{2} + \alpha_e\right)\phi_P^* + \left(\frac{1}{2} - \alpha_e\right)\phi_E^*, \qquad (5.34)$$

originating the expression for α_e,

$$\alpha_e = \frac{1}{2} - \frac{e^{\frac{Pe}{2}} - 1}{e^{Pe} - 1} \qquad (5.35)$$

For determining β_e, Eq. (5.33) results,

$$\Gamma_e^\phi \left.\frac{\partial \phi^*}{\partial \xi}\right|_e = \beta_e \Gamma_e^\phi \frac{\phi_E^* - \phi_P^*}{\Delta \xi}, \qquad (5.36)$$

with the expression for β_e, as

$$\beta_e = Pe \frac{e^{\frac{Pe}{2}}}{e^{Pe} - 1}, \qquad (5.37)$$

in which $\Delta\xi$ was made equal to Peclet, according to the dimensionless variables.

Keeping the expressions for α and β in exponential form entails the same computational difficulties already discussed for the exponential method. In [5] it is proposed the following expressions two avoid exponentials calculation,

$$\alpha_e = \frac{Pe^2}{10 + 2Pe^2} \qquad (5.38)$$

$$\beta_e = \frac{1 + 0.005 Pe^2}{1 + 0.05 Pe^2} \qquad (5.39)$$

recalling that in all relations involving the Peclet number the dimension is Δx.

Substituting the relations for ϕ_e and its derivative in the e face by Eqs. (5.32) and (5.33), and similar equations for ϕ_w and its derivative in the w face into Eq. (5.5), one finds the discretized equation in the usual form,

$$A_P \phi_P = A_e \phi_E + A_w \phi_W, \qquad (5.40)$$

with the following coefficients

$$A_e = -(\rho u)_e \left(\frac{1}{2} - \alpha_e\right) + \frac{\beta_e \Gamma_e^\phi}{\Delta x_e} \tag{5.41}$$

$$A_w = +(\rho u)_w \left(\frac{1}{2} + \alpha_w\right) + \frac{\beta_w \Gamma_w^\phi}{\Delta x_w} \tag{5.42}$$

$$A_P = A_e + A_w \tag{5.43}$$

The 1D mass conservation equation in the form,

$$(\rho u)_e - (\rho u)_w = 0, \tag{5.44}$$

must be used to obtain the coefficients given by Eqs. (5.41) and (5.42).

Figure 5.7 shows the behavior of the coefficient A_e with the velocity. The coefficient will always be positive, regardless of the sign of u. For u equal to zero, only the diffusive part remains in the coefficient, as it should be, while for positive and large u, the coefficient A_e tends to zero, as expected, since the nodal value E should no longer influence the value of the variable on P due to strong advection compared to diffusion. For the coefficient A_w the same analysis applies, i.e., it will be always positive, tending to zero when the velocity is increasing negatively.

Figure 5.8 shows the plot of α and β given by Eqs. (5.35) and (5.37) and by the approximations given by Eqs. (5.38) and (5.39). When the Pe number increases,

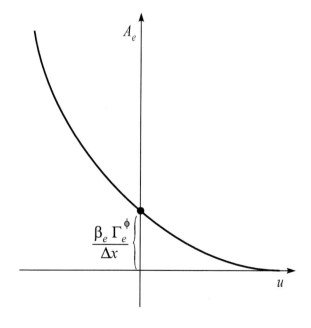

Fig. 5.7 Dependence of A_e on the velocity

5.4 Interpolation Functions for ϕ

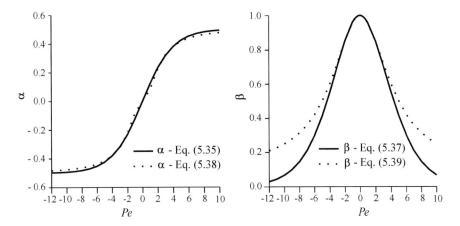

Fig. 5.8 Coefficients α and β

the approximation of the advective term approaches an upstream (UDS) scheme. Schemes of this type, where α and β changes in the calculation domain seeking to "weight" the influences of convection and diffusion, are called hybrid schemes. While it is true that the use of such schemes avoids spatial oscillations and possible divergence of the solution, it is also true that, as the velocities increase, α tends to 1/2, and the so-called numerical diffusion or false diffusion increases.

This is a very practical 1D modified exponential scheme. It is worth remembering, however, that when the velocity increases, soon the coefficient α goes to 0.5 or -0.5, depending on the sign of the velocity. In this situation β goes to zero, but the diffusive term can be kept in the equation with a CDS evaluation with no consequences in the numerical solution.

5.4.2.5 Quick—Quadratic Upwind Interpolation Scheme

The Quick-Quadratic Upwind Interpolation for Convective Kinematics interpolation function [6], is quite widespread and an option in some commercial computational fluid mechanics packages. In this scheme, the idea is to increase the order of approximation of the interpolation function by using a higher order polynomial. Considering Fig. 5.9, if the interest is to calculate the value of ϕ at integration point e, the expressions for positive and negative u are, respectively,

$$\phi_e = -\frac{1}{8}\phi_W + \frac{6}{8}\phi_P + \frac{3}{8}\phi_E \tag{5.45}$$

$$\phi_e = -\frac{1}{8}\phi_{EE} + \frac{6}{8}\phi_E + \frac{3}{8}\phi_P \tag{5.46}$$

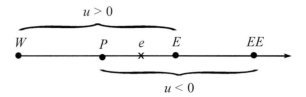

Fig. 5.9 Grid points used in the Quick scheme

The weights of each point in the two previous expressions appear from the weighting of the coefficients of a parabola passing through points W, P and E when u is positive, and EE, E and P when negative. The coefficients in Eqs. (5.45) and (5.46) shows the influence of each point on the evaluation of ϕ at the integration point.

It can be shown, through the Taylor series expansion, that the order of approximation is Δx^3. According to [7], the QUICK approximation is slightly superior than CDS, but both converge with second order error, and considerable differences between the two schemes are rarely observed. Again, it should be noticed that this scheme is prone to numerical oscillations.

5.4.2.6 Deferred Correction or Explicit Correction Scheme

An ever-present concern with the choice of interpolation function is the quality of the resulting matrix of coefficients. If the UDS scheme is used, the system will be always stable with dominant diagonal and no negative coefficients, at least when the donor cell is known. Therefore, it is recommended to work with a matrix with these characteristics but, at the same time, to use higher-order schemes than UDS. The following, proposed by Khosla and Rubin [8], called deferred correction, express the flux as,

$$F_e = F_e^{UDS} + \left(F_e^{CDS} - F_e^{UDS}\right)^o, \tag{5.47}$$

in which the upwind stabilizing part of the flux is treated implicitly and the second order explicitly. In Eq. (5.47), F_e is the approximation of the fluxes, and the super-indexes UDS and CDS indicate the upwind and central differences schemes. Since the second term on the right-hand side is treated explicitly, hence the super-index "o" indicating previous iteration, the linear systems will be solved with the upwind formulation, having the second order correction in the independent term. It is logical that increasing the independent term hurts convergence, but the damage is less than if the equations were solved with the CDS scheme implicitly. At convergence, the upwind evaluations cancel each other and the CDS scheme is the one actually used. In Eq. (5.47), other first-order and high-order schemes can replace the upwind and central schemes, respectively. In other words, the linear system is solved in a stable way using upwind implicitly, tending to a 2nd order when convergence is achieved, what means that numerical oscillation may be present in the solution. There will be no numerical diffusion.

5.4.3 Numerical or False Diffusion

5.4.3.1 Preliminaries

Recapping the effects shown schematically in Fig. 5.3, we learned that if the interpolation function adopted for the advective terms is central differences, there are two risks. The first is to experience divergence of the solution, caused by the use of linear solvers not suitable for the treatment of negative coefficients, and the second is to obtain unrealistic solutions presenting numerical oscillations, since the central differences scheme does not have the ability to dissipate the perturbations inherent to the solution process [9, 10].

The phenomena can be superimposed, that is, the presence of a negative coefficient can give rise to a disturbance that propagates without the possibility of being dissipated. This perturbation can grow and make the solution diverge or it can be limited, establishing a converged solution, but presenting numerical oscillations (Fig. 5.3a). The remedy to the oscillations, while still obtaining a solution with second order accuracy, is to refine the mesh. On the other hand, if the interpolation used is upwind, the scheme is stable, always obtaining a realistic solution, but with high built-in dissipation, as shown in Fig. 5.3b. This dissipation occurs, logically, in the regions of large gradients, often destroying the solution, as is the case of capturing a shock wave, which must be done accurately to identify the real position of the shock.

The use of upwind, accepting some numerical diffusion can, in some circumstances, be beneficial, since for many engineering situations it is better to have a solution, even knowing the inaccuracies, than have none. The gradient smoothing mechanism is equivalent to the physical diffusion process of a property and is therefore called numerical diffusion or false diffusion. How to interpret numerical diffusion has always been controversial in the literature. In the next section it is discussed the existing views and how numerical diffusion is interpreted it in this text.

5.4.3.2 Characterization of Numerical Diffusion

For long, numerical diffusion is explained using Taylor series expansion, demonstrating that when first order schemes, like upwind, are used it produces a term similar to a physical diffusion. Let us term this the mathematical way of demonstrating the appearance of numerical diffusion introduced by a numerical approximation. In the last three decades, when happened the boom of solving the Navier–Stokes equation, considerable work was dedicated in understanding the so-called numerical diffusion. As a result, there are today several interpretations of numerical diffusion, leaving behind the mathematical point of view. And this point of view is very important to

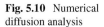

Fig. 5.10 Numerical diffusion analysis

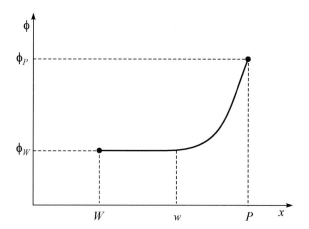

consider, since numerical diffusion is a truncation error introduced by the approximation of the advective term, and Taylor series expansion can do the job of finding this error. Let's see, briefly, how to find the truncation errors when using central difference and upwind schemes. Referring to Fig. 5.10, the calculation of the advective flux at the integration point w is given by

$$A_w = \rho_w u_w \phi_w, \qquad (5.48)$$

in which ϕ_w should be found using ϕ at grid nodes. Consider now a forward and backward Taylor expansion series around w, given by

$$\phi_P = \phi_w + \left(\frac{\partial \phi}{\partial x}\right)_w \left(\frac{\Delta x}{2}\right) + \left(\frac{\partial^2 \phi}{\partial x^2}\right)_w \left(\frac{\Delta x}{2}\right)^2 \frac{1}{2!} - O(\Delta x^3)\ldots \qquad (5.49)$$

$$\phi_W = \phi_w - \left(\frac{\partial \phi}{\partial x}\right)_w \left(\frac{\Delta x}{2}\right) + \left(\frac{\partial^2 \phi}{\partial x^2}\right)_w \left(\frac{\Delta x}{2}\right)^2 \frac{1}{2!} + O(\Delta x^3)\ldots \qquad (5.50)$$

Using a central differencing scheme, ϕ_w can be found using Eqs. (5.49) and (5.50), to give

$$\phi_w = \frac{\phi_P + \phi_W}{2} + O(\Delta x^2), \qquad (5.51)$$

Using UDS for the flux calculation, ϕ_w in Eq. 5.48 should be ϕ_W. The expression for ϕ_w, to find the truncation error can be found from Eq. (5.50), as

$$\phi_w = \phi_W + \left(\frac{\partial \phi}{\partial x}\right)_w \left(\frac{\Delta x}{2}\right) + O(\Delta x^2) \qquad (5.52)$$

Inserting Eqs. (5.51) and (5.52) into Eq. (5.48), one at a time, gives,

5.4 Interpolation Functions for ϕ

$$A_w = \rho_w u_w \phi_w = \rho_w u_w \underbrace{0.5(\phi_P + \phi_W)}_{CDS} + O(\Delta x^2), \qquad (5.53)$$

and

$$A_w = \rho_w u_w \phi_w = \rho_w u_w \left(\phi_W + \left(\frac{\partial \phi}{\partial x} \right)_w \left(\frac{\Delta x}{2} \right) \right) + O(\Delta x^2), \qquad (5.54)$$

which can be written, as

$$A_w = \rho_w u_w \phi_w = \rho_w u_w \underbrace{\phi_W}_{UDS} + \underbrace{\rho_w u_w \left(\frac{\partial \phi}{\partial x} \right)_w \left(\frac{\Delta x}{2} \right)}_{False\ diffusion} + O(\Delta x^2) \qquad (5.55)$$

Equation (5.55) demonstrates that when using an upwind scheme, the approximation carries a first order truncation error. It can be written as,

$$A_w = (A_w)_{UDS} + \Gamma_{false} \left(\frac{\partial \phi}{\partial x} \right)_w \qquad (5.56)$$

As this term contains a first order derivative of ϕ it resembles a diffusion term, called false or numerical diffusion, which is a first order truncation error. Hence, the numerical diffusion is, yes, a truncation error of first order caused by the first order interpolation. As seen in Eq. (5.53) the central differencing scheme doesn't introduce first order errors in the approximation, what means that this scheme doesn't introduce numerical diffusion. It introduces possible numerical oscillation.

The physical approach, always claimed in the developments of schemes in finite volume methods, has also allowed different interpretations of the numerical diffusion. For example, in [3] it is analyzed a 1D problem of the advection of a pulse of ϕ, shown schematically in Fig. 5.11, whose solution is numerically obtained without numerical diffusion and, therefore, without gradient smoothing using the upwind scheme. In this case, the upwind scheme gave the exact solution, and there would be no justification for claiming that the upwind scheme introduces numerical diffusion because of being of first order. The problem is a 1D stream with constant velocity u and the pulse is characterized by two regions with $\phi = 2$ and $\phi = 1$. If the fluid is inviscid there is no physical effect which can disturb this pulse. The differential equation for this problem, therefore, is

$$\frac{\partial}{\partial x}(\rho u \phi) = 0 \qquad (5.57)$$

Using a 1D grid aligned with the flow and integrating Eq. (5.57), the advective term reads,

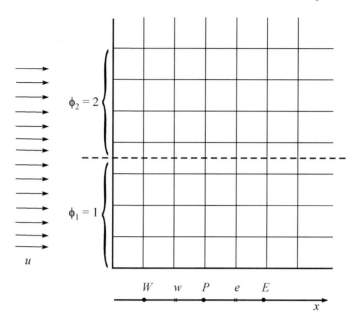

Fig. 5.11 Propagation of a discontinuity in x

$$\frac{\partial}{\partial x}(\rho u \phi) = \rho_e u_e \phi_e - \rho_w u_w \phi_w = 0, \quad (5.58)$$

and using $\phi_e = \phi_P$ and $\phi_w = \phi_W$, what identifies an upwind scheme, and considering the density and the velocity constant, the solution is

$$\phi_P = \phi_W, \quad (5.59)$$

which is the exact solution, that is, maintenance of the pulse of ϕ as x progresses.

Now, consider the same physical problem defined in a 2D domain. A 1D problem becomes 2D in a two-dimensional frame of reference, as shown in Fig. 5.12. The differential equation is, therefore,

$$\frac{\partial}{\partial x}(\rho u \phi) + \frac{\partial}{\partial y}(\rho v \phi) = 0, \quad (5.60)$$

The numerical integration gives,

$$\rho_e u_e \phi_e - \rho_w u_w \phi_w + \rho_n v_n \phi_n - \rho_s v_s \phi_s = 0, \quad (5.61)$$

and using upwind in each direction, the resulting discretized equation reads,

$$\phi_P = \frac{\phi_W + \phi_S}{2} \quad (5.62)$$

5.4 Interpolation Functions for ϕ

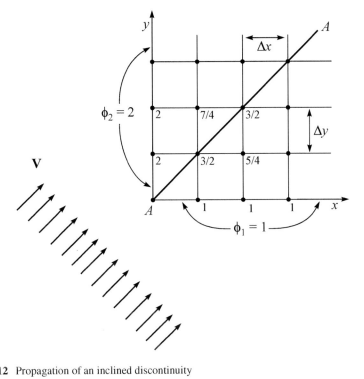

Fig. 5.12 Propagation of an inclined discontinuity

The solution of the problem governed by Eqs. (5.57) and (5.60) should be the same, since the physical problem did not change. Only the coordinate system has changed, making the problem two-dimensional. However, Eq. (5.62) gives a solution that is different from the exact solution given by Eq. (5.59), which is $\phi = 1$ for the points below the $\overline{AA}$ line and $\phi = 2$ for the points above. There is a diffusion of the property from the upper region to the lower region, called numerical diffusion or false diffusion, since there is no physical diffusion predicted in the differential equation.

It is then said that the existence of this numerical diffusion is because the flow is oblique to the meshes and that there is a gradient normal to the flow direction. Why first order upwind approximation gives the exact solution when the grid is aligned with the flow? The mathematical view fails? The next section seeks to clarify these issues.

5.4.3.3 Key Considerations About Numerical Diffusion

The lack of agreement in the literature about the reasons of appearance of numerical diffusion, the mathematical point of view, and the numerical results encountered solving the pulse propagation using upwind motivates the following comments.

(1) First, the reason why Eq. (5.57) has the numerical solution equal to the exact one when using upwind is because an exact interpolation function was used, in this case upwind, which is the solution to the problem itself. It is not because the mesh is aligned with the flow. The rationale should be the reverse. Because the flow is aligned with the grid the 1D upwind interpolation function becomes the analytical solution. And, for any problem, if it would be possible to have the analytical solution as interpolation function there would be no truncation errors and the numerical solution would be equal to the analytical one, regardless of mesh size. When the problem becomes 2D, as given by Eq. (5.60), using the analytical solution of this problem as interpolation function we would have a numerical solution also exact for the problem of Fig. 5.12, with the mesh inclined with respect to the velocity vector.

Of course, the inclination of the velocity vector with the mesh aggravates the numerical diffusion because it is insisted in using 1D interpolation functions in multidimensional problems. But the grid inclination is not the essence of the problem.

(2) To cite one more example, the one-dimensional advection/diffusion problem given by Eq. (5.4) will also have exact numerical solution, no matter the mesh size, if the interpolation function is that of the exponential scheme.

(3) Therefore, if the exact interpolation function is used, that is, obtained from the differential equations to be solved, the numerical solution will be exact, regardless of the mesh size and dimensionality of the problem. Thus, the existence of truncation errors (i.e., inaccurate solution) is directly linked to the nature of the interpolation function employed.

(4) Hence, the use of non-exact interpolation functions to approximate the advective terms generates truncation errors that may be associated with dissipative or non-dissipative behavior. Truncation errors originated via central-difference interpolation are non-dissipative, which produce the so-called numerical oscillations, while those associated with upwind interpolation functions are dissipative and smooth the gradients existing in the domain, producing the so-called numerical diffusion, a phenomenon similar to a physical diffusion, as was seen by the mathematical approach.

(5) Inaccuracies in interpolation functions will always exist and can be generated in several ways. One of them, and critical, is the use one-dimensional interpolation in 2D and 3D problems, which is the case in the problem of Fig. 5.12.

Based on the previous comments, we conclude that the use of non-exact interpolation functions gives rise to truncation errors. Such truncation errors, when associated with the advective terms, can be classified into dissipative and non-dissipative and give rise to numerical diffusion and numerical oscillation, respectively.

It is more consistent, therefore, to define *numerical diffusion as the dissipative truncation errors associated with the advective terms, caused by the fact that the interpolation function is not exact*. It is logical that, in complex two and three-dimensional problems, there will be no possibility of using an exact interpolation functions, because if the solution would exist to be used as interpolation function,

5.4 Interpolation Functions for ϕ

there would be no reason for solving it numerically. Therefore, the solution will always be contaminated with numerical diffusion if the advective terms are approximated by dissipative schemes. It is possible to minimize this by creating interpolation functions as close as possible to the solution of the physical problem to be solved. This will be discussed in a coming section.

The reader may try to conclude that a two-dimensional pure conduction problem, in which one-dimensional central differences are used in each direction independently, which characterizes a non-exact interpolation function, has numerical diffusion. To dispel this doubt, it should always be remembered that numerical diffusion appears only when dissipative schemes are associated to the advective terms. By its turn, numerical oscillation is present only when non-dissipative truncation errors are also associated with the advective terms. In the case of the 2D heat conduction problem truncation errors do exist, because the interpolation function is not exact, but they do not give rise to numerical diffusion neither to numerical oscillations, because there is not advective terms in the equation. This is why there are no difficulties in solving numerically a two or three-dimensional heat conduction problem using central differences.

Therefore, the use of central differences to approximate the advective terms eliminates numerical diffusion, because when using central differences, the truncation errors are non-dissipative, do not give rise to numerical diffusion, and can be eliminated by mesh refinement. An example of this is in the work of Silva [11], who solved the problem of the propagation of the discontinuity given by Eq. (5.60), on the inclined mesh, using one-dimensional central differences as interpolation function of the advective terms, obtaining a result without numerical diffusion, that is, without the smoothing of the discontinuity.

Another ambiguous question in the literature is whether numerical diffusion exists in one-dimensional problems. For this analysis, consider again the one-dimensional advection/diffusion problem given by Eq. (5.4), being approximated using upwind in the advective term and central differences in the diffusive term. Since this interpolation function is not exact and there is a dissipative scheme (upwind) associated with the representation of the advective term, by the definition of this text there is numerical diffusion, yes, for a one-dimensional problem. To prove this, simply solve the problem numerically using upwind and verify that the profile will be more diffusive than that obtained by the exact solution when the advective term predominates.

In two and three-dimensional advective-dominant problems, the numerical diffusion is accentuated, precisely because the use of one-dimensional interpolation functions departs too much from the exact 3D interpolation function.

Among numerical analysts of finite volumes there is a fairly well disseminated concept that numerical diffusion is fundamentally linked to the inclination of the velocity vector with the mesh and not to truncation errors. The non-alignment of the velocity vector with the mesh only makes the one-dimensional interpolation functions more inaccurate, as already emphasized, generating truncation errors that may give rise to numerical diffusion, numerical oscillation, or other errors. Numerical diffusion decreases with mesh refinement, and so do all other truncation errors, if

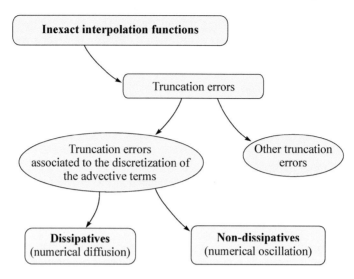

Fig. 5.13 The origin of numerical oscillation and diffusion

the approximation of the differential equation is consistent. Figure 5.13 shows the different truncation errors, showing the origin of numerical diffusion and oscillation.

Finally, it should be emphasized that a good interpolation function is one that comes from the differential equation one is trying to solve. Therefore, transient effects, lateral diffusion, lateral advection, source terms should be considered, if possible, in designing the interpolation function. The outcome of this analysis tells that if it would be possible to use very fine grids, central differential schemes would suffice for the advection as well as diffusion terms. Remember that in this case the cell Peclet or Reynolds numbers would be less than 2 for a Cartesian regular grid, with the same implications for unstructured grids.

In the following section, it is discussed some more interpolation functions that attempt to reduce numerical diffusion by working with grid points in more than one dimension.

5.4.4 Two and Three-Dimensional Interpolation Functions

Since the use of one-dimensional first-order interpolation functions causes numerical diffusion, and this is aggravated using them in 2D and 3D problems, the alternative, besides mesh refinement, is to create functions with the dimensionality of the problem. Only the basic ideas of some of them will be presented, in order to show the procedure of creating a multidimensional interpolation function. The literature on the subject is vast, and it is beyond the scope of this text to review the subject. Again, the developments will be made for a 2D problem, for simplicity, but the idea is the

5.4 Interpolation Functions for ϕ

same for 3D problems, with the corresponding additional work for the computational implementation.

5.4.4.1 SUDS and SWUDS—Skew Weighted Upstream Differencing Scheme

The pioneering ideas in the development of two-dimensional interpolation functions were presented in [12] when it was developed two methods that create interpolation functions aligned with the velocity vector. To develop the scheme along the direction of the velocity vector allows upwind schemes to be applied, and the region which donates the information can be identified, even though it is not being easy to relate this region with grid nodes such that positive coefficients are generated.

Since in most computer codes the mesh is fixed, and even in adaptive codes it is difficult to adapt the mesh to the velocity vector for all regions of the domain, it is natural to write the interpolation function along the velocity vector. In this way, to evaluate, for instance, the value of ϕ on the west face, as shown in Fig. 5.14, the values of the variable in SW and W need to participate in the interpolation function instead of just participating W and P, since it is the southwest region what donates the information. It is easy to see that the one-dimensional interpolation function can be applied along a coordinate s. This is the principle of the SUDS, apply an upstream scheme skewed with the mesh along the velocity direction. By doing this, one of the factors of appearance of numerical diffusion, the dealignment of the velocity vector with the mesh, is eliminated. Other factors must also be taken into consideration, such as diffusion across the flow, source terms etc. The key difficulty is to find ϕ_u which can't be found just averaging the closest grid nodes. The scheme will be a function of 9 points in 2D and will increase the number of non-zeros in the matrix.

As pointed out more than once, the most appropriate interpolation function is the one obtained from a differential equation as close as possible to the equation to be solved. For two-dimensional problems, the two-dimensional advective/diffusive transport equation, without considering the source, pressure, and transient terms, seems to be a good alternative. Taken ρ and u as locally constant, this equation, written for the face w, has the form,

$$(\rho u)_w \frac{\partial \phi}{\partial x} + (\rho v)_w \frac{\partial \phi}{\partial y} = \Gamma_w^\phi \left(\frac{\partial^2 \phi}{\partial x^2} + \frac{\partial^2 \phi}{\partial y^2} \right)_w \quad (5.63)$$

Grouping the advective terms for the s direction,

$$\rho V|_w \frac{\partial \phi}{\partial s}\bigg|_w = \Gamma_w^\phi \left(\frac{\partial^2 \phi}{\partial s^2} + \frac{\partial^2 \phi}{\partial n^2} \right)_w \quad (5.64)$$

in which n is normal to s and $|V|$ is the magnitude of the velocity vector. The analytical solution of Eq. (5.64) is laborious, and a solution which admits a linear variation of ϕ along the normal to the flow is,

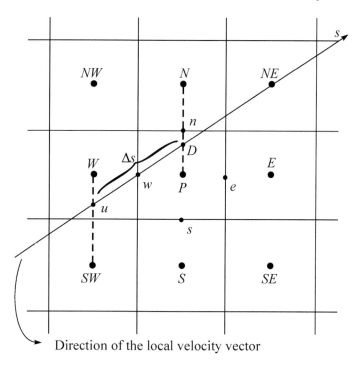

Fig. 5.14 Nodal points for the SUDS scheme

$$\phi = k_1 + k_2 \left(y \frac{u}{|\mathbf{V}|} + x \frac{v}{|\mathbf{V}|} \right) \tag{5.65}$$

in which k_1 and k_2 are constants and (x, y) are measured along the normal to s in the integration point w. The constants are found using two out of the six grid points which interferes in the west face. When the gradient of ϕ normal do the flow is zero the algorithm reduces to a pure upwind. The equations of the coefficients, not shown here, allows to conclude that there is the possibility of negative coefficients, depending on the inclination of the velocity vector, again addressing the problem in finding ϕ_u. This scheme reduces numerical diffusion but, as expected, introduces some complexity to implement. Many schemes based on these ideas have been developed, seeking to eliminate the problem of negative coefficient, as in [13–17]

Another method developed in [12], SWUDS–Skew Weighted Upstream Differencing Scheme, also uses Eq. (5.64) and proposes a solution of the form,

$$\phi = A + B \left(y \frac{u_w}{|\vec{V}|} + x \frac{v_w}{|\vec{V}|} \right) + C e^{ax+by} \tag{5.66}$$

in which a and b are the Peclet number, given by

5.4 Interpolation Functions for ϕ

$$a = \left.\frac{\rho u x}{\Gamma^\phi}\right|_w ; \quad b = \left.\frac{\rho v y}{\Gamma^\phi}\right|_w \tag{5.67}$$

In this scheme there is a balance among advection and diffusion, controlled by the Peclet numbers, therefore, being a weighting differential scheme.

Huget [18], proposed two new schemes based on SUDS, the MSUS (Modified Skew Upstream Scheme) and the MWUS (Mass Weighting Upstream Scheme). They improve the results obtained by SUDS, the latter being quite stable, with no generation of negative coefficients and no oscillations in the converged solution. MWUS dedicates effort in finding the donor cell, the key detail for assuring positive coefficients when applying the interpolation function [19, 20]. A skew scheme related to EbFVM, method to be seen in Chap. 13, can be seen in [21]. Investigating the behavior of interpolation functions and comparing them was a hot topic of research in the last three decades, and the literature is plenty of works which are extensions of the basic ones reported herein. To finalize this section the rationale behind a general interpolation function which works for multidimensional problems will be presented.

5.4.4.2 General Interpolation for ϕ

A general form, applicable to 2D and 3D situation follows the idea of the SWUDS, with a little modification in the form of calculating all other influences than upwind. Consider again Fig. 5.14, in which a Taylor series expansion is performed along s, as

$$\phi_{ip} = \phi_u + \left.\frac{\partial \phi}{\partial s}\right|_u \frac{\Delta s}{2} + \left.\frac{\partial^2 \phi}{\partial s^2}\right|_u \frac{\Delta s^2}{4} + O(\Delta s^3) + \cdots +, \tag{5.68}$$

in which ϕ_{ip} is the value at the integration point treated as e or w in previous developments in Cartesian coordinates.

Two important quantities need to be evaluated in this expression, ϕ_u and $(\partial \phi/\partial s)_u$. The later can be determined with the help of Eq. (5.63) with the advection term written in the s direction, in discrete form, as

$$\left.\frac{\Delta \phi}{\Delta s}\right|_u = \frac{1}{\rho \mathbf{V}}\left[\Gamma^\phi_{ip}\left(\frac{\partial^2 \phi}{\partial x^2} + \frac{\partial^2 \phi}{\partial y^2}\right)_u\right] + B^\phi \tag{5.69}$$

which is similar to Eq. (5.64), whose idea is to use the general conservation equation for ϕ to find its derivative. This equation can be approximated using finite differences, since it is an auxiliary equation and doesn't need to obey the conservation principles. The determination of ϕ_u, since it involves the concept of the donor cell, is the key issue to be considered, due to the difficult of identifying what grid point, or combination of them, donates the information to ϕ_{ip}. This is a matter not easily solved in cell-center methods. For the element-based cell-vertex method it is possible to create schemes in which the donor grid point can be identified. This will be discussed when the Element-based Finite Volume Method will be presented in Chap. 13.

5.4.4.3 Second Order Upwind and Muscl Schemes

The Second Order Upwind (SOU) scheme [22] derives from the general interpolation function just described and can be expressed as

$$\phi_{ip} = \phi_P + (\nabla\phi)_P \cdot \mathbf{r} \tag{5.70}$$

in which ϕ_P and $(\nabla\phi)_P$ are evaluated at the center of the upstream control volume and $\mathbf{r}$ is the vector joining the center of the control volume to the integration point. Based on the previous discussions, this scheme doesn't guarantee positive coefficients because the upstream point (the donor cell) is not chosen based on the mass flow donating the information.

The MUSCL—Monotonic Upstream-Centered Scheme for Conservation Laws [23], blends the central differencing scheme with the SOU scheme, by

$$\phi_{ip} = \theta(\phi_{CD})_{ip} + (1-\theta)(\phi_{SOU})_{ip}, \tag{5.71}$$

which improves the interpolation function compared to the SOU scheme.

5.4.4.4 High Resolution Scheme

In the high-resolution scheme [24], the magnitude of ϕ at the surrounding nodes are kept under control to avoid undershoots and overshoots. The equation,

$$\phi_{ip} = \phi_P + \beta(\nabla\phi)_P \cdot \mathbf{r} \tag{5.72}$$

uses the factor β according to the boundedness principle of Barth and Jesperson [25], in which ϕ_{min} and ϕ_{max} are calculated for the node and its surrounding ones. Then, for each integration point around the node, Eq. (5.72) is solved for β. The value of β used is the minimum of all integration point which surrounds the node. The value of β is also not permitted to exceed 1. This scheme is said to be TVD for a one-dimensional situation [24]. In Eq. (5.72) β and $\nabla\phi)_P$ are taken from the upwind node.

5.5 Conclusions

The purpose of this chapter was to present the key points that should be carefully considered when developing or choosing the interpolation functions and explain their influences on the numerical solution. The main learning point was that the interpolation functions are responsible for the truncation errors of an approximation. In other words, if there would be the possibility of using the exact interpolation functions, the exact numerical solution of the problem would be obtained, regardless of the mesh size.

Furthermore, and perhaps most importantly in the chapter, we classify the truncation errors as dissipative and non-dissipative, defining numerical diffusion as being errors of a dissipative character, associated with the advective terms. Therefore, numerical diffusion does disappear with mesh refinement, although it is possible to find contrary statements in the literature.

It was also pointed out that the central-difference approximations are of a non-dissipative character, and therefore not classified as numerical diffusion errors. Non-dissipative errors associated with the representation of the advective terms give rise to numerical oscillations.

Some widely used one-dimensional interpolation functions and the philosophy embedded in the construction of two and three-dimensional interpolation functions were also shown. A general interpolation function was also presented, suitable for 2D and 3D problems, identifying the crucial terms to be obtained. The literature is vast on this topic; however, as mentioned, it is not the goal of this text to review the literature on the subject. For example, Total Variation Diminishing (TVD) schemes is a good alternative for the development of robust interpolation functions for CFD calculations [26–28].

To end this chapter let's go back to Eq. (5.3) and recognize that both ϕ and $\nabla\phi$ needs interpolation function. It was devoted the whole chapter for the interpolation for ϕ only, with the comment that the interpolation for $\nabla\phi$, besides of being easy to find, poses no difficulties for convergence. It will be discussed the determination of $\nabla\phi$ at the integration points when unstructured grids will be presented in the context of cell-center and cell-vertex methods.

5.6 Exercises

5.1 Explain how numerical oscillations and numerical diffusion appear in the solution of advective-dominant problems.

5.2 Why does mesh refinement eliminate numerical diffusion and numerical oscillation?

5.3 To verify the appearance of numerical oscillations and numerical diffusion, the one-dimensional convection/diffusion problem in the steady-state regime is extremely useful. Using a fully implicit formulation and the finite volume method, solve Eq. (5.4), considering as boundary conditions $\phi = 0$ at $x = 0$ and $\phi = 1$ at $x = 1$, for the following situations:

(a) Central differences for the diffusive and advective term. Gradually increase the velocity and observe the appearance of numerical oscillations in the region of large gradients, that is, near $x = 1$.
(b) Upwind for the advective term and center differences for the diffusive term. Again, increase the velocity and observe that the captured gradient is dissipated, i.e., there is the appearance of numerical diffusion, even in this problem in which the flow is aligned with the grid.
(c) Use as interpolation functions Eqs. (5.26) and (5.27). Vary the number of points of the grid and compare the numerical solution with the exact one given by Eq. (5.25). Why is the solution exact regardless of the mesh size? Are there no more truncation errors?
(d) Now, solve the problem using WUDS with α and β given by Eqs. (5.38) and (5.39). Will the numerical result again be mesh independent? Why?

References

1. Maliska CR (2000) On the physical significance of some dimensionless numbers used in heat transfer and fluid flow. In: 8th Brazilian Congress of engineering and thermal sciences, ENCIT, p 49, Porto Alegre, RS, Brazil. https://abcm.org.br/anais/encit/2000/arquivos/s06/s06p28.pdf
2. Spalding DB (1972) A novel finite-difference formulation for differential expressions involving both first and second derivatives. Int J Numer Methods Eng 4:551
3. Patankar SV (1980) Numerical heat transfer and fluid flow. Hemisphere Publishing Corporation
4. Raithby GD, Torrance KE (1974) Upstream-weighted differencing schemes and their application to elliptic problems involving fluid flow. Comput Fluids 2:191–206
5. Raithby GD, Schneider GE (1980) Prediction of surface discharge jets by a three-dimensional finite-difference model. J Heat Transfer 102(1):138–145
6. Leonard BP (1979) A stable and accurate convection modelling procedure based on quadratic interpolation. Comput Methods Appl Mech.Eng, 59–98
7. Ferziger JH, Peric M (1999) Computational methods for fluid dynamics, 2nd edn. Springer
8. Khosla PK, Rubin SG (1974) A diagonally dominant second-order accurate implicit scheme. Comput Fluids 2:207–209
9. Anderson DA, Tannehill JC, Pletcher RH (1984) Computational fluid mechanics and heat transfer. Hemisphere Pub. Corporation[
10. Pulliam TH (1986) Artificial dissipation models for the Euler equations. AIAA J 24:1931–1940
11. Silva AFC (1991) Finite volume solution of all speed flows (in Portuguese). Ph.D. Thesis, Federal University of Santa Catarina, Florianopolis, Brazil. https://repositorio.ufsc.br/handle/123456789/106327
12. Raithby GD (1976) Skew upstream differencing schemes for problems involving fluid flow. Comput Methods Appl Mech Eng 9:153–164
13. Lillington JN (1981) A vector upstream differencing schemes for problem in fluid flow involving significant source terms in steady-state linear systems. Int J Numer Methods Fluids 1:3–16
14. Hassan YA, Rice JG, Kin JH (1983) A stable mass-flow-weighted two-dimensional skew upwind scheme. Numer Heat Transfer 6:395–408
15. Schneider GE, Raw MJ (1986) A skewed, positive influence coefficient upwinding procedure for control-volume-based finite-element convection-diffusion computation. Numer Heat Transfer 9:1–26

References

16. Marchi CH (1993) High order schemes for the solution of fluid flows without numerical dispersion. Braz J Mech Sci 15(3):231–249
17. Schneider GE (1986) A novel co-located finite difference procedure for numerical computation of fluid flow. In: AIAA PAPER 86-1330, AIAA/ASME IV joint thermophysic and heat transfer conference, Boston, USA
18. Huget RG (1985) The evaluation and development of finite volume approximation schemes for fluid flow and heat transfer predictions. Ph.D. Thesis, University of Waterloo, Waterloo, Canada
19. Hurtado FSV, Maliska CR (2012) A family of positive flow-weighted advection schemes for element-based finite volume methods. Numer Heat Transfer, Part B, Fund: Int J Comput Methodol 62(2–3):113–140
20. Raw MJ (1985) A new control-volume-based finite element procedure for numerical solution of the fluid flow and scalar transport equations. Ph.D. Thesis, University of Waterloo, Canada
21. Rousse DR, Lassue S (2008) A skew upwinding scheme for numerical radiative transfer. https://doi.org/10.1615/ICHMT.2008.CHT.820
22. Ansys Fluent Theory Guide (2021) Ansys, Inc. Canonsburg, PA 15317
23. van Leer B (2006) Upwind and high-resolution methods for compressible flow: from donor cell to residual-distribution schemes. Commun Comput Phys 1:192–206
24. Ansys CFX. Solver Theory Guide (2021) Ansys, Inc. Canonsburg, PA 15137
25. Barth TJ, Jespersen DC (1989) The design and application of upwind schemes on unstructured meshes. In: AIAA Paper No. 89–0366, 27th AIAA aerospace sciences meeting and exhibit, Reno, NV, USA
26. Versteeg HK, Malalasekera W (2007) An introduction to computational fluid dynamics: the finite volume method, 2nd edn. Pearson Education
27. Zhang D, Joang C, Liang D, Cheng L (2015) A review on TVD schemes and a refined flux-limiter for steady-state calculations. J Comp Phys 302:114–154
28. Hirsch C (1991) Numerical computation of internal and external flows, vol 1. Wiley

Chapter 6
Three-Dimensional Advection/Diffusion of ϕ

6.1 Introduction

The previous chapters exhibited all the necessary background to numerically solve an advection/diffusion problem for one-dimensional situations considering the velocity field known. In this chapter, for the sake of completeness, the integration of the equation for a three-dimensional situation is realized. The velocity field continue to be considered as known. There are no novelties in this chapter but the integration of the equation in 3D just to follow the whole process of discretization, applying an interpolation function seen in the previous chapter and writing all coefficients. This helps the student to analyze the coefficients and see the influences of geometry and physics on them.

6.2 Integration of the 3D Equation for ϕ

For a 3D advection/diffusion problem, Eq. (2.64) must be integrated in time and in space as shown in Fig. 6.1, in which, for ease of specifying the dimensions, only the straight lines connecting the centers of the control volumes are shown. It is already known that the six lower case letters identify the interfaces, or integration points, of the control volume centered in P, which is connected to its six neighboring control volumes via coefficients.

The integration of the general equation for ϕ, reads

$$\int_{V,t} \frac{\partial}{\partial t}(\rho\phi)dVdt + \int_{V,t} \frac{\partial}{\partial x}(\rho u\phi)dVdt + \int_{V,t} \frac{\partial}{\partial y}(\rho v\phi)dVdt + \int_{V,t} \frac{\partial}{\partial z}(\rho w\phi)dVdt$$
$$= \int_{V,t} \frac{\partial}{\partial x}\left(\Gamma^\phi \frac{\partial\phi}{\partial x}\right)dVdt + \int_{V,t} \frac{\partial}{\partial y}\left(\Gamma^\phi \frac{\partial\phi}{\partial y}\right)dVdt + \int_{V,t} \frac{\partial}{\partial z}\left(\Gamma^\phi \frac{\partial\phi}{\partial z}\right)dVdt + \int_{V,t} S^\phi dVdt, \quad (6.1)$$

giving,

© The Author(s), under exclusive license to Springer Nature Switzerland AG 2023
C. R. Maliska, *Fundamentals of Computational Fluid Dynamics*, Fluid Mechanics and Its Applications 135, https://doi.org/10.1007/978-3-031-18235-8_6

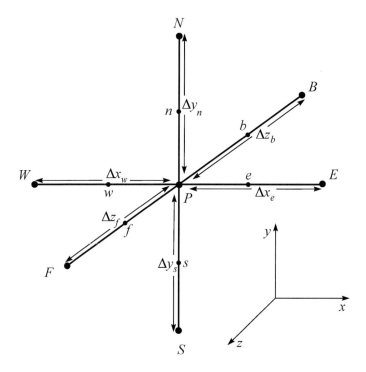

Fig. 6.1 Three-dimensional stencil

$$(M_P\phi_P - M_P^o\phi_P^o)/\Delta t + \dot{M}_e\phi_e^\theta - \dot{M}_w\phi_w^\theta + \dot{M}_n\phi_n^\theta - \dot{M}_s\phi_s^\theta + \dot{M}_f\phi_f^\theta - \dot{M}_b\phi_b^\theta$$
$$= D_1\left.\frac{\partial\phi}{\partial x}\right|_e^\theta - D_1\left.\frac{\partial\phi}{\partial x}\right|_w^\theta + D_2\left.\frac{\partial\phi}{\partial y}\right|_n^\theta - D_2\left.\frac{\partial\phi}{\partial y}\right|_s^\theta + D_3\left.\frac{\partial\phi}{\partial z}\right|_f^\theta - D_3\left.\frac{\partial\phi}{\partial z}\right|_b^\theta + L\left[S^\phi\right]^\theta$$
(6.2)

This equation is still in the general form being possible to obtain the explicit, implicit and fully implicit formulations according to the value of θ. The mass fluxes and the diffusion coefficients at the integration points are given by,

$$\dot{M}_e = \rho u \Delta y \Delta z|_e; \quad \dot{M}_w = \rho u \Delta y \Delta z|_w \tag{6.3}$$

$$\dot{M}_n = \rho v \Delta x \Delta z|_n; \quad \dot{M}_s = \rho v \Delta x \Delta z|_s \tag{6.4}$$

$$\dot{M}_f = \rho w \Delta x \Delta y|_f; \quad \dot{M}_b = \rho w \Delta x \Delta y|_b \tag{6.5}$$

$$D_{1e} = \Gamma^\phi \Delta y \Delta z\big|_e; \quad D_{1w} = \Gamma^\phi \Delta y \Delta z\big|_w \tag{6.6}$$

6.2 Integration of the 3D Equation for ϕ

$$D_{2n} = \Gamma^\phi \Delta x \Delta z\big|_n; \quad D_{2s} = \Gamma^\phi \Delta x \Delta z\big|_s \quad (6.7)$$

$$D_{3f} = \Gamma^\phi \Delta x \Delta y\big|_f; \quad D_{3b} = \Gamma^\phi \Delta x \Delta y\big|_b \quad (6.8)$$

Rigorously, the mass fluxes at the interfaces are also evaluated in θ. However, in the implicit and fully implicit formulation the equation must be linearized, and the mass fluxes will be part of the matrix of coefficients, which are evaluated with the variables of the previous iteration. As the coefficients are updated, when the solution converges both the coefficients and the variable of the equation will be obtained at the same time level. For this reason, the mass flows do not carry the superscript θ. The term $L[S^\phi]^\theta$ indicates the numerical approximation of the term in the brackets, and its linearization is done according to what was already learned in Chap. 3. This term will contain the pressure gradient when ϕ is the components of the velocity vector. Choosing the time interpolation according to what was described in Chap. 3,

$$\phi^\theta = \theta\phi + (1-\theta)\phi^o \quad (6.9)$$

and recalling that the absence of superscript in ϕ represents the time level in which the solution is sought ($t + \Delta t$). For the spatial interpolation one chooses the WUDS scheme, which furnish, for the values of the functions at the interfaces,

$$\phi_e = \left(\frac{1}{2} + \alpha_e\right)\phi_P + \left(\frac{1}{2} - \alpha_e\right)\phi_E \quad (6.10)$$

$$\phi_w = \left(\frac{1}{2} + \alpha_w\right)\phi_W + \left(\frac{1}{2} - \alpha_w\right)\phi_P \quad (6.11)$$

$$\phi_n = \left(\frac{1}{2} + \alpha_n\right)\phi_P + \left(\frac{1}{2} - \alpha_n\right)\phi_N \quad (6.12)$$

$$\phi_s = \left(\frac{1}{2} + \alpha_s\right)\phi_S + \left(\frac{1}{2} - \alpha_s\right)\phi_P \quad (6.13)$$

$$\phi_f = \left(\frac{1}{2} + \alpha_f\right)\phi_P + \left(\frac{1}{2} - \alpha_f\right)\phi_F \quad (6.14)$$

$$\phi_b = \left(\frac{1}{2} + \alpha_b\right)\phi_B + \left(\frac{1}{2} - \alpha_b\right)\phi_P \quad (6.15)$$

and for the gradient at the interfaces,

$$\Gamma^\phi_e \left(\frac{\partial \phi}{\partial x}\right)_e = \beta_e \Gamma^\phi_e \left(\frac{\phi_E - \phi_P}{\Delta x_e}\right) \quad (6.16)$$

$$\Gamma_w^\phi \left(\frac{\partial \phi}{\partial x}\right)_w = \beta_w \Gamma_w^\phi \left(\frac{\phi_P - \phi_W}{\Delta x_w}\right) \quad (6.17)$$

$$\Gamma_n^\phi \left(\frac{\partial \phi}{\partial y}\right)_n = \beta_n \Gamma_n^\phi \left(\frac{\phi_N - \phi_P}{\Delta y_n}\right) \quad (6.18)$$

$$\Gamma_s^\phi \left(\frac{\partial \phi}{\partial y}\right)_s = \beta_s \Gamma_s^\phi \left(\frac{\phi_P - \phi_S}{\Delta y_s}\right) \quad (6.19)$$

$$\Gamma_f^\phi \left(\frac{\partial \phi}{\partial z}\right)_f = \beta_f \Gamma_f^\phi \left(\frac{\phi_F - \phi_P}{\Delta z_f}\right) \quad (6.20)$$

$$\Gamma_b^\phi \left(\frac{\partial \phi}{\partial z}\right)_b = \beta_b \Gamma_b^\phi \left(\frac{\phi_P - \phi_B}{\Delta z_b}\right) \quad (6.21)$$

Substituting the values of ϕ and its derivatives at the integration points in Eq. (6.2), it becomes

$$\begin{aligned}
&\frac{M_P \phi_P}{\Delta t} + \phi_P^\theta \left[\dot{M}_e\left(\frac{1}{2} + \alpha_e\right) - \dot{M}_w\left(\frac{1}{2} - \alpha_w\right) + \dot{M}_n\left(\frac{1}{2} + \alpha_n\right)\right.\\
&\left. - \dot{M}_s\left(\frac{1}{2} - \alpha_s\right) + \dot{M}_f\left(\frac{1}{2} + \alpha_f\right) - \dot{M}_b\left(\frac{1}{2} - \alpha_b\right) - S_P \Delta V\right.\\
&\left. + \frac{D_1 \beta}{\Delta x}\bigg|_e + \frac{D_1 \beta}{\Delta x}\bigg|_w + \frac{D_2 \beta}{\Delta y}\bigg|_n + \frac{D_2 \beta}{\Delta y}\bigg|_s + \frac{D_3 \beta}{\Delta z}\bigg|_f + \frac{D_3 \beta}{\Delta z}\bigg|_b \right]\\
&= \phi_E^\theta\left[-\left(\frac{1}{2} - \alpha_e\right)\dot{M}_e + \frac{D_1\beta}{\Delta x}\bigg|_e\right] + \phi_W^\theta\left[\left(\frac{1}{2} + \alpha_w\right)\dot{M}_w + \frac{D_1\beta}{\Delta x}\bigg|_w\right]\\
&\phi_N^\theta\left[-\left(\frac{1}{2} - \alpha_n\right)\dot{M}_n + \frac{D_2\beta}{\Delta y}\bigg|_n\right] + \phi_S^\theta\left[\left(\frac{1}{2} + \alpha_s\right)\dot{M}_s + \frac{D_2\beta}{\Delta y}\bigg|_s\right]\\
&+ \phi_F^\theta\left[-\left(\frac{1}{2} - \alpha_f\right)\dot{M}_f + \frac{D_3\beta}{\Delta z}\bigg|_f\right] + \phi_B^\theta\left[\left(\frac{1}{2} + \alpha_b\right)\dot{M}_b + \frac{D_3\beta}{\Delta z}\bigg|_b\right]\\
&+ \frac{M_P^o \phi_P^o}{\Delta t} + S_c \Delta V
\end{aligned} \quad (6.22)$$

To help in the derivation of the final discretized equation for ϕ, the mass conservation equation is used. Using $\phi = 1$ in Eq. (6.2) one recovers the mass conservation, as

$$\frac{M_P - M_P^o}{\Delta t} + \dot{M}_e - \dot{M}_w + \dot{M}_n - \dot{M}_s + \dot{M}_f - \dot{M}_b = 0 \quad (6.23)$$

Multiplying Eq. (6.23) by (-1) and adding it inside the brackets which multiplies ϕ_P^θ, it is found

6.3 Explicit Formulation

$$\frac{M_P}{\Delta t}\phi_P + A_P^*\phi_P^\theta = A_e\phi_E^\theta + A_w\phi_W^\theta + A_s\phi_S^\theta + A_n\phi_N^\theta + A_f\phi_F^\theta$$
$$+ A_b\phi_B^\theta + \frac{M_P^o}{\Delta t}\phi_P^o + Sc\Delta V \quad (6.24)$$

with the following coefficients,

$$A_P^* = A_e + A_w + A_n + A_s + A_f + A_b - S_P\Delta V - \frac{M_P}{\Delta t} + \frac{M_P^o}{\Delta t} \quad (6.25)$$

$$A_e = -\left(\frac{1}{2} - \alpha_e\right)\dot{M}_e + \left.\frac{D_1\beta}{\Delta x}\right|_e \;;\; A_w = \left(\frac{1}{2} + \alpha_w\right)\dot{M}_w + \left.\frac{D_1\beta}{\Delta x}\right|_w \quad (6.26)$$

$$A_n = -\left(\frac{1}{2} - \alpha_n\right)\dot{M}_n + \left.\frac{D_2\beta}{\Delta y}\right|_n \;;\; A_s = \left(\frac{1}{2} + \alpha_s\right)\dot{M}_s + \left.\frac{D_2\beta}{\Delta y}\right|_s \quad (6.27)$$

$$A_f = -\left(\frac{1}{2} - \alpha_f\right)\dot{M}_f + \left.\frac{D_3\beta}{\Delta z}\right|_f \;;\; A_b = \left(\frac{1}{2} + \alpha_b\right)\dot{M}_b + \left.\frac{D_3\beta}{\Delta z}\right|_f \quad (6.28)$$

In Chap. 3, it was discussed the general aspects of a numerical approximation using finite volumes in a one-dimensional context, and the various types of formulations regarding the advancement of the solution in time were presented. Now, with the approximate equation for three-dimensional situation, with all coefficients involved, it is wise to return to the subject. It is now pertinent an analysis of the relationship between the different types of transient using explicit and implicit formulations.

6.3 Explicit Formulation

Equation (6.24) written for $\theta = 0$ has the following form, clustering the neighbor coefficients,

$$\frac{M_P\phi_P}{\Delta t} = \phi_P^o\left(\frac{M_P^o}{\Delta t} - A_P^*\right) + \sum A_{nb}\phi_{NB}^o + Sc\Delta V \quad (6.29)$$

in which, in order to guarantee the positivity of the coefficient of ϕ_P^o, it is required that,

$$\Delta t \leq \frac{M_P^o}{A_P^*} \quad (6.30)$$

Simplifying Eq. (6.26) for the 1D case, one finds that the coefficients A_e and A_w reduces to $\Gamma/\Delta x$. Using Eq. (6.25) and applying the restriction given by Eq. (6.30), the maximum time step is given by

$$\frac{\alpha \Delta t}{\Delta x^2} \leq \frac{1}{2}, \quad (6.31)$$

which is the same relation obtained in Chap. 3 given by Eq. (3.24). Equation (6.30), thereby, gives the general form for obtaining the maximum value of Δt that the explicit solution can advance in time.

Satisfying Eq. (6.30) works for 1D and 2D linear problems. The general problem is almost always 3D non-linear and there is not mathematical proof of convergence if the restriction given by Eq. (6.30) is satisfied. Anyway, it is worth to satisfy it in more general cases. It is also clear that the maximum allowed advance in time is different from cell to cell, as shown by Eq. (6.30). Manipulating the value of Δt opens possibilities of advancing the solution in time following either the true or a distorted transient.

6.3.1 True Transient

When the interest is in the true transient the time step (Δt) for advancing the solution in time must be equal for all volumes. This is physically easy to understand, because the transient would be no longer true if after a sweep in the domain the variables are not all evaluated at same time level. So, the maximum possible Δt, is given by,

$$\Delta t|_{for\ all\ cells} = \min\left[\Delta t|_{max}^P\right], \quad (6.32)$$

that is, the maximum possible time step should be equal to the minimum of the maximums Δt allowed for each volume, as Fig. 6.2 tries to illustrate. In doing so, the coefficient of ϕ_P^o will be always positive.

As pointed out in Chap. 3, the explicit formulation does not give rise to a linear system of equations, but rather to a set of equations that are solved one by one (point-by-point). Every time a sweep is performed in the domain, the solution in the new time level is obtained in an extremely fast computational process. This speed, however, is apparent, since the possible values of Δt that can be used and that keep the coefficients positive are rather limited by the criterion of Eq. (6.30). In other words, it can be said that Δt, in general, is limited by the convergence criterion, not by accuracy, and it is unnecessary to perform a mesh refinement study in time, i.e., refining Δt. As already reported in this text, the explicit formulation can be a good choice for very fast transients.

6.3.2 Distorted Transient

When only the steady-state solution is of interest, the maximum Δt possible for each volume can be used, advancing the solution in a distorted way, since each volume

6.3 Explicit Formulation

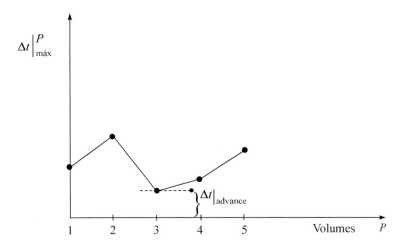

Fig. 6.2 Real transient–maximum time step allowed

advances differently in time. In fact, there is no longer a solution at a given time level, and it would be better saying that the solution is at a given iteration level. Logically, the steady state solution is independent of these intermediate distorted solutions over time. Figure 3.11, Chap. 3, shows alternatives of advancing in time. For a distorted transient, therefore using the maximum possible Δt for each volume, Eq. (6.29) results

$$\frac{M_P \phi_P}{\Delta t} = \sum A_{nb} \phi_{NB}^o + Sc \Delta V \tag{6.33}$$

In general, the explicit formulation is associated with the time coordinate. But that is just the general case. For example, the temperature distribution in the fully developed region between two flat plates is governed by

$$\frac{u}{\alpha} \frac{\partial T}{\partial x} = \frac{\partial^2 T}{\partial y^2}, \tag{6.34}$$

in which u is the velocity along the x axis. This equation can be solved marching explicitly in x in the following manner,

$$T_P^{x+\Delta x} - T_P^x = \frac{\alpha \Delta x}{u} \left[\frac{T_N^x + T_S^x - 2T_P^x}{\Delta y^2} \right], \tag{6.35}$$

or

$$T_P^{x+\Delta x} = T_P^x \left(1 - \frac{2\alpha \Delta x}{u \Delta y^2}\right) + T_N^x \frac{\alpha \Delta x}{u \Delta y^2} + T_S^x \frac{\alpha \Delta x}{u \Delta y^2} \tag{6.36}$$

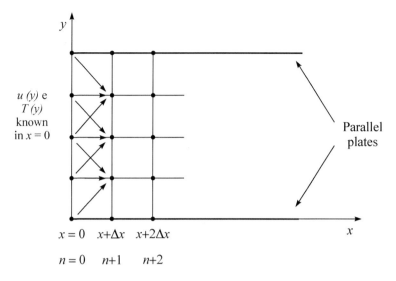

Fig. 6.3 Marching procedure in the x axis

It can be seen that the temperature at point P at position $x + \Delta x$ is obtained as a function of known values at position x. Figure 6.3 illustrates the procedure. Consequently, whenever the process marches in a given coordinate, determining the values of the function without the need for solving linear systems, the formulation is explicit. In such cases, there will always be a restriction on the step size that can be used in the explicit coordinate. In this example, the restriction that Δx must respect for avoiding divergence in the explicit process is

$$\Delta x \leq \frac{u \Delta y^2}{2\alpha}, \tag{6.37}$$

which should be compared with the criterion given by Eq. (6.31) and their similarities recognized.

6.4 Fully Implicit Formulation

Equation (6.24) written for $\theta = 1$ has the following form

$$A_P \phi_P = \frac{M_P^o \phi_P^o}{\Delta t} + \sum A_{nb} \phi_{NB} + Sc \Delta V, \tag{6.38}$$

with

$$A_P = \sum A_{nb} - S_P^\phi \Delta V + \frac{M_P}{\Delta t} \qquad (6.39)$$

Equation (6.38) is the most widely used form for solving fluid mechanics and heat transfer problems numerically. The fully implicit formulation is preferred because of the possibility, one is saying, possibility, of advancing in time with larger Δt. However, using the fully implicit formulation does not mean that time steps of any size can be used since, when a system of equations it is being solved (which is almost always the case), the coupling among the equations can severely limit the time step. Studies refining Δt, to find the solution independent of Δt, for that time level, should always be conducted.

Again, remember that the implicit formulation gives rise to a linear system of equations. When the real transient is of interest, the linear system must be solved accurately for each time level. When adopting a distorted transient, the linear system should be poorly solved at each time level trying to quickly advance to the steady state regime, which is of interest. The other alternative is to make Δt large as possible (may be infinity) and obtain the steady-state solution. Repeating what has been pointed out in previous chapters, it makes no sense (it would be spent an unnecessary computational effort) to solve a distorted transient implicitly using a direct method or a very converged iterative solution of the linear system in each time level, as Fig. 3.11 warns.

As a final remark, it is remembered that when the real transient is sought, either explicitly or implicitly, the initial condition must be the physical condition of the problem. When the distorted transient is adopted (interest in the steady state solution only), the initial condition is an estimative of the variables to start the distorted march in time, or along iterations, better saying.

6.5 Conclusions

This chapter reunited the knowledge of the previous chapters for the discretization of a three-dimensional advection/diffusion equation. To have the discretization in 3D may be helpful for visualizing the coefficients involved in this kind of discretization when constructing a code for fluid flow simulations.

The explicit and fully implicit formulation seen for a 1D were revisited extending to three dimensions, with key comments about the true and distorted transient. For example, even if only the steady state solution is sought, the complexity of the equations system does not allow the use an infinite time step, since there is some explicitness involved, especially when the segregated solution is employed. Therefore, it was again recommended to keep the transient term in the equations. The time coordinate is a relaxation parameter with physics embodied and should be used when distorted transient is followed.

6.6 Exercises

6.1 Obtain Eqs. (6.29) and (6.38) with all coefficients.

6.2 Equation (6.36) is the numerical approximation of Eq. (6.34) when the marching in x is made explicitly. Write a computational algorithm for calculating the temperatures in the domain highlighting the main loops in your code. Is there any linear system to be solved?

6.3 Still considering the problem of flow between parallel flat plates, obtain now the equation for the fully implicit advancing in x. Are there still any restrictions with respect to the size of Δx while advancing the solution in x? Rewrite your computational algorithm to also include the implicit advance in x. Is there now a linear system to be solved in this algorithm?

6.4 For the same problem, consider now the implicit advancing in x with any θ, greater than zero and less than 1, that is

$$\left.\frac{\partial^2 T}{\partial y^2}\right|_{x_P+\theta\Delta x} = (1-\theta)\left(\frac{\partial^2 T}{\partial y^2}\right)_{x_P} + \theta\left(\frac{\partial^2 T}{\partial y^2}\right)_{x_P+\Delta x} \qquad (6.40)$$

Find the relation among θ and Δx such that no oscillations appear, what could provoke divergence during the marching in x. As a special case consider a grid with just three grid nodes in y (and two of them on the plates) with temperatures T_N and T_S. Restarting with the differential equation and approximating only the term $\partial^2 T/\partial y^2$, the following ordinary differential equation is obtained,

$$\frac{d}{dx}(T_P(x)) = -\frac{2\alpha}{u\Delta y^2}T_P(x) \qquad (6.41)$$

Solve this equation to obtain $T_P(x)$. Is there any possibility of oscillation??

Using now the equation for any θ, determine the expression for θ which guarantees the "exact" solution of this simplified problem.

6.5 Solve numerically the two-dimensional advection/diffusion problem shown in Fig. 6.4, where the velocity field V is known. As interpolation function, use the WUDS method. All data of the problem is a free choice.

This is an important problem for preparing your computer code to include, in Chap. 7, the determination of the velocities u and v, which in this problem are considered to be known. Note that the same routine you will develop to calculate the coefficients for T will be used for calculating the coefficients of u and v in Chap.7. Therefore, structure your code so that you reserve local storage for the variables T, u, v, and p, for the coordinates x and y, for the physical properties and for the coefficients.

Make the problem dimensionless and solve for different values of $Pe_{\Delta x}$, $Pe_{\Delta y}$, and a/b ratio. Study the limiting cases and compare, when possible, the numerical solution with the analytical one. For example, for u and v equal to zero, the problem has the exact solution shown in Chap. 10. For insulated

6.6 Exercises

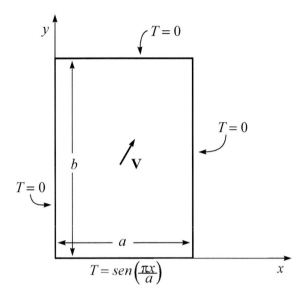

Fig. 6.4 Figure for Problem 6.5

laterals and T constant at $y = 0$, $u = $ constant and $v = 0$, the problem becomes one-dimensional and has exact solution.

To generalize the problem further, consider the plate to be of small thickness ε with a source term as function of temperature, simulating the heat lost by convection by the two surfaces of the plate to an environment at T_∞ with heat transfer coefficient h. Nothing changes from the point of view of the structure of the code, and the resulting model can simulate the cooling of a plate moving with known velocity.

Chapter 7
Finding the Velocity Field—Pressure/Velocity Couplings

7.1 Introduction

For the developments presented so far, the velocity field has been assumed to be known whenever necessary. We are now prepared to study the techniques for obtaining the velocity field, since in any engineering simulation involving fluid flow, the most important and intricate solution is the fluid mechanics one. All other scalars, such as temperature, species concentration, turbulent kinetic energy, turbulent kinetic energy dissipation, and others, require the velocity field. All Computational Fluid Dynamics techniques, as the name suggests, is related to the solution of the mass conservation and momentum conservation equations. Besides the significant non-linearities present in the advection terms, there is a strong, but subtle coupling between pressure and velocity. These two issues are on the kernel of the developments in computational fluid dynamics and has received an enormous amount of research in the last four decades, with lots of publications of articles and textbooks in the field. There are, however, concepts in the numerical machinery for treating the pressure–velocity coupling which takes time to mature for reaching a good deep understanding. This chapter aims to shorten this path.

7.2 System of Equations

A quite general system of partial differential equations which governs the flow of a compressible fluid with heat transfer for many engineering applications is given by the equations of mass, momentum and energy conservation, plus a state equation, given by,

$$\frac{\partial \rho}{\partial t} + \frac{\partial}{\partial x}(\rho u) + \frac{\partial}{\partial y}(\rho v) + \frac{\partial}{\partial w}(\rho w) = 0 \qquad (7.1)$$

$$\frac{\partial}{\partial t}(\rho u) + \frac{\partial}{\partial x}(\rho uu) + \frac{\partial}{\partial y}(\rho vu) + \frac{\partial}{\partial z}(\rho wu) = -\frac{\partial p}{\partial x}$$
$$+ \frac{\partial}{\partial x}\left(\mu \frac{\partial u}{\partial x} - \frac{2}{3}\mu \text{div}\mathbf{V}\right) + \frac{\partial}{\partial y}\left(\mu \frac{\partial v}{\partial x}\right) + \frac{\partial}{\partial z}\left(\mu \frac{\partial w}{\partial x}\right)$$
$$+ \frac{\partial}{\partial x}\left(\mu \frac{\partial u}{\partial x}\right) + \frac{\partial}{\partial y}\left(\mu \frac{\partial u}{\partial y}\right) + \frac{\partial}{\partial z}\left(\mu \frac{\partial u}{\partial z}\right) + B_x \qquad (7.2)$$

$$\frac{\partial}{\partial t}(\rho v) + \frac{\partial}{\partial x}(\rho uv) + \frac{\partial}{\partial y}(\rho vv) + \frac{\partial}{\partial z}(\rho wv) = -\frac{\partial p}{\partial y}$$
$$+ \frac{\partial}{\partial x}\left(\mu \frac{\partial u}{\partial y}\right) + \frac{\partial}{\partial y}\left(\mu \frac{\partial v}{\partial y} - \frac{2}{3}\mu \text{div}\mathbf{V}\right) + \frac{\partial}{\partial z}\left(\mu \frac{\partial w}{\partial y}\right)$$
$$+ \frac{\partial}{\partial x}\left(\mu \frac{\partial v}{\partial x}\right) + \frac{\partial}{\partial y}\left(\mu \frac{\partial v}{\partial y}\right) + \frac{\partial}{\partial z}\left(\mu \frac{\partial v}{\partial z}\right) + B_y \qquad (7.3)$$

$$\frac{\partial}{\partial t}(\rho w) + \frac{\partial}{\partial x}(\rho uw) + \frac{\partial}{\partial y}(\rho vw) + \frac{\partial}{\partial z}(\rho ww) = -\frac{\partial p}{\partial z}$$
$$+ \frac{\partial}{\partial x}\left(\mu \frac{\partial u}{\partial z}\right) + \frac{\partial}{\partial y}\left(\mu \frac{\partial v}{\partial z}\right) + \frac{\partial}{\partial z}\left(\mu \frac{\partial w}{\partial z} - \frac{2}{3}\mu \text{div}\mathbf{V}\right)$$
$$\frac{\partial}{\partial x}\left(\mu \frac{\partial w}{\partial x}\right) + \frac{\partial}{\partial y}\left(\mu \frac{\partial w}{\partial y}\right) + \frac{\partial}{\partial z}\left(\mu \frac{\partial w}{\partial z}\right) + B_z \qquad (7.4)$$

$$\frac{\partial}{\partial t}(\rho c_p T) + \frac{\partial}{\partial x}(\rho u c_p T) + \frac{\partial}{\partial y}(\rho v c_p T) + \frac{\partial}{\partial z}(\rho w c_p T)$$
$$= \frac{\partial}{\partial x}\left(k \frac{\partial T}{\partial x}\right) + \frac{\partial}{\partial y}\left(k \frac{\partial T}{\partial y}\right) + \frac{\partial}{\partial z}\left(k \frac{\partial T}{\partial z}\right) + \beta T \frac{Dp}{Dt} + \mu \Phi \qquad (7.5)$$

$$p = p(\rho, T) \qquad (7.6)$$

Other equations are added to this system depending on the problem under analysis. For example, for a turbulent flow using a RANS approach for turbulence modelling, two new differential equations appear, for the turbulent kinetic energy and for its dissipation, if the $k - \varepsilon$ is employed. Turbulence modelling is a compulsory topic for simulating engineering problems, and the reader finds a good summary of the most used models in [1, 2] among a diverse literature in the field. For multiphase flows, the equations for single phase applies for each phase, that is, for a general multiphase flow model, there are different pressure and velocity fields and different temperatures and mass conservation for each phase. The crucial part of the modelling is to specify how mass, momentum and energy exchange between phases, since there are no boundary conditions in the phase's interfaces, as the grid encompasses all computational domain. It is an impossible task, for example, to model each phase and apply boundary conditions at the interfaces. Therefore, what is learned in this chapter for a single phase applies for multiphase flows [3].

7.2 System of Equations

In the equation system given by Eqs. (7.1–7.6) the major difficulties for the solution are restricted to Eqs. (7.1–7.4), in whose equations the nonlinearities and the coupling between pressure and velocity are contained. Therefore, along this chapter attention will be, mainly, given for this system of equations, recalling that the other scalars can be solved sequentially, in general, without numerical difficulties.

According to what was seen in Chap. 3, each differential equation is represented by a system of linear algebraic equations obtained via discretization. Therefore, a system of systems of algebraic equations should be solved.

The discretization of Eqs. (7.1–7.6), following the numerical procedures of the finite volume method reads,

$$\frac{M_P - M_P^o}{\Delta t} + \dot{M}_e - \dot{M}_w + \dot{M}_n - \dot{M}_s + \dot{M}_f - \dot{M}_b = 0 \tag{7.7}$$

$$A_p u_P = A_e u_E + A_w u_W + A_n u_N + A_s u_S \\ + A_f u_F + A_b u_B - L[p^u]\Delta V + B_P^u \tag{7.8}$$

$$A_p v_P = A_e v_E + A_w v_W + A_n v_N + A_s v_S \\ + A_f v_F + A_b v_B - L[p^v]\Delta V + B_P^v \tag{7.9}$$

$$A_p w_P = A_e w_E + A_w w_W + A_n w_N + A_s w_S \\ + A_f w_F + A_b w_B - L[p^w]\Delta V + B_P^w \tag{7.10}$$

$$A_p T_P = A_e T_E + A_w T_W + A_n T_N \\ + A_s T_S + A_f T_F + A_b T_B + B_P^T \tag{7.11}$$

$$p = p(\rho, T) \tag{7.12}$$

7.2.1 About Segregated and Simultaneous Solution

In these equations $L[.]$ represents the numerical approximation of the term in the brackets, in this case the pressure gradient term. It is warned that the coefficients, although without different notations, may not be the same. For simplicity, it is not identified in the coefficients to which conservation equation they belong, as they appear throughout the text almost always multiplying the variable of the equation considered, precluding misunderstandings.

When trying to solve these equations, the first decision to be made is about the nature of the solution, if segregated or simultaneous, the latter also called coupled. It is preferred to call simultaneous instead of coupled because even solving in a segregated (iterative) manner, at the end of the process, the equations are solved coupled. The segregated approach solves each linear system independently and, consequently, each variable must have its evolutive equation for the iterative solution. It could be said that each variable "owns" one equation.

On the other hand, the simultaneous solution creates a single matrix involving all coefficients of all equations and solving all unknowns simultaneously. In this case, no variable "owns" any equation, and it can be said that all variables share the whole system. The simultaneous solution avoids having special algorithm for solving the segregated pressure–velocity coupling (SPVC)[1] between variables, which is considered automatically. The nonlinearities remain, which requires updating the matrix of coefficients and solve it again until a predefined tolerance is attended. When the solution is segregated, non-linearities and coupling are the reasons for the iterations, while, when it is simultaneous, the reason is just because of the non-linearities. The iteration one is talking about are not relate to the iterative solution of the linear system, but of the whole system of equations.

If the alternative chosen is the simultaneous solution, the linear algebraic system becomes much bigger and should be solved with a robust iterative solver, perhaps accelerated by some multigrid algorithm. It shouldn't be solved with a traditional direct solver, because this requires the storage of all coefficients, including the zeros, and the resulting size of the matrix is phenomenal. To have a glance on this, imagine a three-dimensional incompressible flow with heat transfer being solved with a mesh of 50,000 volumes, which is not a too large grid. There are 250,000 unknowns, originating a matrix with 62.5×10^9 entries, of which only 0.0028% are non-zeros. There is no point in trying to invert this matrix, unless a robust sparse matrix handling method is used, as commented in Chap. 4.

Because of the size of the linear system, the segregated approach is still largely used and solves each linear system with an iterative solver supported by some accelerating technique, in general, multigrid. The iterative solvers require only the storage of the non-zero coefficients. Opting for the segregate solution the coupling between pressure and velocity (SPVC) appears as a difficulty, opposed to the simultaneous solution, in which this coupling is already considered solving all equations in a single equation system.

In the segregated formulation, the nature of the flow, if compressible or incompressible, has strong effects on the solution procedure. Because of this importance, the following section is dedicated to present the compressible and incompressible formulation in the context of segregated solutions.

---

[1] The usual coupling referred in the literature due to the solution of incompressible flows in a segregated approach will be called in this text SPVC, to differentiate from the CPVC coupling due to the co-located grid.

7.3 Segregated Formulation. Incompressibility

Using the segregated approach, if the fluid is compressible, that is, pressure can be found through the state equation, $p = f(\rho, T)$, all partial differential equations possess its evolutive variable, namely, density found from the mass conservation equations and velocities from the equations of motion. Still using the segregated alternative, the situation changes drastically when density is constant or a function of temperature only. Flows of gases with little or no pressure variation and flows of liquids with or without pressure variation fit into this class of problems. In this case, density is no longer a variable and turns to be a physical property, constant or function of temperature only, found through $\rho = f(T)$. Pressure, therefore, no longer has an evolutive equation, and the mass conservation equation no longer has an evolutive variable. In some way, mass conservation needs to be used for finding pressure.

Consider a three-dimensional flow with heat transfer where there are five equations to solve: conservation of mass, an equation of motion in each coordinate direction, and the energy equation. The unknowns are specific mass, pressure, temperature, and the three components of the velocity vector.

If ρ has considerable variation with p, then the equation of state, relating ρ to temperature and pressure, is the relation employed for the problem closure. The equation of state is then the evolution equation for pressure, while the continuity equation is for specific mass (density). This formulation, in which all dependent variables have their evolution equation, is called the **compressible formulation**. The most important class of problems using this formulation is that of high-speed gas flows. In these problems, it is also common to use the simultaneous solution of the equations, where the nature of the linearization avoids the need of updating the coefficient matrix [4, 5].

In principle, any compressible problem can be solved using the procedure given below by advancing the solution from time t to time $t + \Delta t$. The values of the variables at time t are known, either through initial conditions (true transient) or through an estimate that starts the iterative process (distorted transient). The steps are:

(1) Calculate ρ at instant $t + \Delta t$ using the mass conservation equation;
(2) Find the temperature from the energy equation;
(3) Calculate the pressure using the equation of state;
(4) Calculate the velocities (u, v, w) using the equations of motion for each direction;
(5) Back to step 1 and advance the solution for a new time interval until it reaches the steady state regime according to a tolerance (true transient) or until convergence is achieved (distorted transient). It should be also considered in this iterative cycle the nonlinearities of the conservation equations, expressed in the coefficients of the algebraic equations. The iterative steps can keep the coefficients fixed and only the variables are updated during a given number of iterations, or coefficients and variables can advance simultaneously. The appropriate procedure depends on the problem at hand.

If density does not change significantly with pressure, but does vary considerably with temperature, the problem could, rigorously, still be defined as compressible. However, the equation of state $p = f(\rho, T)$ can no longer be used as an equation for the determination of p, because small errors made in the calculation of ρ, via the mass conservation equation, may produce large errors in p calculated via the relation $p = f(\rho, T)$. If this pressure field, which may contain large errors, is introduced into the equations of motion for obtaining the velocities, and those velocities substituted into mass conservation equation for calculating ρ for the next time interval, the errors in ρ will be augmented and serious instabilities will occur in the numerical solution of the system of equations.

As density no longer depends on p, it seems logical that the equation of state be used to calculate ρ, depending only on T, that is, $\rho = f(T)$, in which T is determined through the energy conservation equation. The difficulty that arises is clear. In doing so, the equation of state becomes an equation for ρ, and pressure does not have an evolutive equation, its influence appearing only through its gradient in the equations of motion. It is easy to see that it is not enough to isolate p from one of the equations of motion, since the gradients in the three directions must be combined to determine the pressure. This is the difficulty: extracting p from the equations of motion in such a way that the velocities obtained satisfy the conservation of mass. The mass conservation equation, by its turn, does not serve as an evolution equation for any variable, and becomes only a restriction that must be obeyed by the velocity field.

The challenge, thus, is to determine a pressure field that, when inserted into the equations of motion, results in a velocity field that satisfies the mass conservation equation. In other words, the fact that ρ does not vary with p introduces a great difficulty in dealing with the coupling between pressure and velocity. When $\rho = cte$ or $\rho = f(T)$ the same numerical treatment applies, and this formulation is called **incompressible**. As a reminder, note that if the system of equations is solved simultaneously, the segregated pressure–velocity coupling problem would not exist.

Fundamentally, the procedure for advancing the solution from time t to $t + \Delta t$ is as follows:

(1) Provide the initial values of the dependent variables;
(2) Calculate T, using the energy equation;
(3) Calculate ρ with $\rho = f(T)$;
(4) Calculate p. An algorithm for this should be used. Developing these algorithms is the main task in the pressure–velocity coupling treatment;
(5) Calculate the components of the velocity vector, using the equations of motion.
(6) Check whether velocities satisfy the mass conservation equation. If they don't go back to item 4 and recalculate the pressure. Iterate through items 4-5-6 until the mass conservation equation is satisfied. The error in the mass conservation is the input for correcting the pressure;
(7) Since temperature depends on velocities, go back to item 2 and start again;
(8) After convergence, advance to a new time level until the steady state regime is achieved or the desired simulation time is reached.

7.4 Variable Arrangement on the Grid

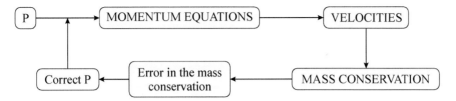

Fig. 7.1 Segregated pressure–velocity coupling (SPVC) loop

Figure 7.1 pictures the iterative process inside steps 4-5-6. The key issue in the iterative loop 4-5-6, is the creation of a special algorithm to treat the SPVC. It also seems logical that the equation of conservation of mass should be transformed into an equation in which the variable pressure appears. By doing so, when determining p, the conservation of mass will be satisfied. A few methods that use this philosophy will be described in Sect. 7.6. They are the well-known SIMPLE-like methods, which are schemes for treating the SPVC due to the segregated approach of solving the system of equations.

From now on, when it is referred to the system of equation being solved it is meant the (u, v, w, p) system. Besides the SPVC coupling, depending on the arrangement of the variables on the grid adopted, another type o coupling may be created. This is now addressed.

7.4 Variable Arrangement on the Grid

There are several possible alternatives for the arrangements of the variables on the grid, but two of them deserves be detailed. The variables may be stored all together using the same control volume for the balances, named the co-located arrangement, or they may have different control volumes, the staggered arrangement. They have marked differences related to algorithms implementation and to the stability of the pressure fields. By now it is wise to anticipate that the co-located arrangement was used up to the 60s, always facing pressure instabilities. The definitive elimination of pressure wiggles came with the clever idea of Harlow and Welch [6], creating the staggered grids which completely eliminates the instabilities. However, by the 80s, with the need of solving more complex engineering problems in irregular geometries and in 3D conditions, the staggered grid could no longer be used due to its cumbersome implementation in unstructured grids. The co-located grid and its related difficulties, thus, came back to the forefront and, nowadays, all schemes used in commercial software employ co-located arrangement supported by algorithms developed to suppress the pressure oscillation issue. This will be discussed in Sect. 7.5.

7.4.1 Co-located Grid Arrangement

The co-located arrangement has just one control volume defined on the grid with all variables sharing this volume. It is a very convenient arrangement helping the discretization process considerably. The same control volume for all variables eliminates many calculations, as fluxes of all properties are calculated at the same point at the same face of the control volume. Variable indexing, geometric data storage, areas calculations in only 6 faces instead of 24 in 3D Cartesian meshes greatly simplifies the construction and computational implementation of the algorithm and the computer code. Implementation of a simultaneous solution are also not so difficulty with co-located grids. Figure 7.2 depicts this arrangement for a 2D situation. But, as a rule, there is not always gains, as the co-located arrangement causes pressure oscillations, an undesirable and critical behavior of the solution.

The principal reason for the appearance of oscillations is because this arrangement fails in promoting a strong coupling between pressure and velocity. Mass conservation equation is in the center of this issue. In a finite volume formulation, it is necessary to calculate the mass flow at the interfaces of all control volumes. As can be seen in Fig. 7.2, there are no velocities available at the interfaces of the control volumes. It is tempting to calculate the interface velocity by a linear interpolation of nodal velocities, but it is well-known that this approach does not promote the desired coupling because a consistent pressure gradient must appear as driving force for the advecting velocities. Any interpolation of velocity which doesn't involve pressure in a consistent way will not work. A possible reason for this decoupling and appearance of the checkerboard pressure field [7] is now presented using a 1D problem as example, as seen in Fig. 7.3 [8]. Considering a steady state problem for simplicity, the mass and momentum conservation are, respectively,

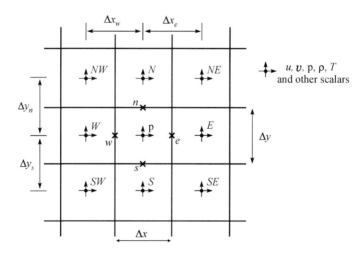

Fig. 7.2 Co-located grid arrangement

7.4 Variable Arrangement on the Grid

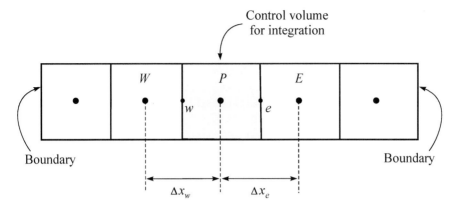

Fig. 7.3 One-dimensional control volume

$$\frac{\partial u}{\partial x} = 0 \tag{7.13}$$

$$\frac{\partial}{\partial x}(\rho u u) = -\frac{\partial p}{\partial x} + OT, \tag{7.14}$$

in which OT means other terms, like diffusion and possible source terms.

Integrating Eq. (7.13), one gets

$$u_e - u_w = 0, \tag{7.15}$$

and assuming a linear interpolation using nodal velocities,

$$u_e = \frac{1}{2}(u_P + u_E) \tag{7.16}$$

$$u_w = \frac{1}{2}(u_P + u_W) \tag{7.17}$$

Inserting Eqs. (7.16) and (7.17) into Eq. (7.15), it reads

$$u_E - u_W = 0 \tag{7.18}$$

The integration of the momentum equation gives,

$$\dot{m}(u_E - u_W) = A(p_e - p_w) + OT, \tag{7.19}$$

or

$$\dot{m}(u_E - u_W) = \frac{A}{2}(p_E - p_W) + OT \qquad (7.20)$$

in which A, a constant, is the cross-sectional area of the duct. Inserting Eq. (7.18), into Eq. (7.20), the pressure gradient balances the diffusion and possibly a source term, disappearing its coupling with the momentum variation in the control volume, which is the dominant term to promote the coupling between velocity and pressure (mass and momentum).

Therefore, having the velocities at the integration points not dependent of a pressure gradient, as shown in Eqs. (7.16) and (7.17), it is a potential risk for decoupling. Besides that, in the momentum equation for the velocity at node p, according to Eq. (7.20), the pressure at this node doesn't appear, with a clear sign that this velocity is not coupled with the pressure, originating the checkerboard pressure fields just described [7].

It is possible to see another sign of decoupling by recognizing that, independent of the stabilization scheme created, we always will have a set of velocity that satisfy mass and a set that satisfy momentum. The advecting velocities, which satisfy mass, are not variables of the problem (unknowns). Those velocity fields must be tightly coupled through a unique pressure field for stabilization. It seems logical, therefore, that the best scheme is the one in which the velocity that satisfies mass is the same as the one which satisfies momentum, eliminating the need of any interpolation. This is, precisely, the staggered grid arrangement. The coupling due to the co-located arrangement will be named in this text CPVC (Co-located Pressure–Velocity Coupling) and can be avoided naturally using the staggered arrangement. It is due solely because of the co-located arrangement.

If the staggered grid cannot be used as a remedy for the pressure oscillations due to its geometrical complexity, the idea is to create stabilization schemes, permitting the use of co-located grids [9–11]. The strategies for calculating the advecting velocity in co-located arrangement are "seen" as stabilization schemes and they will be seen just after the staggered arrangement is presented.

7.4.2 Staggered Grid Arrangement

It is important to keep in mind that CPVC happens because of the discretization, since there are no velocities at the interfaces for calculating the mass flow. The staggered grid [6] avoids this problem, since velocities are located where they are needed for the mass balance (Fig. 7.4). In this case, the same set of velocities that satisfy mass satisfy momentum. It is a perfect arrangement. If it is thought as a stabilization scheme, it is the best one, since it is the unique in which the advecting velocities are also unknows of the problem. Pressure is located staggered with respect to velocities creating a strong coupling, avoiding the checkerboard pressure field. Perhaps, one could say that the creation of this grid arrangement was one of the greatest contributions in the last five decades in numerical fluid flow.

7.5 Co-located PV Coupling (CPVC) Methods

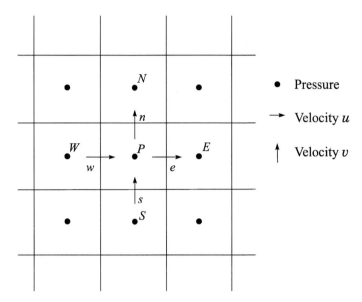

Fig. 7.4 Staggered grid arrangement

Again, there is not always gains. Even for curvilinear coordinate systems the implementation is cumbersome, and in the 80s the staggered grid was practically abandoned. It was a turning point in fluid mechanics computation. Methods for promoting the pressure–velocity coupling for co-located grids started to be developed, since this type of arrangement is the standard nowadays. The current growth of methods using unstructured meshes (in which it is even more complicated to implemented staggered grids) practically eliminated the use of this type of arrangement. There are recent developments using 2D unstructured meshes with staggered grids aiming to have a better felling of the difficulties and trying to overcome them with better programming and new tools [12–14].

For domains in which it is possible to define a Cartesian box inside, perhaps it would be possible to use staggered Cartesian grids joining the irregular boundary using unstructured grid. But, again, the flexibility will be hampered by the geometry.

7.5 Co-located PV Coupling (CPVC) Methods

Since almost all major commercial CFD software employs co-located grids, it is a significant topic to be treated. To start this Section, it is necessary to have a clear understanding what is meant by "pressure velocity coupling", since this denomination is encountered in two distinct situations and raises misunderstanding among beginners. This topic was a major focus in CFD research in the last three decades but, in general, the literature doesn't leave it clear enough, sometimes conveying to the reader that this coupling happens only in segregated solutions.

When co-located grids are used there will be always the CPVC issue, doesn't matter if the solution is segregated or simultaneous. This is the problem caused by the co-located discretization, which causes the advecting velocity not be available at the control volume interfaces to perform the mass balance, as seen in Sect. 7.4.

The other situation is when the solution is segregated, that is, the linear systems for each differential equation are solved iteratively, briefly discussed previously. In this case the coupling is due to the solution procedure. Even using staggered grids, which eliminates the CPVC, there will be the need of methods to deal with this inter-equation coupling, since the equations are solved one-by-one. We will call this coupling as "segregated pressure–velocity coupling", or SPVC.

When the equations are solved in a segregated manner with co-located arrangement, both couplings are present, that is, one needs a SIMPLE-like method to treat the SPVC, and some Rhie and Chow-like method (RC-like) or PIS-Physical Influence Scheme, for treating the CPVC. When solving in a segregated manner with a staggered grid, one just needs to deal with the SPVC.

The coming section is devoted to present two general approaches [9, 11, 15] to deal with the CPVC, ideas which are followed by many works available in the literature. There is nothing new, but just variations over the same ideas of the pioneering developments reported in this text.

7.5.1 Rhie and Chow-Like Methods

One of the pioneering works trying to overcome the co-located PV coupling is reported in [9]. It was a great contribution to numerical fluid flow literature because in the 80s there was an enormous engineering demand for solving complex 3D problems. Staggered grid was not on the game due to its implementation complexity, so, co-located grid was of compulsory use. To overcome the CPVC it is just obtaining an interface velocity which is coupled with a local consistent pressure gradient. The natural choice is to obtain a pseudo momentum equation based on the neighbor nodal equations. This pseudo momentum equation, of course, will not take part on the linear system, since it is not an unknown, but will bring to the interface velocity all the required pressure connections for a good coupling.

7.5.1.1 1D Representation

The RC idea has many variants, and the following developments can be found in [8]. According to Fig. 7.5, what is needed is the velocity u_e at the interface of the control volumes P and E. This velocity will enter the mass conservation, but it is not an unknown of the problem. It should be obtained as function of the neighbor velocities and pressure. The idea is to use the discretized equations for u_P and u_E, and based

7.5 Co-located PV Coupling (CPVC) Methods

on some average, to find a pseudo-velocity u_e. The discretized momentum equation for the control volume centered at E is

$$(A_P)_E u_E = \left(\sum_{nb} A_{nb} u_{nb}\right)_E - \left(\frac{\partial p}{\partial x}\right)_E \Delta V + B_E. \tag{7.21}$$

in which ΔV is the volume of the control volume centered in E. For simplicity, the following notation is adopted,

$$\hat{u}_E = \frac{1}{(A_P)_E}\left[\left(\sum_{nb} A_{nb} u_{nb}\right)_E + B_E\right] \tag{7.22}$$

and

$$d_E = -\frac{\Delta V}{(A_P)_E} \tag{7.23}$$

With these definitions Eq. (7.21) can be written as,

$$u_E = \hat{u}_E + d_E \left(\frac{\partial p}{\partial x}\right)_E \tag{7.24}$$

A similar equation can be written for the control volume centered at P, as

$$u_P = \hat{u}_P + d_P \left(\frac{\partial p}{\partial x}\right)_P \tag{7.25}$$

Equations (7.24) and (7.25) are the momentum equations discretized for control volumes E and P. Let's now imagine a control volume centered at e. Its discretized equation would have the same form, as

$$u_e = \hat{u}_e + d_e \left(\frac{\partial p}{\partial x}\right)_e \tag{7.26}$$

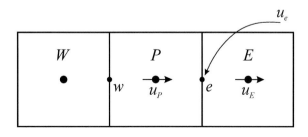

Fig. 7.5 One dimensional Rhie and Chow-like scheme

The terms in Eq. (7.26) are not known and can be determined by averaging the terms of the equations written for E and P, that is,

$$\hat{u}_e = \frac{1}{2}(\hat{u}_P + \hat{u}_E) \qquad (7.27)$$

$$d_e = \frac{1}{2}(d_E + d_P) \qquad (7.28)$$

and for the pressure at the interface e,

$$\left(\frac{\partial p}{\partial x}\right)_e = \frac{p_E - p_P}{(\Delta x)_e} \qquad (7.29)$$

It should be observed that the last equation is an important movement in this derivation, since it is employed the consistent pressure gradient that drives u_e. One could have used the average of the pressure gradients evaluated in E and P but this would keep the decoupling. Recall that the interface velocity u_e is found through some average of the neighbors discretized momentum equations, being each term averaged in a chosen manner. Pressure gradient has no average, it is chosen as the consistent gradient, exactly to produce the required coupling. Substituting Eqs. (7.27–7.29) in Eq. (7.26), it results

$$u_e = \frac{1}{2}(\hat{u}_P + \hat{u}_E) + \frac{1}{2}(d_P + d_E)\left(\frac{p_E - p_P}{(\Delta x)_e}\right) \qquad (7.30)$$

Keeping Eq. (7.30) in this form, when expanding $\hat{u}_P$ and $\hat{u}_E$ the stencil will be too large, involving too many surrounding nodal velocities. This can be simplified rewriting Eqs. (7.24) and (7.25), as

$$\hat{u}_E = u_E - d_E\left(\frac{\partial p}{\partial x}\right)_E$$

$$\hat{u}_P = u_P - d_P\left(\frac{\partial p}{\partial x}\right)_P, \qquad (7.31)$$

and, putting back those equations in Eq. (7.30), one gets

$$u_e = \frac{1}{2}(u_P + u_E)$$
$$+ \frac{1}{2}(d_P + d_E)\left\{\left(\frac{\partial p}{\partial x}\right)_e - \frac{1}{2}\left[\left(\frac{\partial p}{\partial x}\right)_P + \left(\frac{\partial p}{\partial x}\right)_E\right]\right\} \qquad (7.32)$$

Inspecting Eq. (7.32), the first term in the right-hand side is the averaging of the neighbor velocities, the alternative that could have been chosen, but as known, it

7.5 Co-located PV Coupling (CPVC) Methods

doesn't produce the required coupling. The remaining terms is a pressure correction which stabilizes the solution.

Repeating this procedure for the interface velocity u_w and applying the mass conservation equation for this 1D problem given by Eq. (7.15), it results

$$\frac{u_E - u_W}{2\Delta x} + \frac{A\Delta x^3}{4\dot{m}} \left(\frac{-p_{WW} + 4p_W - 6p_P + 4p_E - p_{EE}}{\Delta x^4} \right) = 0, \qquad (7.33)$$

The leading term in Eq. (7.33) is a CDS approximation of the first order velocity derivative, with would be the CDS approximation of the mass conservation equation without the stabilization term. The continuous operator represented by the finite difference of Eq. (7.33) is

$$\left(\frac{\partial u}{\partial x} \right)_P + \frac{A\Delta x^3}{4\dot{m}} \left(\frac{\partial^4 p}{\partial x^4} \right)_P = 0, \qquad (7.34)$$

which is the mass conservation for 1D added to a fourth order pressure derivative which acts as a stabilization term, avoiding pressure decoupling which would occur if only the CDS scheme would represent the mass conservation equation. It is reported in [8] that using Eq. (7.33) with an upwind approximation for the advection terms in the momentum equation there will be no decoupling at all in the pressure or velocity fields. Minor overshoot and undershoot may occur in regions with high gradients, but with small magnitudes. This stabilization term, in fact, alters the mass conservation equation, but it disappears quickly compared to the velocity derivative as the grid is refined, recovering the original mass conservation equation. In other words, the term reduces at a rate of Δx^3 compared to the second order approximation of the velocity derivative. not compromising the accuracy of the solution.

7.5.1.2 Multidimensional Representation

Following the ideas just described, the interface velocity for multidimensional problems are easily found. The mass conservation equation for a control volume of any shape is

$$\sum_{ip} (\rho \mathbf{V}.\mathbf{n} dA)_{ip} = 0 \qquad (7.35)$$

The momentum conservation equation for a nodal velocity including the transient term can be written as

$$(A_P)_P u_P = \left(\sum_{nb} A_{nb} u_{nb} \right)_P - \Delta V \left(\frac{\partial p}{\partial x} \right)_P + \left(\rho \frac{\Delta V}{\Delta t} \right)_P u_P^o + B_P \qquad (7.36)$$

This is a general equation written for a generic control volume P and they are the equations for the control volumes that shares the ip interface. The term $(A_P)_P$ should be interpreted as the central coefficient of the discretized momentum equation located at the control volume centered in P, and $(A_{nb})_P$ are the connecting coefficients of the momentum equation centered at P with its neighbors. Dividing Eq. (7.36) by $(A_P)_P$,

$$u_P = \hat{u}_P + d_P \left(\frac{\partial p}{\partial x}\right)_P - c d_P u_P^o, \tag{7.37}$$

with

$$c = \frac{\rho}{\Delta t} \tag{7.38}$$

$$d_P = -\frac{\Delta V}{(A_P)_P} \tag{7.39}$$

$$\hat{u}_P = \frac{\left[\left(\sum_{nb} A_{nb} u_{nb}\right)_P + B_P\right]}{(A_P)_P} \tag{7.40}$$

Again, following what was done for the 1D case, one assumes that the interface velocity u_{ip} has the same form of Eq. (7.37), giving,

$$u_{ip} = \hat{u}_{ip} + d_{ip}\left(\frac{\partial p}{\partial x}\right)_{ip} - c d_{ip} u_{ip}^o, \tag{7.41}$$

in which all terms with a subscript ip will be an averaging of the values from the equations written for the nodal velocities. To reduce the stencil Eq. (7.37) is written in terms of $\hat{u}_{ip}$, as

$$\hat{u}_{ip} = \overline{\left(u_P - d_P\left(\frac{\partial p}{\partial x}\right)_P + c d_P u_P^o\right)}, \tag{7.42}$$

recalling that u_P, for example, is an average of nodal velocities which shares the interface ip. Inserting Eq. (7.42) in Eq. (7.41), results in

$$u_{ip} = \overline{\left[u_P - d_P\left(\frac{\partial p}{\partial x}\right)_P + c d_P u_P^o\right]} + d_{ip}\left(\frac{\partial p}{\partial x}\right)_{ip} - c d_{ip} u_{ip}^o \tag{7.43}$$

Approximating $\overline{d_P\left(\frac{\partial p}{\partial x}\right)_P}$ by $d_{ip}\left(\frac{\partial p}{\partial x}\right)_P$, Eq. (7.43) becomes

7.5 Co-located PV Coupling (CPVC) Methods

$$u_{ip} = \bar{u}_P + d_{ip}\left[\left(\frac{\partial p}{\partial x}\right)_{ip} - \overline{\left(\frac{\partial p}{\partial x}\right)_P}\right] - cd_{ip}(u_{ip}^o - \bar{u}_P) \qquad (7.44)$$

Inspecting Eq. (7.44) it is clear that the interface velocity that enter in the mass conservation equation is an average of the surrounding nodal velocities plus a pressure correction which can be controlled by the value o d_{ip}. The same approach is now applied for the finding v_{ip} and w_{ip}, and when all components of velocity are introduced into mass conservation equation, one obtains an equation that involves nodal velocities and pressures. Together with the three momentum conservation equations it forms the system of four equations and four unknowns that can be solved simultaneously. The mass flow $\dot{m}_{ip}$, which are obtained with the three components of velocity at the integration points, are required by all other transport equations. The segregated and the simultaneous solution will be discussed in coming sections with possible iterative schemes.

7.5.2 PIS—Physical Influence Scheme

What was possible to infer from the discussions made so far, is that the velocities at the interfaces of the control volumes, not existing at these locations due to the co-located arrangement, must involve for their determination a consistent pressure gradient. By consistent pressure gradient it is meant the one which is the driving force of the velocity and compatible with the mesh size.

The natural choice for doing this is to create a pseudo momentum equation at the interface using the discretized momentum equation of the control volumes that share the same interface. This procedure brings into the interface velocity all the physical ingredients which helps in promoting the coupling between pressure and velocity. This is the philosophy of the so-called RC-like schemes presented in the previous section.

The PIS-like schemes follow the same idea of introducing all physical effects, especially pressure, to obtain the velocity at the integration points. This is done by applying the partial differential equation at the integration point and discretizing it to obtain the velocity at this interface. To exemplify, the 1D problem [11], discussed when demonstrating the coupling difficulties is employed. The discretization of the mass conservation gives,

$$u_e - u_w = 0 \qquad (7.45)$$

Instead of using the linear interpolation as,

$$u_e = \frac{1}{2}(u_P + u_E) \qquad (7.46)$$

$$u_w = \frac{1}{2}(u_P + u_W) \tag{7.47}$$

which will produce conditions for uncoupling, u_e and u_w are found using the corresponding momentum equation, which, for this 1D case reads,

$$\rho u \frac{du}{dx} = -\frac{dp}{dx} + \mu \frac{d^2 u}{dx^2} \tag{7.48}$$

Applying this equation to the integration point e (see Fig. 7.5), and working term-by-term, one obtains

$$\rho u \left(\frac{du}{dx}\right)_e \approx \rho u_e^o \frac{(u_e - u_P)}{\Delta x / 2} \tag{7.49}$$

$$\left(\frac{dp}{dx}\right)_e \approx \frac{p_E - p_P}{\Delta x}, \tag{7.50}$$

which is the consistent pressure gradient for driving u_e. For the last term,

$$\left(\mu \frac{d^2 u}{dx^2}\right)_e \approx \mu \frac{u_P + u_E - 2u_e}{(\Delta x / 2)^2} \tag{7.51}$$

This is just an auxiliary equation for finding u_e, therefore, it can be written in a non-conservative form. Introducing Eqs. (7.49–7.51) into Eq. (7.48) one gets the velocity at the integration point "e", expressed as

$$u_e = u_P \left(\frac{P_e + 2}{P_e + 4}\right) + u_E \left(\frac{2}{P_e + 4}\right) - \frac{\Delta x}{2\rho u_e^o (1 + 4/P_e)} \left(\frac{p_E - p_P}{\Delta x}\right) \tag{7.52}$$

having as a parameter the Peclet number,

$$Pe = \frac{\rho u_e^o \Delta x}{\mu} \tag{7.53}$$

Limiting Eq. (7.52) for Peclet tending to zero and infinity, one obtains, respectively,

$$u_e = (1/2)(u_P + u_E) + \frac{\Delta x^2}{8\mu}(p_P - p_E) \tag{7.54}$$

$$u_e = u_P + \frac{\Delta x}{2\rho u_e^o}\left(\frac{p_P - p_E}{\Delta x}\right) \tag{7.55}$$

In the above expressions if the pressure term is removed, the approximation for the velocity at the integration point reduces to the usual central differencing and upwind schemes.

The same procedure can be done for finding u_w. Inserting them into the mass conservation equation, Eq. (7.45), it is obtained the mass conservation equation which contains the pressure gradients for stabilization, that is, avoiding decoupling. The expressions for Pe tending to zero and to infinity reads, respectively,

$$\frac{u_W - u_E}{2\Delta x} + \frac{\Delta x^2}{8\mu}\left(\frac{p_E + p_W - 2p_P}{\Delta x^2}\right) = \frac{u_W - u_E}{2\Delta x} + \frac{\Delta x^2}{8\mu}\left(\partial^2 p/\partial x^2\right)_P = 0 \quad (7.56)$$

$$\frac{u_W - u_P}{\Delta x} + \frac{\Delta x}{2\rho u_e^o}\left(\frac{p_E + p_W - 2p_P}{\Delta x^2}\right) = \frac{u_W - u_P}{\Delta x} + \frac{\Delta x}{2\rho u_e^o}\frac{\partial^2 p}{\partial x^2} = 0 \quad (7.57)$$

As in the RC-like methods, the PIS algorithms work in the same way, adding, based on physical principles embodied in the momentum equations, a stabilization term for avoiding decoupling. Comments done with respect to this stabilization term was already presented when the RC method was discussed. Of interest is to say that this stabilization is required irrespective if the solution of system of equations is simultaneous or segregated. These methods will be always required when the grid is co-located. There are several works in the literature bringing modifications to speed up this type of algorithms. They have all the same basic idea.

7.6 Segregated PV Coupling (SPVC) Methods

Bringing a bit of history concerning the solution of momentum and mass conservation equations, the technique used, perhaps around five decades ago, was to solve the equations in terms of the stream function and vorticity. With this practice, a two-dimensional incompressible flow problem, whose unknowns are the components of the velocity vector and pressure, is reduced to a problem of two unknowns: the stream function and the vorticity, and the pressure disappears from the formulation. Getting rid of pressure means that pressure oscillations would not appear, of course. It looks as a very attractive technique, but its advantages (only apparent) are just for two-dimensional problems.

For three-dimensional problems, the methodology does not apply, because there is no definition of the stream function for three dimensions. There is the definition of an equivalent variable, the potential function, but this makes the problem, whose unknowns are, primitively, the three components of the velocity vector and the pressure, in a problem with six unknowns, three components of the potential function and three components of the vorticity. The apparent advantage of the two-dimensional case is lost. Even in the 2D case, one considerable drawback is the need to provide boundary conditions for the vorticity, a variable with no easy physical interpretation, and not known as primitive boundary condition of the problem. Another drawback is the coupling between the stream function and the vorticity, via boundary conditions, making the iterative process quite unstable and of slow convergence. This approach is no longer used, and rarely appears in the literature a problem solved with this technique.

Nowadays all effective numerical schemes use the primitive variables, that is, pressure and velocity, adopting methods to deal with the coupling between them, with emphasis on the pressure–velocity couplings that appears due to the segregated nature of the solution of the system of differential equations.

This section is devoted to review the SPVC methods, that is, the coupling appearing when the equation system is solved in a segregated manner and when density is constant or a function of temperature only. It is not an exaggeration to remember that this coupling appears irrespective if the layout is co-located or staggered. If it is staggered, the velocities at the interface of a control volume for mass balance are already available, and there will be no need of a RC-like or PIS-like method. If it is co-located, it will.

The SPVC will be described considering that the velocities at the interfaces are known. For completeness, in Sects. 7.6.7 and 7.6.8, for two SPVC methods, SIMPLEC and PRIME, it will be shown how the velocities at the interfaces are determined if the grid is co-located. Perhaps, it would be didactic remember that inside a SPVC method, there will be the CPVC to be treated, if a co-located arrangement is used.

There are currently several methods to deal with the SPVC. The goal of all of them is to create an equation for pressure that allows the iterative solution process to proceed, while watching closely and carefully on the mass conservation equation, which is not a working equation but just a restriction to be obeyed.

Before start describing some methods, it is advisable to make clear the physical mechanism that must be observed when developing algorithms to deal with this coupling. The correct solution of a momentum transport problem will be obtained when the pressure field introduced in the Navier–Stokes equations generates velocities that satisfy the mass conservation equation.

The discretized system of equations of interest are Eqs. (7.7–7.12), with the last equation changed to $\rho = f(T)$ or ρ constant. Other scalars may need to be solved, such as mass concentration, turbulent kinetic energy, turbulent kinetic energy dissipation etc., but in what follows only the momentum and mass conservation equations will be considered. It is in this system in which all possible numerical difficulties are hidden.

Segregated pressure–velocity coupling methods are losing their importance as computer resources grows and new tools for the solution of large linear systems become available for the engineers. This availability permits to solve the pressure and velocity simultaneously, avoiding, therefore, the need for segregated pressure–velocity coupling methods. As these methods are still used in some commercial applications, and they are a rich source of thinking and learning about the behavior of the pressure–velocity equation system, it is important to keep the topic in this book, which intends to be a text for learning the fundamentals of CFD.

7.6 Segregated PV Coupling (SPVC) Methods

In the construction of any iterative procedure, the algorithm which feeds back the information is crucial for achieving good rate of convergence. In the case of the coupling in question, the mass residue, calculated using the mass conservation equation is the fundamental input to indicate how the new pressure field should be changed. At the same time, this new pressure field must, together with the new velocities, satisfy the equations of motion. There are several ways to "adjust" the variables during the iterative process until the solution of the problem is attained, when all conservation equations involved will be satisfied. Harlow and Welch [6], Chorin [16, 17], Amsden and Harlow [18], Patankar and Spalding [19] were the precursors in the development of methods to treat the pressure–velocity coupling.

Based on Chorin's ideas, many others were developed, in which, two of them, one because of its pioneering role and the other due to its widespread utilization in the scientific community will be presented herein. Once again, it is didactic to make clear that the creation of an algorithm for this purpose is to obtain an equation to advance pressure in an efficient way, since pressure does not possess one equation for evolution. As mentioned, with the enormous advances in computer memory and speed, these methods are almost falling in disuse, since the simultaneous solution is becoming preferable because of its robustness. There are commercial applications in the market which already offer to the user only the simultaneous solution approach, while others offer simultaneous as well as segregated.

7.6.1 Chorin's Method

Just to mention one of the precursors of the methods for treating the SPVC for incompressible flows, few words are devoted to one of the methods developed by Chorin [16, 17]. Two methods were advanced by him. The first one [16] was conceived for problems where only the steady-state solution was of interest. As the intermediate solutions along time are distorted, he used the concept of artificial compressibility, a widely used strategy in the past, by which the flow is treated as compressible, and the compressibility disappears when the steady state solution is obtained. In this way, the compressible formulation can be used, in which all variables have its evolutive equation avoiding, therefore, the SPVC. Chorin's second method [17] can solve problems following the real transient. To exemplify, consider the equation of motion for the x direction written in the form,

$$\frac{\partial}{\partial t}(\rho u) + \frac{\partial p}{\partial x} = F_x u, \qquad (7.58)$$

in which ρ is constant and all terms non explicitly written are grouped in the term $F_x u$. In a specified time level u is considered known and, for the time level $t + \Delta t$ the following equation is solved,

$$\frac{\partial}{\partial t}(\rho u) = F_x u \tag{7.59}$$

As the solution didn't consider the pressure gradient, this velocity is called u^*. Recognizing that F_x can be decomposed into a zero-divergence and a zero-rotational vectors, and that the zero-rotational vector must be $\partial p/\partial x$, the relations between u and u^* and v and v^*, are

$$\rho u = \rho u^* - \Delta t \frac{\partial p}{\partial x} \tag{7.60}$$

$$\rho v = \rho v^* - \Delta t \frac{\partial p}{\partial y} \tag{7.61}$$

Since the pressure is not known, it must be determined such that conservation of mass is satisfied. The following iterative scheme was proposed by Chorin,

$$p^{k+1} = p^k - \lambda D \tag{7.62}$$

in which D is the numerical approximation of the mass conservation equation, or the error in satisfying mass, λ is a relaxation parameter, and k is the iterative level within the time interval at which the solution is being obtained. When it converges, D goes to zero, and the pressure has also converged. The iterative cycle for this method can be written as:

1. Obtain u^* from Eq. (7.59).
2. Correct u and v (and w if it is 3D) using Eqs. (7.60) and (7.61).
3. Calculate p using Eq. (7.62).
4. Iterate between items 2 and 3 until determining the velocities and pressure within the desired accuracy.
5. Move to a new time level.

Based on Chorin's method, described extremely brief, methods with high impact on numerical fluid flow simulation in the last decades were developed. This method introduced the two-steps solution procedure present in almost all SPVC methods, that is, correcting the velocities and, as second step, correcting pressure.

7.6.2 SIMPLE—Semi Implicit Linked Equations

This method, widely used until recently, which historically was a turning point in numerical fluid flow calculations, and from which many others were derived, was developed by Patankar and Spalding [19]. In this procedure the pressure is also written as the sum of the best estimate of the available pressure, p^*, plus a correction p', which is calculated to satisfy the mass conservation, that is, $p = p^* + p'$.

7.6 Segregated PV Coupling (SPVC) Methods

In virtually all methods for treating pressure–velocity coupling, the calculation sequence involves two distinct steps, as presented in Chorin's method: in the first, the velocities are corrected to satisfy the equation of conservation of mass; in the second, the pressure is advanced to complete the iterative cycle. Equations (7.60) and (7.61), which correct the estimated velocity field, are called the velocity correction equations, and the way they are obtained significantly influences the convergence rate of the iterative process.

In the SIMPLE method, the velocity correction equations are obtained from the equations of motion. If a pressure field p^* is introduced in Eqs. (7.8–7.10) they become,

$$A_p u_P^* = A_e u_E^* + A_w u_W^* + A_n u_N^* + A_s u_S^* + A_f u_F^*$$
$$+ A_b u_B^* - \left(\frac{\Delta p}{\Delta x}\right)^* \Delta V + B^u \quad (7.63)$$

$$A_p v_P^* = A_e v_E^* + A_w v_W^* + A_n v_N^* + A_s v_S^* + A_f v_F^*$$
$$+ A_b v_B^* - \left(\frac{\Delta p}{\Delta y}\right)^* \Delta V + B^v \quad (7.64)$$

$$A_p w_P^* = A_e w_E^* + A_w w_W^* + A_n w_N^* + A_s w_S^* + A_f w_F^*$$
$$+ A_b w_B^* - \left(\frac{\Delta p}{\Delta z}\right)^* \Delta V + B^w \quad (7.65)$$

When the correct pressure field, which produces the correct velocity fields, is introduced in the momentum equations, they read,

$$A_p u_P = A_e u_E + A_w u_W + A_n u_N + A_s u_S$$
$$+ A_f u_F + A_b u_B - \left(\frac{\Delta p}{\Delta x}\right) \Delta V + B^u \quad (7.66)$$

$$A_p v_P = A_e v_E + A_w v_W + A_n v_N + A_s v_S$$
$$+ A_f v_F + A_b v_B - \left(\frac{\Delta p}{\Delta y}\right) \Delta V + B^v \quad (7.67)$$

$$A_p w_P = A_e w_E + A_w w_W + A_n w_N + A_s w_S$$
$$+ A_f w_F + A_b w_B - \left(\frac{\Delta p}{\Delta z}\right) \Delta V + B^w \quad (7.68)$$

Subtracting Eqs. (7.63–7.65) from Eqs. (7.66–7.68), considering the source terms not functions of pressure, and neglecting the differences $(u - u^*)$, $(v - v^*)$ and $(w - w^*)$, the velocity correction equations are,

$$u_P = u_P^* - \frac{\Delta V}{A_P}\frac{\Delta p'}{\Delta x} \qquad (7.69)$$

$$v_P = v_P^* - \frac{\Delta V}{A_P}\frac{\Delta p'}{\Delta y} \qquad (7.70)$$

$$w_P = w_P^* - \frac{\Delta V}{A_P}\frac{\Delta p'}{\Delta z} \qquad (7.71)$$

in which, again, it is warned that the A_P coefficients are different in each correction equation. The rigorous notation would be A_P^u, A_P^v, A_P^w, but the super indexes are not carried because there are no chances for confusion.

Following the 2-steps procedure, one needs to find the pressure correction field p'. Now, the mass conservation equation enters into play. Substituting the velocity correction equations given by.

$$u_e = u_e^* - d_e^u(p'_E - p'_P) \qquad (7.72a)$$

$$u_w = u_w^* - d_w^u(p'_P - p'_W) \qquad (7.72b)$$

$$v_n = v_n^* - d_n^v(p'_N - p'_P) \qquad (7.72c)$$

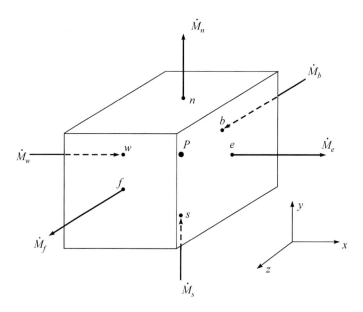

Fig. 7.6 Three-dimensional control volume

7.6 Segregated PV Coupling (SPVC) Methods

$$v_s = v_s^* - d_s^v(p'_P - p'_S) \tag{7.72d}$$

$$w_f = w_f^* - d_f^w(p'_F - p'_P) \tag{7.72e}$$

$$w_b = w_b^* - d_b^w(p'_P - p'_B) \tag{7.72f}$$

into the mass conservation equation,

$$\dot{M}_e - \dot{M}_w + \dot{M}_n - \dot{M}_s + \dot{M}_f - \dot{M}_b = 0 \tag{7.73}$$

the equation for p' is obtained, as

$$A_P p'_P = A_e p'_E + A_w p'_W + A_s p'_S + A_n p'_N \\ + A_f p'_F + A_b p'_B - \nabla.\mathbf{V}^*, \tag{7.74}$$

in which $\nabla.\mathbf{V}^*$ is obtained using Eq. (7.73) applied to the velocity field $\mathbf{V}^*$. The coefficients of Eq. (7.79) are given by

$$A_e = \left((\Delta y \Delta z) d_e^u\right)_e \tag{7.75a}$$

$$A_w = \left((\Delta y \Delta z) d_w^u\right)_w \tag{7.75b}$$

$$A_n = \left((\Delta x \Delta z) d_n^v\right)_n \tag{7.75c}$$

$$A_s = \left((\Delta x \Delta z) d_s^v\right)_s \tag{7.75d}$$

$$A_f = \left((\Delta x \Delta y) d_f^w\right)_f \tag{7.75e}$$

$$A_b = \left((\Delta x \Delta y) d_b^w\right)_b \tag{7.75f}$$

while the d_e^u coefficient is written as,

$$d_e^u = \left(\frac{(\Delta y \Delta z)}{A_P}\right)_e \tag{7.76}$$

with all others d coefficients easily obtained by comparison. Boundary conditions for p' will be discussed soon, later in this chapter.

After the solution of the linear system for p', the velocities which satisfy mass can be obtained using Eq. (7.72). At this point the velocity field satisfy the mass conservation equation. The next step is to correct the pressure by

$$p = p^* + p' \tag{7.77}$$

For the new iterative cycle, p^* is made equal to the new p obtained with Eq. (7.77), and a new velocity field is calculated using the momentum equations. This procedure continues until convergence is attained within prescribed tolerances.

At this point, it is important to remember that if the staggered grid arrangement is used, that is, pressures are located in the centers of the mass conservation control volumes and velocities at the surfaces of this control volume, the velocities given by Eq. (7.72) are already available where they are needed for mass conservation balance.

This means that the CPVC is already treated. When the co-located arrangement is used, a way to calculate the velocities at the interfaces of the pressure control volume should be sought, that is a CPVC method should be devised.

The SIMPLE method has limitations, especially with respect to the rate of convergence. The advantages and limitations of this method, and others similar can be found in [20], where a comparative analysis of several methods for treating pressure–velocity coupling is performed. In the literature it is reported several articles with the same goal as in [20]. Briefly, it can be advanced that the velocity correction equations, Eq. (7.72), have a strong simplification when deriving them, the reason for requiring a strong under relaxion in p', as

$$p = p^* + \alpha p' \tag{7.78}$$

in order to obtain convergence of the system of equations. The complete iterative cycle for solving the pressure–velocity coupling using the SIMPLE method is, as follows:

1. Estimate the velocity and pressure fields p^*.
2. Calculate the coefficients of the momentum equations for u, v, and w.
3. Solve the momentum equations, Eqs. (7.63–7.65) using p^*, obtaining u^*, v^* and w^*. This action involves the solution of the linear system of algebraic equations for each variable.
4. Solve Eq. (7.74) and obtain p'. A linear system needs to be solved.
5. Correct the stars velocities using Eq. (7.72), obtaining the velocity field that satisfies the mass conservation equation.
6. Calculate p using Eq. (7.77) or Eq. (7.78).
7. Solve the conservation equations for other variables such as temperature, mass concentration, turbulence etc.
8. Do $p = p^*$ and start over at step (2) until convergence.

In item 7, if the values of these variables have no influence on the flow, for example, in forced convection heat transfer with constant physical properties, these equations can be solved after the flow has been calculated.

An important detail of pressure–velocity coupling methods that use velocity correction equations is to remember that the solution of the problem does not depend on the type of the correction equations, since these are auxiliary equations and not

7.6 Segregated PV Coupling (SPVC) Methods

part of the system of equations being solved. Their influence is on the rate of convergence. For this reason, it is advisable to have correction equations that originates from the equations to be solved, that is, bringing the physics into the correction equations. The SIMPLE method was derived for steady state problems, therefore, time was not in the equations, which is equivalent in having an infinite time step. This means that the correction applied in the velocities corresponds to an infinite time step, which results in a large correction, which needs to be underrelaxed.

As commented in another chapter, there are two reasons why iterations are necessary when solving the system of equations in a segregated manner. The first is because the equations are coupled to each other, and the second is to take nonlinearities into account. In the procedure described above, the iterative cycle shown is playing both roles simultaneously, that is, it solves the coupling and at the same time advances the coefficients, bringing the effects of nonlinearity into each iterative cycle.

Another possible, and widely adopted, procedure is to create two iterative cycles, one dealing with the pressure–velocity coupling for a fixed set of coefficients, and another one recalculating the coefficients. This is equivalent in doing an internal iteration loop between items 3 and 6 and, after this step has converged, calculating item 2.

One of the advantages of the SIMPLE method is that it is not necessary to solve a linear system to determine the pressure. However, the rate of convergence is slow. An analysis of the convergence difficulties and reasons for that can be found in [20].

7.6.3 SIMPLER—Simple-Revisited

Aiming to improve the convergence characteristics of the SIMPLE method, it was revisited, originating the SIMPLE-Revisited, or, simply, SIMPLER [7]. In this method the pressure is no longer found through Eq. (7.78), but a linear system is found for this task. The rationale behind, largely used in the past, is to play with the momentum conservation equations to extract pressure, since the pressure gradient is present in all of them. The key movement is how to reunite the momentum equations to extract pressure? In SIMPLER this is done using the mass conservation equation, that is, the momentum equations are introduced into mass conservation equation to obtain a discretized equation for pressure determination.

To this end, the momentum equations are written as,

$$u_P = \hat{u}_P - \frac{\Delta V}{A_P} \frac{\Delta p}{\Delta x} \tag{7.79}$$

$$v_P = \hat{v}_P - \frac{\Delta V}{A_P} \frac{\Delta p}{\Delta y} \tag{7.80}$$

$$w_P = \hat{w}_P - \frac{\Delta V}{A_P} \frac{\Delta p}{\Delta z} \tag{7.81}$$

It is important to observe the similarities of Eqs. (7.69–7.71) and Eqs. (7.79–7.81), keeping in mind that the former are just velocity correction equations created for correcting the velocities such they conserve mass, while the later are the momentum equations themselves written in a compact form. The hat velocities are

$$\hat{u}_P = \frac{A_e u_E + A_w u_W + A_n u_N + A_s u_S + A_f u_F + A_b u_B + B^u}{A_P} \quad (7.82)$$

$$\hat{v}_P = \frac{A_e v_E + A_w v_W + A_n v_N + A_s v_S + A_f v_F + A_b v_B + B^v}{A_P} \quad (7.83)$$

$$\hat{w}_P = \frac{A_e w_E + A_w w_W + A_n w_N + A_s w_S + A_f w_F + A_b w_B + B^w}{A_P} \quad (7.84)$$

Note that the hat velocities are the momentum equations without the pressure term. Our goal now is to determine the pressure that is present in the three equations of motion. One way, among many, to isolate the pressure from these three equations, is to substitute it into the mass conservation equation. Using the equation of mass conservation not only does the task but it is a robust way of doing it. It is worth noting that in this step there is no need to satisfy the conservation of mass. This was already done in the previous step, when the velocities u^*, v^* and w^* were corrected with the p', which is still used for correcting the velocities, but no longer for updating pressure.

Equations (7.79–7.81), written for the integration points of the mass conservation control volume read

$$u_e = \hat{u}_e - d_e^u(p_E - p_P) \quad (7.85a)$$

$$u_w = \hat{u}_w - d_w^u(p_P - p_W) \quad (7.85b)$$

$$v_n = \hat{v}_n - d_n^v(p_N - p_P) \quad (7.85c)$$

$$v_s = \hat{v}_s - d_s^v(p_P - p_S) \quad (7.85d)$$

$$w_f = \hat{w}_f - d_f^w(p_F - p_P) \quad (7.85e)$$

$$w_b = \hat{w}_b - d_b^w(p_P - p_B) \quad (7.85f)$$

Introducing them into the mass conservation equation, Eq. (7.73) one obtains an equation for calculating the pressure, as

$$A_P p_P = A_e p_E + A_w p_W + A_n p_N + A_s p_S$$

7.6 Segregated PV Coupling (SPVC) Methods

$$+ A_f p_F + A_b p_B - \nabla . \widehat{\mathbf{V}} \qquad (7.86)$$

Equation (7.86) calculates the pressure for the next iteration level. The technique used to obtain the pressure is similar the one used for calculating p', with Eq. (7.85) being similar to Eq. (7.72). However, let's keep in mind that Eq. (7.85) are not velocity correction equations. The velocity field were already corrected and satisfies mass conservation, an operation performed in the item of velocity correction with p'. The solution procedure for the SIMPLER method is:

1. Estimate the velocity and pressure fields p^*;
2. Repeat items 2 through 5 of the SIMPLE procedure;
3. Calculate $\hat{u}_e$, $\hat{u}_w$, $\hat{v}_n$, $\hat{v}_s$, $\hat{w}_f$ and $\hat{w}_b$;
4. Calculate p using Eq. (7.86)
5. Solve the conservation equations for other variables, such as temperature, mass concentration etc.
6. Go back to item 2 and iterate until convergence.

In this procedure, two Poisson equations, one for p and one for p', are solved. However, advancing p toward convergence is faster and safer.

As a reminder, in the SIMPLE method in three dimensions, to solve the fluid flow problem, one has in each iterative cycle, the solution of four linear systems, for u, v, w and p'). In the SIMPLER method, the solution of five linear systems is required, the four just mentioned plus the one for p.

7.6.4 PRIME—Pressure Implicit Momentum Explicit

In this method [21, 22], the main motivation was to perform both steps (velocity correction and pressure calculation) at once. This can be achieved by using Eq. (7.86) not only for the pressure calculation but also to use the pressures obtained for the velocity correction, making it unnecessary to obtain the p' field, which in the methods SIMPLE and SIMPLER serve to correct the velocity field. That is, the equations of motion, Eq. (7.85), are also used as velocity correction equations. These equations, repeated for completeness, are

$$u_e = \hat{u}_e - d_e^u(p_E - p_P) \qquad (7.87a)$$

$$u_w = \hat{u}_w - d_w^u(p_P - p_W) \qquad (7.87b)$$

$$v_n = \hat{v}_n - d_n^v(p_N - p_P) \qquad (7.87c)$$

$$v_s = \hat{v}_s - d_s^v(p_P - p_S) \qquad (7.87d)$$

$$w_f = \hat{w}_f - d_f^w(p_F - p_P) \qquad (7.87e)$$

$$w_b = \hat{w}_b - d_b^w(p_P - p_B) \qquad (7.87f)$$

As pointed out, the equations for correcting the velocities can be any, and, therefore, can be the momentum equations themselves, as proposed by the PRIME method.

It is important to note that in the PRIME method, since the velocity correction equations are the equations of motion themselves, the pressure field that corrects the velocities is the pressure solution. This is why it is said that the velocity correction and the pressure determination steps are done together. The algorithm of the PRIME method then presents the following steps

1. Estimate the velocity fields.
2. Calculate the coefficients of the equations of motion for u, v, and w.
3. Calculate the velocities $\hat{u}_e$, $\hat{u}_w$, $\hat{v}_n$, $\hat{v}_s$, $\hat{w}_f$ and $\hat{w}_b$ at the integration points of the control volume for pressure (mass conservation). Remember that these are not obtained from the solution of linear systems, but from the algebraic expressions, Eqs. (7.82–7.84) already presented in the SIMPLER method.
4. Calculate p by solving Eq. (7.86).
5. Correct the velocities using Eq. (7.87). This is not done in the SIMPLER method.
6. Solve for other scalars that may exist in the model (temperature, mass concentration, turbulence etc.)
7. Go back to item 2 and iterate until convergence.

Some important details of this method should be pointed out. There is no solution of implicit linear systems for advancing the velocities. They are advanced during the iterative cycle in a similar way to the Jacobi method when calculating the hat velocities. The word Explicit in the name of the method comes from this fact. Explicit in this case does not mean that the formulation is explicit, but rather that the linear systems for the velocities are being solved iteratively in a similar way to the Jacobi method.

Only one Poisson equation is solved at each iteration cycle, and the method is extremely simple to implement and robust. It is highly stable and allows larger time steps to be used than similar methods. On the other hand, since the solution of the equations of motion is performed with a point-by-point iterative method, convergence, in principle, is slower and CPU time consuming may be higher [23].

Some results, using the PRIME method and its extension to all speed flows and co-located grids, can be seen in Chap. 15.

7.6.5 SIMPLEC—Simple Consistent

The SIMPLEC method [24] has identical procedure as the SIMPLE method, with a minor difference in the equations for correcting the velocities. In the SIMPLEC method, the differences $u - u^*$, $v - v^*$, $w - w^*$ are not neglected as done in the SIMPLE method. This looks like of a minor modification, but which has important consequences in convergence. Taking the velocity u as an example, consider Eqs. (7.63) and (7.66) written in the form,

$$A_P u_P^* = \sum A_{nb} u_{NB}^* - \left(\frac{\partial p}{\partial x}\right)^* \Delta V + B^u \quad (7.88)$$

$$A_P u_P = \sum A_{nb} u_{NB} - \left(\frac{\partial p}{\partial x}\right) \Delta V + B^u \quad (7.89)$$

Subtracting Eq. (7.88) from Eq. (7.89) one gets

$$A_P u_P' = \sum A_{nb} u_{NB}' - \left(\frac{\partial p'}{\partial x}\right) \Delta V \quad (7.90)$$

in which $u_P' = u_P - u_P^*$. In the SIMPLE method $(u_{NB} - u_{NB}^*)$ are neglected and this is a first order difference. In the SIMPLEC for having a more robust equation the neglecting differences is of second order. Subtracting $\sum A_{nb}(u_P')$ in both sides of Eq. (7.90) and rearranging, one gets

$$u_P'\left(A_P - \sum A_{nb}\right) = \sum A_{nb}(u_{NB}' - u_P') - \left(\frac{\partial p'}{\partial x}\right) \Delta V \quad (7.91)$$

The term which is now neglected are the differences in the prime velocities, and the velocity correction equation for the u variable becomes

$$u_P = u_P^* - \frac{\left(\frac{\Delta p'}{\Delta x}\right) \Delta V}{A_P - \sum A_{nb}}, \quad (7.92)$$

or

$$u_P = u_P^* - d_P^u \Delta p', \quad (7.93)$$

with

$$d_P^u = \frac{\Delta y \Delta z}{\left(A_P - \sum A_{nb}\right)} \quad (7.94)$$

Notice that the difference between SIMPLEC and SIMPLE is only in the expression of d_e^u, in whose denominator now appears the difference between A_P and $\sum A_{nb}$, and not just A_P, as in the SIMPLE method. This avoids the severe under-relaxation in p, necessary in the SIMPLE method to achieve convergence.

The expressions for the velocities at the interfaces for the SIMPLEC method are therefore the same as in the SIMPLE method with the modified d. The whole procedure is identical. This modification, while seeming simple, deserves an analysis. Going back to Chap. 6, when the equations for three dimensions were developed, the following expression for the central coefficient was obtained,

$$A_P = \sum A_{nb} - S_P \Delta V + \frac{M_P}{\Delta t} \tag{7.95}$$

Considering no source terms, Eq. (7.92) can be written as,

$$u_P = u_P^* - \frac{\Delta x \Delta y \Delta z}{M_P}\left(\frac{\Delta p'}{\Delta x}\right)(\Delta t) \tag{7.96}$$

in which the time step acts a relaxation parameter. In the original SIMPLE the equations were deduced for steady state which implies $\Delta t = \infty$, which means that the correction to be applied in u_P^* is very large, requiring then a sub-relaxation in p' represented by the parameter α in Eq. (7.78). Having the time step in the velocity correction, the correction applied will be consistent with the time step used in the simulation. It is worth to reinforce that the coordinate time must be always kept in the discretization of the equations. It is not a good practice, as many textbooks do, to teach numerical techniques dividing in two topics, one for steady state and another one for transient, since the coordinate time is extremely important in fluid flow problems, even when the only interest is the steady state solution, because of the difficulties, if not the impossibility, of solving the system of equations using infinite time step. The steady state is part of the transient and there is no numerical reason for treat them separated.

7.6.6 PISO—Pressure Implicit with Split Operator

This scheme [25] is a combination of the SIMPLE and PRIME methods, as very well described in [26], taking advantage of the implicitness in the solution of the momentum equations of the former, and the stability of the latter in advancing the pressure field. In the SIMPLE method, the solution of the momentum equation to obtain the star velocity, solving for p' and correcting the velocities to satisfy mass conservation, is done in each iteration cycle, as seen in items 2, 3, 4, and 5 of the solution procedure for SIMPLE (Sect. 7.6.2). This step in the PISO scheme is called predictor and is done only once. Having all velocities satisfying mass conservation via

7.6 Segregated PV Coupling (SPVC) Methods

SIMPLE, the PRIME method enters in action. The hat velocities, given by Eqs. (7.82–7.84), which means the explicit solution of the momentum equation are obtained. In the sequence, Eq. (7.86), having as source term the divergence of the hat velocities is solved for pressure.

With this pressure field the hat velocities are corrected using Eqs. (7.85a–7.85f). This PRIME loop can be done more than once, and then back to item 2 of the SIMPLE procedure. The combination of these two methods speeds up the solution convergence and is stable, as reported in the literature.

All SPVC methods seen in this Sect. 7.6 can use staggered or co-located grids, and they were demonstrated considering staggered grids, that is, the velocities at the interfaces of the control volume for mass conservation were known. For the sake of completeness, the SIMPLEC and the PRIME method are now demonstrated for co-located grids, situation in which the velocities are not known at the interfaces and a RC-like or PIS-like should be applied.

7.6.7 SIMPLEC for Co-located Grids

When the colocalized arrangement is used, all variables have the same control volume for the balance of the properties. The difficulty that arises was intensively described in previous sections, which is the determination of a velocity at the integration points, that is, at the interface of the control volume. Somehow, these velocities must be obtained, and the process of obtaining them will have a marked influence on the robustness of the solution, and the strategies were discussed in Sect. 7.5 in which the CPVC was discussed. Those techniques are now applied in conjunction with methods that solves the SPVC and, as commented before, when solving the system of equations in a segregated manner and using co-located grids both, the CPVC and the SPVC appear.

Therefore, for this section it would suffice to say that the interface velocity for the mass balance is found using one of the techniques shown in Sect. 7.5, or a similar one, since the remaining of the procedure is exactly the same. Perhaps, it would be didactic to write few words about, just for completeness. Inspecting the coefficients of the equation for p', given by Eq. (7.75) it is observed that the parameters d_i are required at the interfaces. They do not exist at the interfaces, because the equations of motion are not being solved there. The source term for the equation for p' requires the velocities (superscript star) at the interfaces. It must be devised, therefore, an efficient way to calculate u^*, v^* and the d_i parameters at the interfaces, the job that should be done by a CPVC method. Besides the methods presented in Sect. 7.5, it is reported here an alternative method akin to the one presented in [9]. Following the methodology developed in [10], which belong to the family of the RC methods, the discretized equations for the volumes P and E, for the velocity u, for example, are given by

$$(A_P)_P u_P^* = \sum (A_{nb} u_{NB}^*)_P + \frac{M_P^o u_P^o}{\Delta t} - \left(\frac{\Delta p*}{\Delta x}\right)_P \Delta V + (S_P^{u*})_P \Delta V \quad (7.97)$$

$$(A_P)_E u_E^* = \sum (A_{nb} u_{NB}^*)_E + \frac{M_E^o u_E^o}{\Delta t} - \left(\frac{\Delta p*}{\Delta x}\right)_E \Delta V + (S_P^{u*})_E \Delta V \quad (7.98)$$

These equations are written for the star velocities since these are the calculated velocities from the linear system for the momentum equations using the available pressure field p^* which form the source term of the equation for p' in the SIMPLEC (or SIMPLE) method. The procedure adopted is to average the momentum equations to obtain a pseudo momentum equation at the interface (integration points) keeping a consistent pressure gradient. The resulting equation for u_e^* is

$$u_e^* = \frac{(A_P)_e}{(A_P)_P + (A_P)_E} \quad (7.99)$$

in which

$$(A_P)_e = \sum (A_{nb} u_{NB}^*)_P + \sum (A_{nb} u_{NB}^*)_E + (S_P^{u*})_P \Delta V$$
$$+ (S_P^{u*})_E \Delta V + \frac{(M_P^o + M_E^o) u_e^o}{\Delta t} - 2 \frac{(p_E^* - p_P^*)}{\Delta x} \Delta V \quad (7.100)$$

Similarly, the other velocities are calculated, and the divergence of the estimated velocities is computed. This divergence is the source term of the equation for p'. The values of d_i that are needed in Eq. (7.72) are obtained by an arithmetic average between the values of d_i in P and in E. The iterative cycle is the same as in the SIMPLE and SIMPLEC methods for the staggered arrangement. It differs only in one small detail. In the staggered arrangement, when the velocities are corrected, they become the most recent estimates of the velocity field. In the collocated case, when they are corrected, originating velocities that satisfy mass conservation, the nodal velocities, which are the true variables of the problem, must also be updated to start the next iterative cycle. Two ways are possible. Either correct the nodal velocities with the given p', or, from the corrected velocities at the integration points, determine the new nodal velocities through an average. The first alternative proved to be more efficient when the SIMPLEC method was employed. This observation is probably valid for the SIMPLE-like methods, since for the PRIME method the second alternative was the most efficient. This is, probably, because the correction equations in PRIME are the equations of motion themselves.

This method [10] using SIMPLEC was applied for all speed flows solving incompressible as well as compressible flow with Mach number up to 3.0.

7.6.8 PRIME for Co-located Grids

First, it should be remembered that the velocity correction equations don't change when changing from staggered to co-located grids, and for the PRIME method they are the momentum equations. The procedure, as before, is to introduce Eq. (7.87) into the mass conservation equation resulting in an equation for determination of p. To this end, the velocities $\hat{u}$ and $\hat{v}$ (considering a 2D problem) and the parameters d, should be evaluated at the interfaces. The velocities $\hat{u}$ and $\hat{v}$ form the divergent of $\hat{\mathbf{V}}$, which in turn is the source term for the equation for p. Since there are no velocities neither coefficients stored at the interfaces, an average must be chosen to find them at the integration points (interfaces). The nature of this averaging is important, especially for the PRIME method, which uses as correction equations the equation of motion itself. Taking the velocity $\hat{u}_e$ as example, one has

$$u_e = \hat{u}_e - \overline{d}^u_e (p_E - p_P), \qquad (7.101)$$

in which $\hat{u}_e$ is determined by

$$\hat{u}_e = (1/2)(\hat{u}_P + \hat{u}_E), \qquad (7.102)$$

or

$$\hat{u}_e = (1/2)\left\{\left[\left(\sum A_{nb} u_{NB} + B^u\right)/A_P\right]_P + \left[\left(\sum A_{nb} u_{NB} + B^u\right)/A_P\right]_E\right\} \qquad (7.103)$$

in which B^u includes all possible source terms, except the pressure gradient term and $\overline{d}^u_e$ is

$$\overline{d}^u_e = (1/2)\left((\Delta V / A_P \Delta x)_P + (\Delta V / A_P \Delta x)_E\right) \qquad (7.104)$$

Recall that with co-located grids, for avoiding the coupling difficulties, the main idea is to create pseudo equations at the interfaces. In the PRIME method, since the correcting equations are the proper momentum equations, a RC-like method is automatically embodied, as can be appreciated in Eq. (7.103). The final correcting equations is

$$u_e = \hat{u}_e - (1/2)\left(\overline{d}^u_P + \overline{d}^u_E\right)(p_E - p_P) \qquad (7.105)$$

Similarly, it is possible to determine $\hat{u}_w$, $\hat{v}_n$, $\hat{v}_s$ and the other two components in the z direction in the three-dimensional case. Knowing all these velocities at the interfaces and substituting them into the mass conservation equation, an equation like Eq. (7.86) is obtained. Once the pressure field is known, the velocity correction equations, Eq. (7.87), are used to determine the velocities that satisfy mass conservation. To obtain the velocities at the centers of the control volumes, it is enough to average the neighboring velocities that satisfy the mass conservation equation. There are other engineering applications using the PRIME method, as phase change, subsonic and supersonic flows, which demonstrates the good stability of the method. In all those problems the strong point of PRIME was its stability, being, for the problems tested, independent of the size of the time step, which allows the user to use any time interval without concern for convergence. The drawback is the poor rate of convergence, which deserves further studies for improving this.

There are many other SIMPLE-like methods for dealing with segregated pressure–velocity coupling available in the literature. The purpose of this section is to present the philosophy behind the coupling methods, not a review of them.

Following, let's have a look on the boundary conditions, since in order to solve the equations for p and p' it is necessary to specify the corresponding boundary conditions.

7.7 Boundary Conditions for p and p'

The boundary conditions for p and p' in SIMPLE and SIMPLER methods, respectively, as recommended in [19], are obtained from Eqs. (7.69–7.71) and Eqs. (7.79–7.81), respectively. When velocities are prescribed at the boundaries, there is no corrections for these velocities, which is achieved by enforcing zero gradients of p and p' normal to the surfaces. This procedure is easily applicable when using orthogonal coordinate system, otherwise, it is not recommended, because in non-orthogonal systems, curvilinear or unstructured, the gradient of p normal at the boundary cannot be expressed as a function of only two grid points, becoming a cumbersome implementation. Such a possibility has been investigated in [21] and proved to be inadequate. And, Neumann boundary condition, as is the case, requires that the physics be respected. Zero gradients at the boundaries means no mass fluxes through the boundaries and, if during the iterative procedure, even small errors in mass conservation for each control volume appears, there will be no convergence. In incompressible flows, since the level of the pressure doesn't matter, this difficult in having zero pressure gradients in all boundaries can be circumvented by removing one equation of the linear system setting the pressure to some value.

A general and simple alternative to the problem was proposed in [21], for non-orthogonal curvilinear grids, also presented later in [24] for Cartesian coordinates.

The idea is to apply for the boundary volumes the same procedure adopted for the internal volumes, that is, apply the mass conservation equation, respecting the existing boundary condition. Consider Fig. 7.7 with a prescribed velocity at the right boundary. Applying the mass conservation equation for the control volume centered at P, it reads

$$\dot{M}_e - \dot{M}_w + \dot{M}_n - \dot{M}_s = 0, \tag{7.106}$$

7.7 Boundary Conditions for p and p'

and only the internal mass fluxes will be substituted by the correcting equations. Since the mass flow $\dot{M}_e$ is known by boundary condition it will be in the B term and the discretized equation will be of the form for the control volume centered at P,

$$A_P p'_P = A_w p'_W + A_n p'_N + A_s p'_S + B \tag{7.107}$$

Hence, the boundary condition is incorporated in the approximate equation for the boundary volume. This procedure does not depend on the coordinate system used or the shape of the control volume, it is general, satisfies the balances also for the boundary volumes and does not increase the number of equations of the linear system, because it has no fictitious points.

Another advantage of this procedure, and very important from the point of view of convergence, is the exact satisfaction of the equation of conservation of mass, because the prescribed velocities enter directly into the equation for that volume. The method of specifying the gradient of p' equal to zero creates a Neumann problem with known difficulties for the solution.

A boundary condition of prescribed pressure in a flow entering the domain may be also possible. In this case, the treatment will depend on the type of variable arrangement that is being used. Considering the staggered arrangement, Fig. 7.8 shows a control volume where, on the west face, the pressure is prescribed and equal to p_f. The approximate equation for the volume P is obtained, as always, by doing a mass balance. The equations of correcting the velocities u_e, v_n and v_s are the same as for internal control volume, while for u_w it reads,

$$u_w = u^*_w - \frac{2\Delta V}{A_P} \frac{\Delta p'_P}{\Delta x} \tag{7.108}$$

since p'_f is equal to zero because p_f is prescribed and equal to p^*_f and should not be corrected. It should be noted that the velocity u_w is unknown and therefore there

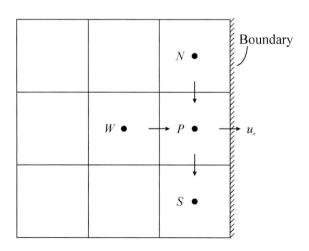

Fig. 7.7 Prescribed velocity at east face

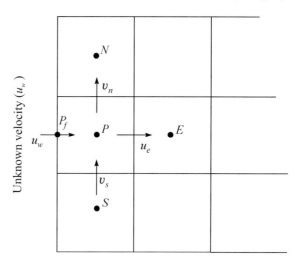

Fig. 7.8 Prescribed pressure at the west face

must be an equation for it in the linear system such that u_w^* can be determined. For an incompressible problem, it should not be forgotten that the pressure at the outlet must be specified, because the pressure difference in that case establishes the mass flow.

Just as a reminder, the pressure gradient in the equation for the momentum conservation for u_w is, in this case, $(p_f - p_P)$. If the PRIME method is being used, the correction equation for u_w is

$$u_w = \hat{u}_w - \frac{2\Delta V}{A_P} \frac{(p_P - p_f)}{\Delta x} \qquad (7.109)$$

in which $\hat{u}_w$ is found using its definition, given by Eq. (7.82).

7.8 Simultaneous Solution and the Couplings

As seen before, the main coupling existing in the conservation equations when simulating engineering problems involving fluid flow is between velocity and pressure, that is, u, v, w, p variables with the three equations from the momentum conservation plus mass conservation. When the solution is segregated, to solve this coupling, which we are calling SPVC, to distinguish from the CPVC, one needs to employ a SIMPLE-like method just described. Until three decades ago the segregated solution was almost mandatory due to the lack of memory of the computers at that time. The simultaneous solution eliminates the need of the SIMPLE-like methods because the coupling is considered when the momentum conservation equations and mass are solved simultaneously. This translates into robustness of the numerical solution. Nowadays, commercial applications can be installed on personal computers solving

7.8 Simultaneous Solution and the Couplings

engineering problems of importance, and the simultaneous solution became a viable alternative. Modern commercial applications for CFD computations even no longer offer the alternative of solving the equations in a segregated form, and some others offer both possibilities, segregated and simultaneous solution. It looks like that very soon the segregated solution will no longer be used, and all the SIMPLE-like methods will no longer be needed.

We have also seen that the important part of an algorithm to solve the segregated pressure–velocity coupling was to transform the mass conservation equation into an equation for pressure using the correcting equations for velocity. By doing this, all velocities disappear as active variables in the mass conservation equation and only pressure remains as the active variable. This form of the mass conservation equation is not suitable for solving the system simultaneously. In the same way, the mass conservation equation in its original form, in which only velocities appear is also not adequate, since pressure is missing. That is, a good system of equations to be solved simultaneously is when all variables are present in all equations. To exemplify this, consider the following system of algebraic equations,

$$a_1 x + a_2 y + a_3 z = a$$
$$b_1 x + b_2 y + b_3 z = b$$
$$c_1 x + c_2 y + c_3 z = c \qquad (7.110)$$

whose linear system is,

$$\begin{bmatrix} a_1 & a_2 & a_3 \\ b_1 & b_2 & b_3 \\ c_1 & c_2 & c_3 \end{bmatrix} \begin{bmatrix} x \\ y \\ z \end{bmatrix} = \begin{bmatrix} a \\ b \\ c \end{bmatrix} \qquad (7.111)$$

Inspecting Eq. (7.111) one sees that there are no zero values on the diagonal, what helps the solution of the linear system. This simple example of an algebraic linear system can be extended to the system of partial differential equations of interest. The equation system for solving the velocity and pressure fields is given by,

$$\frac{\partial \rho}{\partial t} + \frac{\partial}{\partial x}(\rho u) + \frac{\partial}{\partial y}(\rho v) + \frac{\partial}{\partial w}(\rho w) = 0 \qquad (7.112)$$

$$\frac{\partial}{\partial t}(\rho u) + \frac{\partial}{\partial x}\left(\rho u u\right) + \frac{\partial}{\partial y}\left(\rho v u\right) + \frac{\partial}{\partial z}\left(\rho w u\right) = -\frac{\partial p}{\partial x}$$
$$+ \frac{\partial}{\partial x}\left(\mu \frac{\partial u}{\partial x} - \frac{2}{3}\mu \mathrm{div} \mathbf{V}\right) + \frac{\partial}{\partial y}\left(\mu \frac{\partial v}{\partial x}\right) + \frac{\partial}{\partial z}\left(\mu \frac{\partial w}{\partial x}\right)$$
$$+ \frac{\partial}{\partial x}\left(\mu \frac{\partial u}{\partial x}\right) + \frac{\partial}{\partial y}\left(\mu \frac{\partial u}{\partial y}\right) + \frac{\partial}{\partial z}\left(\mu \frac{\partial u}{\partial z}\right) + B_x \qquad (7.113)$$

$$\frac{\partial}{\partial t}(\rho v) + \frac{\partial}{\partial x}\left(\underline{\rho u v}\right) + \frac{\partial}{\partial y}\left(\underline{\rho v v}\right) + \frac{\partial}{\partial z}\left(\underline{\rho w v}\right) = -\frac{\partial p}{\partial y}$$

$$+ \overline{\frac{\partial}{\partial x}\left(\mu\frac{\partial u}{\partial y}\right)} + \frac{\partial}{\partial y}\left(\mu\frac{\partial v}{\partial y} - \frac{2}{3}\mu\text{div}\mathbf{V}\right) + \overline{\frac{\partial}{\partial z}\left(\mu\frac{\partial w}{\partial y}\right)}$$

$$+ \overline{\frac{\partial}{\partial x}\left(\mu\frac{\partial v}{\partial x}\right)} + \frac{\partial}{\partial y}\left(\mu\frac{\partial v}{\partial y}\right) + \frac{\partial}{\partial z}\left(\mu\frac{\partial v}{\partial z}\right) + B_y \qquad (7.114)$$

$$\frac{\partial}{\partial t}(\rho w) + \frac{\partial}{\partial x}\underline{(\rho uw)} + \frac{\partial}{\partial y}\underline{(\rho vw)} + \frac{\partial}{\partial z}\underline{(\rho ww)} = -\frac{\partial p}{\partial z}$$

$$+ \overline{\frac{\partial}{\partial x}\left(\mu\frac{\partial u}{\partial z}\right)} + \overline{\frac{\partial}{\partial y}\left(\mu\frac{\partial v}{\partial z}\right)} + \frac{\partial}{\partial z}\left(\mu\frac{\partial w}{\partial z} - \frac{2}{3}\mu\text{div}\mathbf{V}\right)$$

$$+ \frac{\partial}{\partial x}\left(\mu\frac{\partial w}{\partial x}\right) + \frac{\partial}{\partial y}\left(\mu\frac{\partial w}{\partial y}\right) + \frac{\partial}{\partial z}\left(\mu\frac{\partial w}{\partial z}\right) + B_z \qquad (7.115)$$

These equations are valid for a variable viscosity μ and compressible fluids. Taking the u-momentum equation as example, when μ is constant and the fluid is incompressible, after some algebraic manipulation, the velocities v and w disappear from the right hand side of Eq. (7.113). Similarly, u and w disappear from Eq. (7.114), and u and v from Eq. (7.115).

Considering that the underlined terms in Eq. (7.113), due to the linearization process, will be part of the matrix coefficients, Eq. (7.113) contains only u as active variable. In Eq. (7.114) only v and in Eq. (7.114) only w will be active velocities. A look at the mass conservation equation reveals that pressure is absent. In this situation the final matrix for the simultaneous solution of these equations will have the following structure

$$\begin{bmatrix} A^{uu} & 0 & 0 & A^{up} \\ 0 & A^{vv} & 0 & A^{vp} \\ 0 & 0 & A^{ww} & A^{wp} \\ A^{pu} & A^{pv} & A^{pw} & 0 \end{bmatrix} \begin{bmatrix} u \\ v \\ w \\ p \end{bmatrix} = \begin{bmatrix} B^u \\ B^v \\ B^w \\ B^p \end{bmatrix} \qquad (7.116)$$

It is apparent that the system (7.116) doesn't fill the conditions of having all variables in all equations, but the most crucial problem is the zero on the diagonal due to the lack of pressure in the mass conservation equation. This system is possible to be solved, of course, but it is not in a preferable form. The absence of a term related to pressure in the mass conservation equations demonstrates the delicate coupling when the fluid is incompressible, since mass conservation is not an active actor in the system, but just a restriction that needs to be satisfied. Unfortunately, the mass conservation equation doesn't help in finding the velocities, but just enforces that the velocities must satisfy itself. Not all equations work as a team to find the velocities.

The condition of the system can be improved leaving on the momentum equations the terms with an overbar [11, 15], even if the fluid is incompressible and μ constant, since they will be of no effect when the solution is converged. During the iterations they will help in promoting a discretization in which all velocity components will appear in all momentum equations.

7.8 Simultaneous Solution and the Couplings

When the grid arrangement is co-located, the zero entry in the diagonal will be automatically substituted by a nonzero because it will be necessary to apply a RC-like method, or PIS-like, to calculate the velocity at the interfaces. This stabilization algorithm will always involve the pressure, what inserts a nonzero entry in the diagonal. Since in all practical applications the grid is co-located, these zero entries in the diagonal will not exist. It would exist if the grid were staggered, but this arrangement is out of consideration in the simulation practice.

This issue was already discussed, but now it becomes clear the help on the conditioning of the matrix if the corresponding coefficient for pressure appears at the diagonal. With these two actions the matrix will have the following form,

$$\begin{bmatrix} A^{uu} & A^{uv} & A^{uw} & A^{up} \\ A^{vu} & A^{vv} & A^{vw} & A^{vp} \\ A^{wu} & A^{wv} & A^{ww} & A^{wp} \\ A^{pu} & A^{pv} & A^{pw} & A^{pp} \end{bmatrix} \begin{bmatrix} u \\ v \\ w \\ p \end{bmatrix} = \begin{bmatrix} B^u \\ B^v \\ B^w \\ B^p \end{bmatrix} \quad (7.117)$$

with all entries now non-zeros. The A^{ij} are 4×4 matrices, as depicted in Fig. 7.9 in which a 1D problem is recognized by the tridiagonal blocked matrix. This linear system can be also written using the Δu, Δv, Δw and Δp as unknowns with the independent term being the residues of each equation. The solution of this linear system of equations can be solved using an iterative base solver with a multigrid accelerator. Since the matrix $[\mathbf{A}]$ contain the variables to be calculated, iterations are required accounting for the non-linearities.

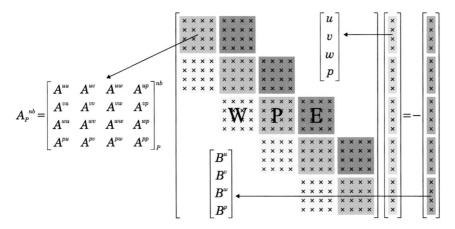

Fig. 7.9 Matrix for the simultaneous solution

7.9 A Note on Boundary Conditions

In Chap. 3, the boundary conditions for pure diffusion problems with the possible boundary conditions, Dirichlet Neumann and Robin, were presented. In Sect. 7.6 the boundary conditions for p and p' were presented in the context SPVC treatment. In this section, it is completed the procedures for applying the boundary conditions by also including problems where there is fluid flowing in and out of the control volumes.

Again, the boundary conditions can be applied realizing balances for the boundary volumes considering the existing condition at that boundary, thus, incorporating the boundary events in the approximate equations.

Fig. 7.10 shows a boundary control volume centered in p. For a 2D problem, considered for simplicity, the conservation equation for ϕ is,

$$\frac{\partial}{\partial t}(\rho\phi) + \frac{\partial}{\partial x}(\rho u\phi) + \frac{\partial}{\partial y}(\rho v\phi) = \frac{\partial}{\partial x}\left(\Gamma^\phi \frac{\partial \phi}{\partial x}\right) + \frac{\partial}{\partial y}\left(\Gamma^\phi \frac{\partial \phi}{\partial y}\right) + S^\phi \quad (7.118)$$

The integration of this equation will require the advective and diffusive fluxes at the four faces of the control volume. In finite volume methods the fluxes are the natural boundary conditions, and the integration of Eq. (7.118) reveal those fluxes and, since the boundary condition is at the w boundary, only the underlined fluxes need to be specified, as in Eq. (7.119)

$$\cdots + (\rho u\phi)_e \Delta y - \underline{(\rho u\phi)_w \Delta y} + \cdots = \Gamma^\phi \frac{\partial \phi}{\partial x}\bigg|_e \Delta y - \underline{\Gamma^\phi \frac{\partial \phi}{\partial x}\bigg|_w \Delta y} + \cdots \quad (7.119)$$

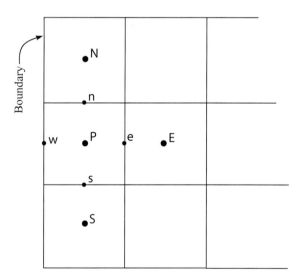

Fig. 7.10 Boundary conditions at west boundary

7.9 A Note on Boundary Conditions

Only the fluxes at the integration point w will be involved with the boundary condition at that face. The other internal fluxes will be the equate using the interpolation functions, and the resulting equation for control volume P will contain the boundary condition embodied. The underlined advective and diffusive fluxes by unit of area at face w are given by, respectively,

$$A''_w = \rho u \phi \big|_w \tag{7.120}$$

$$D''_w = \Gamma^\phi \frac{\partial \phi}{\partial x}\bigg|_w \tag{7.121}$$

For the boundary conditions being exemplified, and valid for all other boundaries. Three conditions are of interest.

7.9.1 Impermeable Boundary—ϕ Prescribed

In this case the advective flux is equal to zero, since there is no mass flow crossing the boundary, and the diffusive flux is determined by,

$$\Gamma^\phi \frac{\partial \phi}{\partial x}\bigg|_w = \Gamma^\phi_w \frac{(\phi_P - \phi_w)}{\Delta x / 2} \tag{7.122}$$

which must be introduced in the discretized equation, Eq. (7.119), with all other terms to obtain the discretized equation in the form,

$$A_P \phi_P = A_e \phi_E + A_n \phi_N + A_s \phi_S + B \tag{7.123}$$

7.9.2 Impermeable Boundary—Flux of ϕ Prescribed

In an impermeable boundary the advective flux is zero and the value of the prescribed diffusive flux is substituted directly in the discretized equation, Eq. (7.119). When ϕ is a component of the velocity vector, as the boundary is impermeable, the boundary conditions will be of zero velocity. If ϕ is temperature, one can have prescribed temperature or prescribe heat flux, since it is not possible to have both conditions applied, since it would be an overspecification of boundary conditions. The final equation looks like Eq. (7.123).

7.9.3 Inflow and Outflow Boundary Conditions

Initially, let's consider mass entering the domain, in which it is possible to have a constant velocity or a distribution of velocity giving the specified mass flow. This means to have a specified advection flux of ϕ, giving by $(\dot{m}\phi)_w$. This boundary condition has subtleties. It is always reported, including in commercial software manuals, that the diffusive flux can also be calculated using the prescribed ϕ_w. This would be done using a first order approximation using the prescribed ϕ_w and the closest internal grid point.

However, if ϕ is prescribed at the boundary (ϕ_w), physically, it required zero diffusive flux, because, on the contrary, this diffusive flux would change the value of ϕ and it would become unknown at the boundary. Therefore, when there is an inlet flow with ϕ_w prescribed, it should be automatically specified the diffusive flux equal to zero at that face. If there are doubts about the specification or not of ϕ_w in a boundary, it is because the boundary of the domain is not well chosen.

Physics should again guide for specifying the boundaries of the calculation domain, since this is one of the crucial problems in simulation, because a problem is almost always coupled to some other physics, which would require expanding the doERENmain. The clever analyst will choose the small domain possible in which it is possible to devise some physically consistent boundary conditions. Mathematically, the flux is

$$\rho u \phi |_w = prescribed, \qquad (7.124)$$

it will require

$$\Gamma^\phi \frac{\partial \phi}{\partial x}\bigg|_w = 0 \qquad (7.125)$$

For the case of mass exiting the domain, if no information is given at the outlet, what is in general what happens, the most used boundary conditions is the so-called locally parabolic boundary condition, hence, the diffusive flux should be set to zero. The downstream boundary condition is not needed, since the coefficient A_e will be zero (imagining the outgoing boundary volume to the right side of the domain), since an upwind type approximation should be used. If, for some reason, the value of ϕ at the boundary is required after the solution, it should be extrapolated from internal values. Remember that ϕ, in this discussion, can be the components of the velocity vector or temperature. For either of these cases, the form of the approximate equation for the boundary volume will be of the type of Eq. (7.123).

7.9.4 General Comments About Boundary Conditions

In the recent past the solution of fluid flow problems were in its majority done using segregated solution. The literature, then, gives emphasis in explaining boundary conditions for incompressible flows and, most of the times, considering staggered grids. Not too much is found about prescribed pressure boundary conditions and of pressure and velocity boundary condition in the same problem. Mixing the conditions, if the physical characteristics are not observed, can bring serious convergence implications, as well as can represent physically unrealistic situations. For these reasons, the remainder of this section will attempt to clarify some of these issues considering cell-center methods with co-located variables.

7.9.5 Incompressible Flows

For incompressible flows, only the pressure gradient has influence on the solution, and the pressure level is of no importance. For example, for incompressible flow in a pipe of length L and cross section A, only the pressure difference between the inlet and outlet is sufficient to determine the mass. Any constants that are added to the pressure values will not change the mass flow.

Therefore, for incompressible flows, if the inlet and outlet pressures are specified, the velocity cannot also be prescribed. If the pressure and velocity are prescribed at the inlet, pressure can't be prescribed at the outlet, because two values for the mass flow would be specified.

A widely used combination is that of a prescribed velocity at the inlet and a locally parabolic condition at the outlet, without setting any condition for the pressure. Since the solution is iterative, the pressure level adjusts automatically. If desired, one can also preset the value of the pressure at a point in the domain, which is equivalent to choosing one of the infinite solutions for the pressure that satisfy the equations of motion. Another condition would be specifying pressure at the inlet and outlet and let the mas flow to satisfy the prescribed pressure difference.

The way the parabolic condition is mathematically written, is to impose the null derivative of the variables at the output. It is worth recalling that prescribing null derivative is equivalent to prescribed null diffusive flow of the property. This boundary condition becomes more real the more fully developed the flow is. Care should be exercised in avoid locally parabolic boundary condition in regions with recirculation. This is not correct, because in these regions there are important diffusive effects that are not considered with this condition. The name says "boundary conditions" and, logically, if in the boundary chosen the values of the variables or fluxes are not known, it is because the boundary of the calculation domain were not well chosen. Figure 7.11 shows a problem indicating that it is not convenient to choose the boundary of the domain in regions in which diffusive effects dominates.

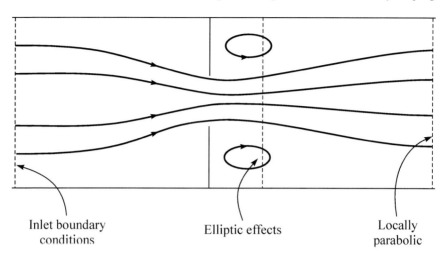

Fig. 7.11 Choosing the boundary of the domain

7.9.6 Compressible Flows

In this section, it is considered compressible flows when density varies significantly with pressure. When it varies only with temperature, the problem may be considered incompressible from the numerical point of view, and what was already studied applies. For compressible flows, the relationship between density, pressure, and temperature, given by the equation of state, must be satisfied. For internal compressible flows, as in nozzles, the boundary conditions must be given according to the nature of the flow at the inlet and outlet.

For subsonic inlet and supersonic outlet flows, the inlet boundary conditions should be of prescribed total pressure and total temperature, with the outlet locally parabolic. Note that, in this case, there must be a velocity update process at the inlet, since the mass flow is not known but determined with the solution of the problem.

For supersonic flows at the inlet and outlet, one can prescribe the static pressure, velocity, and total temperature at the inlet. At the outlet, the locally parabolic condition can be employed.

For external supersonic flows, free flow conditions are prescribed (Mach number and temperature) at the domain boundaries, while locally parabolic conditions are used at the domain outlet. It is also common, when methodologies designed for supersonic flows are employed, to apply the exit conditions along the characteristics.

7.10 Conclusions

The mail goal of this chapter was to present the techniques for the determination of velocity and pressure fields using finite volume methods. It is in the solution of the fluid mechanics problem, that is, the velocity and pressure determination, that the intricate numerical difficulties lie. The first of these, is the treatment of the SPVC resulting from the segregated process of solving the conservation equations for incompressible flows. It was tried to let clear the differences between the co-located PV coupling (CPVC) and the segregated PV coupling (SPVC), issue that is not normally clarified in the literature.

For the segregated solution the philosophy of SIMPLE-like methods was emphasized, pointing out that, besides the SPVC, the CPVC also appears when the grid is co-located. It was also demonstrated that the staggered grid arrangement avoids the CPVC, since the velocity which enters the mass conservation equation is the proper variable of the problem and is available at the interfaces of the control volume. In this arrangement the velocity field which satisfies momentum conservation also satisfies mass conservation. This doesn't happen when the arrangement is co-located. Boundary conditions for p and p' in the context of SIMPLE-like methods, as well as for other variables were considered.

Most of the methods discussed are the ones intensively used and embedded in the commercial applications used by CFD analysts in solving their engineering problems.

7.11 Exercises

7.1 The velocity correction equations in a segregated solution can be of any type, since they do not influence the final solution, just the convergence process. For a two-dimensional incompressible advective/diffusive problem, one possibility is to have the following correcting equations,

$$u = u^* + \frac{\partial \varphi}{\partial x} \qquad (7.126)$$

$$v = v^* + \frac{\partial \varphi}{\partial y} \qquad (7.127)$$

Obtain the equation for φ. What are the boundary conditions for φ when the velocity and the derivative of the velocity are prescribed as boundary condition?

7.2 Show that the velocity correction in Exercise 7.1 does not alter the vorticity of the velocity field.

7.3 Write the expressions for $\overline{d}_e^u$, $\overline{d}_e^u$, $\overline{d}_n^v$, $\overline{d}_s^v$, $\overline{d}_b^w$ and $\overline{d}_f^w$ for the SIMPLEC and PRIME methods for the co-located and staggered arrangement of variables.

7.4 Find the equation for pressure employing the PRIME method.

7.5 Considering Fig. 7.8 in which the pressure is prescribed at the boundary w. Find the equation for p' for the control volume P using the SIMPLEC method.

References

1. Wilcox DC (2006) Turbulence modeling for CFD, 3rd edn. DCW Industries, Inc.
2. Versteeg HK, Malalasekera W (2007) An introduction to computational fluid dynamics, 2nd edn. Pearson/Prentice Hall
3. Yeoh GH, Tu J (2010) Solution method for multiphase flows. Computational techniques for multiphase flows, pp 95–242. https://doi.org/10.1016/B978-0-08-046733-7.00003-5
4. Beam RM, Warming RF (1978) An implicit factored scheme for the compressible navier-stokes equations. AIAA J 16:393–402
5. Pullian TH, Stege JL (1980) Implicit finite difference simulation of three-dimensional compressible flow. AIAA J 18:159–167
6. Harlow FH, Welch JE (1965) Numerical calculation of time-dependent viscous incompressible flow of fluid with free surface. Phys Fluids 8:2182–2189
7. Patankar SV (1980) Numerical heat transfer and fluid flow. Hemisphere Publishing Corporation
8. CFX-TASCflow computational fluid dynamics software. Theory documentation, Version 2.10
9. Rhie CM (1981) A numerical study of the flow past an isolated airfoil with separation. Ph.D. thesis, University of Illinois, Urbana-Champaign
10. Marchi CH, Maliska CR (1994) A nonorthogonal finite volume method for the solution of all speed flows using co-located variables. Numer Heat Transf 26:293–311
11. Schneider GE, Raw MJ (1987) Control volume finite-element method for heat transfer and fluid flow using colocated variables-computational procedure. Numer Heat Transf 11:363–390
12. Hwang YH (1995) Calculation of incompressible flow on a staggered triangular grid, Part I: mathematical formulation. Numer Heat Transf Part B Fundam 27(3):323–336
13. Peters S, Maliska CR (2017) A staggered grid arrangement for solving incompressible flows with hybrid unstructured meshes. Numer Heat Transf Part B Fundam 71(1):50–65
14. Peters S, Honório HT, Maliska CR (2020) The physical influence scheme applied to staggered unstructured grids for solving fluid flows problems. Numer Heat Transf Part B Fundam. https://doi.org/10.1080/10407790.2020.1787037
15. Raw MJ (1985) A new control-volume-based finite element procedure for numerical solution of the fluid flow and scalar transport equations. Ph.D. thesis, University of Waterloo, Canada
16. Chorin AJ (1967) A numerical method for solving incompressible viscous flow problems. J Comp Phys 2:12–26
17. Chorin AJ (1971) Numerical solution of the Navier-Stokes equations. Math Comput 22:745–762
18. Amsden AA, Harlow FH (1970) The SMAC method: a numerical technique for calculating incompressible fluid flows. Los Alamos Scientific Laboratory, LA-4370, Los Alamos, EUA
19. Patankar SV, Spalding DB (1972) A calculation procedure for heat, mass and momentum transfer in three-dimensional parabolic flows. Int J Heat Mass Transf 15:1787–1806
20. Raithby GD, Schneider GE (1979) Numerical solution of problems in incompressible fluid flow: treatment of the velocity-pressure coupling. Numer Heat Transf 2:417–440
21. Maliska CR (1981) A solution method for three-dimensional parabolic fluid flow problems in nonorthogonal coordinates. Ph.D. thesis, University of Waterloo, Waterloo, Canada
22. Maliska CR, Raithby GD (1983) Calculating three-dimensional fluid flows using nonorthogonal grids. In: Laminar and turbulent flows. Pineridge Press, pp 656–666
23. van Doormaal JP (1985) Numerical methods for the solution of incompressible and compressible fluid flows. PhD. thesis, University of Waterloo, Waterloo

References

24. van Doormaal JP, Raithby GD (1984) Enhancements of the simple method for predicting incompressible fluid flow. Numer Heat Transf 7:147–163
25. Issa RI (1982) Solution of implicit discretized fluid flow equations by operator splitting. Mechanical engineering report, FS/82/15, Imperial College, London
26. Moukalled F, Mangani L, Darwish M (2016) The finite volume method in computational fluid dynamics. Springer International Publishing, Switzerland
27. Peric M, Kessler R, Scheuerer G (1988) Comparison of finite volume numerical methods with staggered and colocated grids. Comput Fluids 16:389–403

Chapter 8
All Speed Flows Calculation—Coupling $P \to [V - \rho]$

8.1 Introduction

Historically, numerical methods for fluid flow have been developed covering two major classes of flows: low speed (low Mach number) and high speed (high Mach number). Low speed flows are, in general, related to determination of pressure drop in ducts or convection heat transfer problems, where the interest is in determining the heat exchange among solid–fluid interfaces. In low speed flows of gases, with small pressure variations, density is considered constant or is determined as a function of temperature only. They all fit in the class of incompressible flows.

High-speed flows are generally compressible, related to aerodynamics and aerothermodynamics problems, with large influence of pressure on density. Among these two limits there are the transonic flows, in which the physical behavior of the flow is mixed. Perhaps, due to these limits with distinct applications, the developments of numerical methods for fluid flows were done in, let's say, different schools of research, the "compressible school" and the "incompressible school", the former devoted to the solution of high speed flows (supersonic) while the later for low speed flows (subsonic).

For compressible flows, therefore, the compressible formulation seen in Chap. 7 can be used in a segregated or simultaneous approach for the solution of the equations system. The difficulty encountered with incompressible flows in which the pressure is no longer determined using the equation of state does not exist in the compressible formulation. For incompressible flows, if it is solved in a segregated manner, a SIMPLE-like method is used to deal with the segregated pressure velocity coupling (SPVC). The target is to have methods working for all speed flows.

Long ago both schools started to develop methods for all speed flows, but each one keeping their numerical roots. In the subsonic (incompressible) school the route followed was to extend the known SPVC methods for incompressible flows to account for the effects of density variations with pressure. It was learned that the CPVC appears when the velocity is not available at the boundaries of the control volume. Therefore, if a staggered grid is used for incompressible flows, there is no need to deal with the co-located pressure–velocity coupling. In all speed flows, density must, together with the velocity, be available at the interfaces to compute the mass flow rate. Therefore, even using a staggered grid, it will be needed an scheme (like a PIS method) to find density at the interfaces.

This chapter is devoted to present an all speed flow scheme which extends segregated pressure–velocity coupling methods designed for incompressible flows to compressible flows.

8.2 Pressure–Velocity and Pressure-Density Coupling

High speed flows are related, in general, to supersonic aerodynamics, whose main interest is the determination of the pressure coefficients on the surface to calculate the drag and lift forces of aerodynamic profiles in the aerospace industry.

For low-speed flows, the basic methodologies consider that density is not a function of pressure. Logically, it is not possible to solve supersonic flow, in which shock waves may be present, with this methodology, because in this case density changes significantly with pressure. Consequently, the methods using the incompressible formulation, described in Chap. 7, are not directly applicable to high-speed compressible flows.

On the other hand, the basic methodology of the methods for high-speed flows (compressible flows) considers density as a variable and pressure found by the equation of state. Needless to say, these methods are not applicable for low-speed flows, in which the pressure cannot be determined by an equation of state due to numerical instability reasons.

Another way to see the reasons for the limitation of the two methodologies in their limits, is to analyze the product $\rho \mathbf{V}$ in the mass conservation equation. As in any numerical discretization in which one wants to obtain a linear system of algebraic equations, any product of variables must be linearized. For low-speed flows, in which density doesn't change with pressure, the active variable in the mass conservation equation is the velocity, and density can be put in the coefficient. This means that the velocity will be the variable to be modified for satisfying mass conservation. Pressure variations changes velocities, not densities. Density was put in the coefficients, with its value from the previous iteration level. The linearization of $\rho \mathbf{V}$ in this form can't solve high speed flows.

The other linearization possible is when velocity is incorporated into the coefficients and density becomes the active variable in the mass conservation equation, situation that will work only for high-speed flows.

Based on the above, it seems, therefore, that what characterizes the formulation to be able to solve compressible and/or incompressible flows is the way the mass flux is linearized. There are currently two lines of cutting-edge research seeking to extend the methodologies to all speed flows. One which uses the numerical basis developed for compressible flows and the extension devises to contemplate all speed flows. The other considers the opposite and is the subject of the present chapter.

8.2.1 Linearization of the Mass Flow

Let's consider the application of the mass conservation equation to a control volume in which the mass flow at east (e) face is taken as example,

$$\dot{M}_e = \rho_e u_e A \tag{8.1}$$

This term can be linearized as

$$\dot{M}_e = \rho_e^* u_e A \tag{8.2}$$

in which the superscript (*) means evaluated at the previous iteration level. This linearization considers that the density is not active in the mass conservation equation and is, thus, suitable for solving incompressible flows only. The linearization as,

$$\dot{M}_e = \rho_e u_e^* A \tag{8.3}$$

assumes that density will be the active variable that will change when pressure changes for satisfying mass conservation. None of the linearization alternatives can solve all speed flows. A Newton–Raphson linearization as,

$$\dot{M}_e = \rho_e^* u_e \Delta y + \rho_e u_e^* \Delta y - \rho_e^* u_e^* \Delta y \tag{8.4}$$

keeps both density and velocity active in the mass conservation equation. An interesting analysis can be done inspecting Eq. (8.4). When the incompressible limit is considered, the ρ_e in the second term in the RHS becomes ρ_e^*, therefore, cancelling with the last term and recovering the linearization of Eq. (8.2), that is, for the incompressible limit. For high speed flows u_e in the first term in the RHS is no longer active, and becomes u_e^*, cancelling with the third term, recovering the linearization for high speed flows, Eq. (8.3). The same linearization as Eq. (8.4) is applied for the remaining interfaces and then introduced in the mass conservation equation.

When co-located grid is used both u_e and ρ_e are needed at the interfaces. For both variables a type of RC-like, or PIS scheme should be used. For the u_e velocity the schemes were already discussed in Chap. 7. How to find ρ_e is now addressed. In Eq. (8.4) one is using the subscript e, but in fact it is a integration point ip when dealing with unstructured grids.

The general interpolation function seen in Chap. 5 when interpolation functions were discussed, is a PIS-like method and can be used for determining the values of a function at the integration points. Repeating Eq. (5.68) for a generic integration point (ip), which could be e, one gets

$$\phi_{ip} = \phi_u + \left.\frac{\partial \phi}{\partial s}\right|_u \frac{\Delta s}{2} + \frac{\partial^2 \phi}{\partial s^2}\frac{\Delta s^2}{4} + O(\Delta s^3) + \ldots + \tag{8.5}$$

in which s is the streamwise direction. For the u component of the velocity vector, Eq. (8.5) becomes

$$u_{ip} = u_u + \left.\frac{\partial u}{\partial s}\right|_u \frac{\Delta s}{2} + \frac{\partial^2 u}{\partial s^2}\frac{\Delta s^2}{4} + O(\Delta s^3) + \ldots +, \qquad (8.6)$$

and for the density,

$$\rho_{ip} = \rho_u + \left.\frac{\partial \rho}{\partial s}\right|_u \frac{\Delta s}{2} + \frac{\partial^2 \rho}{\partial s^2}\frac{\Delta s^2}{4} + O(\Delta s^3) + \ldots + \qquad (8.7)$$

In Eqs. (8.6) and (8.7) $(\partial u/\partial s)_u$ and $(\partial \rho/\partial s)_u$ are found from the momentum equations and mass conservation equation, respectively. For $\partial \rho/\partial s$, expanding the derivatives in the mass conservation equation, it results

$$\left.\frac{\partial \rho}{\partial s}\right|_u = -\frac{\rho}{V}\frac{\partial u}{\partial x}, \qquad (8.8)$$

in which $(\partial u/\partial x)$, considering 1D, is the divergence of the velocity vector [15].

As discussed in Chap. 5, the determination of u_u and ρ_u is crucial for obtaining positive coefficients. These upstream values should be represented as functions of the nodal ones. Knowing the velocities and densities at the integration points the mass flow can be calculated with Eq. (8.4) and the simultaneous solution of the equation can be done. In this Section it will be demonstrated the application of this linearization in the context of a segregated solution, therefore using a SVPC, in this case, the SIMPLEC and PRIME.

The proposition of this linearization for solving all speed flows was made, more than five decades ago, in [1, 2] but the idea, apparently, did not prosper. More recently, developments for Cartesian grids were made [3] and for curvilinear coordinate systems in [4–10].

8.3 Two-Dimensional All Speed Flow Discretization

To demonstrate the application of the methodology, a 2D problem in Cartesian coordinates using a segregated solution will be used, that is u, v and p will be solved sequentially. It is worth remembering that the developments depend on the arrangement of variables used. To our intent here, it doesn't matter, since it is assumed that there are velocities at the interfaces of the control volume for mass conservation, or by the use of the staggered arrangement, or by the application of a PIS-like method, as in Chap. 6, if the arrangement is co-located. Equation (8.6) can also be used as alternative. Density, independently of the arrangement used, are never available at the control volume interfaces, since it is always located with the other scalars, like

8.3 Two-Dimensional All Speed Flow Discretization

temperature, concentration, kinetic and dissipation of turbulent energy among others. Application of Eq. (8.7) is, then, required.

The mass conservation equation integrated in space and time, reads

$$\frac{M_P - M_P^o}{\Delta t} + \dot{M}_e - \dot{M}_w + \dot{M}_n - \dot{M}_s = 0, \tag{8.9}$$

in which

$$\begin{aligned}
M_P &= \rho_P \Delta V \\
M_P^o &= \rho_P^o \Delta V \\
\dot{M}_e &= (\rho u)_e \Delta y \\
\dot{M}_w &= (\rho u)_w \Delta y \\
\dot{M}_n &= (\rho v)_n \Delta x \\
\dot{M}_s &= (\rho v)_s \Delta x
\end{aligned} \tag{8.10}$$

in which e, w, n and s are the integration points (ip) for the Cartesian control volume under use. Using a one-dimensional upwind approximation for the density at interfaces

$$\rho_e = \left(\frac{1}{2} + \gamma_e\right)\rho_P + \left(\frac{1}{2} - \gamma_e\right)\rho_E \tag{8.11}$$

$$\rho_w = \left(\frac{1}{2} + \gamma_w\right)\rho_W + \left(\frac{1}{2} - \gamma_w\right)\rho_P \tag{8.12}$$

$$\rho_n = \left(\frac{1}{2} + \gamma_n\right)\rho_P + \left(\frac{1}{2} - \gamma_n\right)\rho_N \tag{8.13}$$

$$\rho_s = \left(\frac{1}{2} + \gamma_s\right)\rho_S + \left(\frac{1}{2} - \gamma_s\right)\rho_P \tag{8.14}$$

in which γ takes the values $1/2$ and $-1/2$, depending on whether the velocity is positive or negative. An upwind scheme is used for the density in order to ensure the positivity of the coefficients of the mass conservation equation. Neglecting the term $(\partial \rho/\partial s)_u$ in Eq. (8.7) it recovers the upwind scheme used in Eqs. (8.11–8.14). See Chap. 5 for similarities among γ and α.

The procedure which now follows is identical to the one already done for the treatment of pressure–velocity coupling in incompressible flows, that is, one must replace the linearized mass fluxes which appear in the mass conservation equation by the correcting equations for velocity and for density, since now the coupling is among pressure and velocity/density.

Initially, substituting Eq. (8.4) and the similar ones for the other three faces in Eq. (8.9), it is obtained the mass conservation equations with the linearized flows in the form,

$$m_P^\rho \rho_P + m_e^\rho \rho_E + m_w^\rho \rho_W + m_n^\rho \rho_N + m_s^\rho \rho_S + m_e^u u_e + m_w^u u_w + m_n^v v_n + m_s^v v_s = b^c \tag{8.15}$$

in which

$$m_P^\rho = \frac{\Delta V}{\Delta t} + \left(\frac{1}{2} + \gamma_e\right) u_e^* \Delta y - \left(\frac{1}{2} - \gamma_w\right) u_w^* \Delta y$$
$$+ \left(\frac{1}{2} + \gamma_n\right) v_n^* \Delta x - \left(\frac{1}{2} - \gamma_s\right) v_s^* \Delta x \tag{8.16}$$

and

$$m_e^\rho = \left(\frac{1}{2} - \gamma_e\right) u_e^* \Delta y \tag{8.17}$$

$$m_w^\rho = -\left(\frac{1}{2} + \gamma_w\right) u_w^* \Delta y \tag{8.18}$$

$$m_n^\rho = \left(\frac{1}{2} - \gamma_n\right) v_n^* \Delta x \tag{8.19}$$

$$m_s^\rho = -\left(\frac{1}{2} + \gamma_s\right) v_s^* \Delta x \tag{8.20}$$

$$m_e^u = \left[\left(\frac{1}{2} + \gamma_e\right) \rho_P^* + \left(\frac{1}{2} - \gamma_e\right) \rho_E^*\right] \Delta y \tag{8.21}$$

$$m_w^u = -\left[\left(\frac{1}{2} + \gamma_w\right) \rho_W^* + \left(\frac{1}{2} - \gamma_w\right) \rho_P^*\right] \Delta y \tag{8.22}$$

$$m_n^v = \left[\left(\frac{1}{2} + \gamma_n\right) \rho_P^* + \left(\frac{1}{2} - \gamma_n\right) \rho_N^*\right] \Delta x \tag{8.23}$$

$$m_s^v = -\left[\left(\frac{1}{2} + \gamma_s\right) \rho_S^* + \left(\frac{1}{2} - \gamma_s\right) \rho_P^*\right] \Delta x \tag{8.24}$$

$$b^c = \frac{\Delta V}{\Delta t} \rho_P^o + m_e^u u_e^* + m_w^u u_w^* + m_n^v v_n^* + m_s^v v_s^* \tag{8.25}$$

Next, the values of ρ, u and v in Eq. (8.15) should be substituted by relations which contains the pressure, or pressure correction, transforming the mass conservation equation in an equation for pressure or for pressure correction. This step depends on the method employed for the segregated PV coupling (SPVC). These relations are now derived.

8.3.1 Velocity Relations as Function of p' - SIMPLEC

The expressions that relate velocities as a function of pressure (or p') are the well-known velocity correction equations already seen in Chap. 7. For the SIMPLE or SIMPLEC method, they are obtained by subtracting the equation of motion written for p^* from the equation of motion written for p and neglecting the differences $(u - u^*)$ and $(v - v^*)$. The resulting equations for the two-dimensional case, already seen in Chap. 7, and repeated here, are

$$u_e = u_e^* - \overline{d}_e^u (\Delta p')_e \tag{8.26}$$

$$u_w = u_w^* - \overline{d}_w^u (\Delta p')_w \tag{8.27}$$

$$v_n = v_n^* - \overline{d}_n^v (\Delta p')_n \tag{8.28}$$

$$v_s = v_s^* - \overline{d}_s^v (\Delta p')_s \tag{8.29}$$

8.3.2 Density Relations as Function of p' - SIMPLEC

Similar equations which corrects the velocity must be found for correcting the densities. The procedure used for finding the correcting equations for the velocities involved the momentum conservation equation. It seems logical that, for finding a correction equation for density, it is used the equation of state. Linearizing the state equation as function of pressure, one obtains

$$\rho = C^\rho p + B^\rho, \tag{8.30}$$

and writing this equation when a pressure field $p*$ is applied, results

$$\rho^* = C^\rho p^* + B^\rho \tag{8.31}$$

Subtracting Eq. (8.31) from Eq. (8.30), it results

$$\rho = \rho^* + C^\rho p', \tag{8.32}$$

in which $p' = p - p*$. Equation (8.32) is the relation to be used for correcting the density as function of pressure. The density at the grid nodes can be calculated as,

$$\rho_P = \rho_P^* + C_P^\rho p_P' \tag{8.33}$$

$$\rho_E = \rho_E^* + C_E^\rho p_E' \tag{8.34}$$

$$\rho_W = \rho_W^* + C_W^\rho p_W' \tag{8.35}$$

$$\rho_N = \rho_N^* + C_N^\rho p_N' \tag{8.36}$$

$$\rho_S = \rho_S^* + C_S^\rho p_S' \tag{8.37}$$

Substituting Eqs. (8.26–8.29) and (8.33–8.37) into Eq. (8.15), the mass conservation equation written in terms of p' results,

$$A_P p_P' = A_e p_E' + A_w p_W' + A_n p_N' + A_s p_S' + b^{p'}, \tag{8.38}$$

in which the coefficients are,

$$A_P^p = m_P^\rho C_P^\rho + m_e^u \overline{d}_e^u - m_w^u \overline{d}_w^u + m_n^v \overline{d}_n^v - m_s^v \overline{d}_s^v \tag{8.39}$$

$$A_e^p = -m_e^\rho C_E^\rho + m_e^u \overline{d}_e^u \tag{8.40}$$

$$A_w^p = -m_w^\rho C_W^\rho - m_w^u \overline{d}_w^u \tag{8.41}$$

$$A_n^p = -m_n^\rho C_N^\rho + m_n^v \overline{d}_n^v \tag{8.42}$$

$$A_s^p = -m_s^\rho C_S^\rho - m_s^v \overline{d}_s^v \tag{8.43}$$

$$b^{p'} = \frac{\Delta V}{\Delta t} \rho_P^o - m_P^\rho \rho_P^* - m_e^\rho \rho_E^* - m_w^\rho \rho_W^* - m_n^\rho \rho_N^* - m_s^\rho \rho_S^* \tag{8.44}$$

in which, for example,

$$\overline{d}_e^u = \left(\frac{\Delta y}{A_P^u}\right)_e \tag{8.45}$$

The parameter given by Eq. (8.45) was given by Eq. (7.76) in Chap. 7. There it carries Δz because it was derived for 3D problems, and with no overbar because the arrangement was staggered. Here we left it with the overbar, and it should be chosen according to the grid arrangement.

Equation (8.38) has the same nature as the one already presented in last chapter including now the influences of the coupling among pressure and density, allowing, therefore, all speed flows to be solved. Knowing the initial fields of u, v, p, T and ρ at $t = 0$, a possible solution procedure [11, 12], is

8.3 Two-Dimensional All Speed Flow Discretization

1. The variables u, v, p and ρ are estimated for the time level $t + \Delta t$;
2. With the estimated variables the coefficients of the momentum equations in the x and y directions are calculated;
3. Calculate the coefficients for the p' equation
4. Do $p^* = p$ and solve the linear systems for determining the u^* and v^* fields;
5. Calculate the $b^{p'}$ (source term for p');
6. Solve the p' equation and correct the u^* and v^* velocities using Eqs. (8.26–8.29). Correct the density ρ^* using Eqs. (8.33–8.37). Calculate the new pressure field by $p = p^* + p'$;

    Up to this point, conservation of mass has been satisfied for a given set of coefficients.
7. Return to item 3 and iterate until convergence. Iterative cycles are still needed for the calculation of the momentum equation, for the solution of the energy equation and taking account the transient;
8. Return to item 2 and iterate until convergence.

    So far, the density is that estimated in item 1 and corrected only for the variation in the pressure field by the approximate expression, Eq. (8.32);
9. The coefficients of the energy equation are calculated, and a new temperature field is determined.
10. Using the equation of state, a new field of density is calculated.
11. Return to item 2 and iterate until convergence.

    The solution of the problem at $t + \Delta t$ has been obtained so far.
12. Consider the solution at $t + \Delta t$ as an initial field for the new time level.
13. Return to item 2 and iterate until the steady stated is reached.

This transient could be distorted using the strategies already discussed in previous chapters to save computer time and having control of the iterative process through the time step size when only the steady state is of interest.

8.3.3 Velocity/Density Relations as Function of p-PRIME

Again, let us consider that the velocities are available at the interfaces of a volume for conservation of mass. This availability may be due to the staggered arrangement or by some CPVC method, topic already exhaustively discussed and presented. Again, considering a two-dimensional situation, for simplicity, the velocity correction equations are those of Chap. 7, and when written for the four faces of the control volume for mass conservation, are given by

$$u_e = \hat{u}_e - \overline{d}_e^{u}(p_E - p_P) \tag{8.46}$$

$$u_w = \hat{u}_w - \overline{d}_w^{u}(p_P - p_W) \tag{8.47}$$

$$v_n = \hat{v}_n - \overline{d}_n^v(p_N - p_P) \tag{8.48}$$

$$v_s = \hat{v}_s - \overline{d}_s^v(p_P - p_S) \tag{8.49}$$

while the equation for density as function of pressure is the state equation, given by

$$\rho = C^\rho p + B^\rho \tag{8.50}$$

Since the interest is in transforming the mass conservation equation into an equation for pressure, substitute the expressions for the density as a function of p (Eq. (8.50)) and the components of velocity into the mass conservation equation. For the PRIME method, the expressions given by Eqs. (8.46–8.49) are the momentum equations themselves. Substituting them into the already linearized mass conservation equation, Eq. (8.4), it is obtained an equation for pressure in the form

$$A_p p_P = A_e p_E + A_w p_W + A_n p_N + A_s p_S + B \tag{8.51}$$

It is left to the reader, as exercise, the determination of the coefficients of Eq. (8.51). A possible iterative procedure follows:

1. Considering the problem as transient, the distribution of u, v, p, T are known at time t with ρ calculated by the equation of state. If a distorted transient is employed, these distributions are not initial conditions but estimated fields;
2. Estimate the distribution of u, v, p, T, with ρ again calculated by the equation of state, at time level $t + \Delta t$.
3. Calculate the coefficients and source terms for the momentum equations for u and v.
4. Calculate $\hat{u}$ and $\hat{v}$ at the interfaces of the control volume for conservation of mass. It is important to repeat that these velocities contain all terms of the momentum conservation equations except the pressure gradient.
5. Calculate the coefficients and the source term for the pressure equation. Solve Eq. (8.51) and obtain p through the solution of a linear system.
6. Correct the velocities at the interfaces using Eqs. (8.46–8.49).

    At this point one has velocities and densities that satisfy mass conservation. The coefficients have not yet updated. If the arrangement is staggered, the correction of the velocity at the interface will correct the variable itself. If it is co-located, the velocity correction at the interface will not correct the variable itself, because it is located in the center of the control volume. The velocity at the center should be averaged using the corrected velocities at the interface and not applying the correction process to the velocities at the center of the volume.
7. Calculate ρ using the equation of state, Eq. (8.50).
8. Return to item 3 and iterate until convergence.

    At this point, the distribution of u, v and T that satisfy the mass and momentum equations for a given temperature field is known.

9. Calculate the coefficients for the energy equation. Solve the linear system and get T. Return to item 3 and iterate until convergence.
10. Return to item 2 and advance to a new time level.

8.4 Conclusions

The development of methodologies for the solution of subsonic, transonic, and supersonic flows that employ a single numerical model, as described in this section, helps considerably the development of computer codes for general flows. This is a convenient way to attack the problems, because often there are flows with all three velocity regimes simultaneously. In these cases, the method must be versatile in order to obtain the solution. The method presented here, developed from the concept of keeping velocity and density active in the mass conservation equation, through a special linearization of the mass flow, has been applied to a series of practical problems with low and high Mach numbers, representing the limits of the methodology, always with very good results. The transonic region requires a more refined grid than the subsonic and supersonic regimes for getting results of same quality. The procedure can be applied to segregated as well as simultaneous solutions, and the SIMPLEC and PRIME methods applied in the context of the former approach.

The set of equations to be solved when fluid flow is coupled with heat transfer, is very rich in couplings and nonlinearities, like the velocity/temperature coupling [13, 14]. Many other ways of treating such couplings and nonlinearities can be devised, and the reader is motivated to exercise his knowledge by seeking to visualize methods with this goal. This all-speed flow methodology was used for solving subsonic, transonic and supersonic flows over a blunt body, and few results are reserved for Chap.15.

8.5 Exercises

8.1. Proposing the following corrections for velocity and density,

$$u = u^* + u'$$
$$\rho = \rho^* + \rho' \tag{8.52}$$

obtain Eq. (8.4).

8.2. Obtain the mass conservation equation, Eq. (8.15), and set $\gamma = 0$ in Eqs. (8.11–8.14) and observe the signal of the coefficients. Which is the minimum value, in modulus, for γ to get positive coefficients?

8.3. Another way to obtain the equation for p', Eq. (8.38), is to directly substitute into the mass conservation equation, Eq. (8.9), the expressions for u, v and ρ as functions of p', without using the linearization given by Eq. (8.4). The result will be an equation in which terms with products of p' appear, therefore, nonlinear. Propose a linearization for this equation.

8.4. The boundary conditions with inlet and outlet mass flow in the domain are the most difficult to apply in a practical problem. Specify the boundary conditions that result in physically consistent problems for the following situations:

a. Incompressible internal flow in which the pressure difference between the inlet and outlet is prescribed;
b. Same, except that the mass flow rate at the inlet is known;
c. Internal flow, subsonic at the inlet, becoming supersonic in the domain and back to subsonic at the outlet;
d. Subsonic internal flow at the inlet and supersonic at the outlet;
e. Supersonic external flow over a blunt body.

8.5. Consider the nonlinear product $\mathbf{V}T$ in the energy equation. Linearize it as $\mathbf{V}^*T + \mathbf{V}T^* - \mathbf{V}^*T^*$ and obtain the discretized energy equation.

8.6. Obtain the coefficients for Eqs. (8.38) and (8.51) for an internal control volume. Consider now a prescribed velocity as boundary condition and obtain the coefficients for the boundary volume in this case.

References

1. Amsden AA, Harlow FH (1970) The smac method: a numerical technique for calculating incompressible fluid flows. Los Alamos Scientific Laboratory-LA-4370, Los Alamos, EUA
2. Patankar SV (1971) Calculation of unsteady compressible flows involving shocks. London Rept. UF/ TN/A/4, Imperial College, England
3. Van Doormaal JP (1985) Numerical methods for the solution of incompressible and compressible fluid flows. Ph.D. Thesis, University of Waterloo, Waterloo, Canada
4. Silva AFC, Maliska CR (1988) A segregated formulation in finite volumes for compressible and incompressible flows using boundary-fitted coordinates. (in Portuguese) II ENCIT-Encontro Nacional de Ciências Térmicas, Águas de Lindoia, Brasil, pp 11–14, abcm.org.br/anais/encit/1988/ii-encit-88.pdf
5. Maliska CR, Silva AFC (1989) A boundary-fitted finite volume method for solution of compressible and/or incompressible fluid flows using both velocity and density correction. VII Int. Conference on Finite Element in Flow Problems. University of Alabama Press, EUA, pp. 405–412
6. Karki KC, Patankar SV (1989) Pressure based calculation procedure for viscous flows at all speeds in arbitrary configurations. AIAA J 27:1167–1174
7. Marchi CH, Maliska CR (1994) A nonorthogonal finite volume method for the solution of all speed flows using co-located variables. Numer Heat Transfer Part B 26:293–311
8. Demirdzic I, Lilek Z, Peric M (1993) A co-located finite volume method for predicting flows at all speeds. Int. J. Numer. Meth. Fluids 16:1029
9. Darbandi M, Schneider GE (1988) Comparison of pressure-based velocity and momentum procedures for shock tube problem. Numer Heat Transfer Part B 33:287
10. Moukalled F, Darwish M (2001) A high-resolution pressure-based algorithm for fluid flow at all speeds. J Comput Phys 168(1):101–133
11. Maliska CR, Silva AFC (1987) High speed flows code developments. Part II. Internal Report IAE/CTA-SINMEC/EMC/UFSC, RT-87-5, Florianópolis, SC, Brasil

12. Silva AFC (1991) A procedure in finite volume method for the solution of all speed flows (in Portuguese). Ph.D. Thesis, Federal University of Santa Catarina, Florianópolis, Brasil, https://repositorio.ufsc.br/handle/123456789/106327
13. Galpin PF, Raithby GD (1986) Numerical solution of problems in incompressible fluid flow:treatment of the temperature-velocity coupling. Numer Heat Transfer 10:105–129
14. Sheng Y, Shoukri M, Sheng G, Wood P (1998) A modification of the SIMPLE method for buoyancy-driven flows. Numer Heat Transfer Part B 33:65–78
15. CFX-TASCflow (2000) Theory documentation, V. 2.10

Chapter 9
Two and Three-Dimensional Parabolic Flows

9.1 Introduction

Strictly speaking, it can be said that all three-dimensional problems of practical interest are elliptic, requiring the complete solution of the Navier–Stokes equations. There are, however, many problems, among them the three-dimensional flow inside ducts and high-momentum free jets, in which, as discussed in Chap. 2, diffusion effects can be neglected in one direction. The advection is, therefore, dominant and the problem can be treated as parabolic in that direction.

The parabolic approximation allows a marching problem in that direction, and for that reason, there is no need of downstream boundary conditions. All physical effects that convey information opposite to the flow can't exist if the problem is to be treated parabolically. Physically, the conditions that must prevail for the parabolic approximation to be valid are:

1. Existence of a predominant flow direction;
2. Negligible diffusion of momentum, mass, heat etc. in the predominant flow direction;
3. The downstream pressure must not influence the upstream flow conditions. This condition introduces a decoupling of the pressure fields. Since the pressure effects are elliptic, it is necessary to decouple the pressure field in the parabolic direction from the pressure field in the other directions, so that the equation of motion in the predominant (parabolic) direction is decoupled from the equations of motion in the other directions.

Condition (1) means that there must be no separation or flow recirculation in the predominant direction, or in other words, the velocity must always be positive. With respect to secondary flow (transverse to the parabolic direction), it is important to keep in mind that the parabolic approximation does not impose any restrictions, allowing it to be analyzed completely.

The decoupling of the pressure field is obtained by writing the three-dimensional pressure field, as

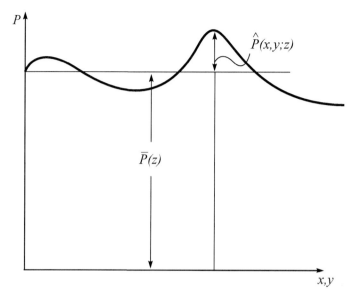

Fig. 9.1 Pressure decoupling in parabolic flows

$$p(x, y, z) = \hat{p}(x, y; z) + \overline{p}(z) \qquad (9.1)$$

in which $\overline{p}(z)$ and $\hat{p}(x, y; z)$ represent the average and the local variation of the pressure at a given section z. Figure 9.1 illustrates Eq. (9.1). The pressure in the cross section, $\hat{p}$, is function of x and y only, but in the notation $\hat{p}(x, y; z)$, the z coordinate is kept to clarify that $\hat{p}(x, y; z)$ is a two-dimensional pressure field, but that it changes with z, while the solution marches, plane-by-plane along z.

It is important to note that, since the pressure fields are decoupled, a method to treat the pressure–velocity coupling in the parabolic direction is required.

The iterative schemes for two and three-dimensional problems now follows. The equations will be written for steady state and incompressible flow. These assumptions bring, absolutely no prejudice to the generality of the subject discussed in this section.

9.2 Two-Dimensional Parabolic Flows

After the parabolic assumptions are done, this flow, considering constant fluid properties, has the following mathematical model,

$$\frac{\partial}{\partial x}(\rho u u) + \frac{\partial}{\partial y}(\rho v u) = -\frac{\partial \overline{p}}{\partial x} + \frac{\partial}{\partial y}\left(\mu \frac{\partial u}{\partial y}\right) \qquad (9.2)$$

9.2 Two-Dimensional Parabolic Flows

$$\frac{\partial}{\partial x}(\rho uv) + \frac{\partial}{\partial y}(\rho vv) = -\frac{\partial \hat{p}}{\partial y} + \frac{\partial}{\partial y}\left(\mu \frac{\partial v}{\partial y}\right) \qquad (9.3)$$

$$\frac{\partial}{\partial x}(\rho u) = -\frac{\partial}{\partial y}(\rho v) \qquad (9.4)$$

in which it was assumed that $(\partial \hat{p}/\partial x) \ll (\partial \overline{p}/\partial x)$, and x is the parabolic direction in this 2D problem. The decoupling of the pressure fields according to Eq. (9.1) is,

$$p(x, y) = \hat{p}(y; x) + \overline{p}(x), \qquad (9.5)$$

This decoupling introduced a new unknown, $\overline{p}$, and, as $\partial \overline{p}/\partial x$ is unknown a new equation should be added to the system to get the solution. External and internal flows can be solved with this equation system. It is possible to show that Eq. (9.3) is of second order and can be discarded. If there is interest in obtaining the $\hat{p}$ for some reason, Eq. (9.3) can be used after the velocities are calculated.

9.2.1 External Two-Dimensional Parabolic Flows

The classical boundary layer flow fits in this category. Considering that the pressure variation normal to the flow is small, since the thickness of the boundary layer is of order of 10^{-3} m for usual flow velocities, the pressure gradient inside the boundary layer can be calculated using Bernoulli's equation, as

$$\frac{d\overline{p}}{dx} = -\rho_\infty u_\infty \frac{du_\infty}{dx} \qquad (9.6)$$

Discarding Eq. (9.3) the system of equations reduces to

$$\frac{\partial}{\partial x}(\rho uu) + \frac{\partial}{\partial y}(\rho vu) = \rho_\infty u_\infty \frac{du_\infty}{dx} + \frac{\partial}{\partial y}\left(\mu \frac{\partial u}{\partial y}\right) \qquad (9.7)$$

$$\frac{\partial}{\partial x}(\rho u) = -\frac{\partial}{\partial y}(\rho v) \qquad (9.8)$$

which can be solved for finding u and v, since the (du_∞/dx) is known. The reader is realizing that there are nothing new in this derivation, except that one is explicitly saying that in parabolic (boundary layer) flows the pressure is decoupled, allowing to have a parabolic problem in the predominant direction and an elliptic problem in the transversal direction. The numerical solution is, therefore, starting at the leading edge of the plate ($x = 0$) marching in the x direction solving a 1D elliptic problem in y for each x position, up to a desired downstream position. In each x position the elliptic problem can be solved segregated or simultaneously.

9.2.2 Internal Two-Dimensional Parabolic Flows

Firstly, it is instructive to recognize that the internal flow between two flat plates, for example, is a boundary layer problem that occurs in two plates that form the duct. Many authors like to classify as boundary layers only external flows, which is not totally correct, since a boundary layer is a thin viscous flow interacting with an inviscid flow. In external boundary layer one knows the velocity of this external flow, while in internal, it interacts with an inviscid flow (acting in the center of the duct) which is unknown. The entrance length of the duct is an internal boundary layer flow.

For internal problems, therefore, there is no longer the possibility of equating the pressure gradient in x with the flow outside the boundary layer, since the latter also varies with x and is unknown. The gradient of $\overline{p}$ cannot, therefore, be eliminated, leaving as unknowns u, v and $\overline{p}$. Using the mathematical model represented by Eqs. (9.2) and (9.4), one needs a third equation to closure the model. The closure of the problem is obtained through the conservation of global mass, since, for internal flows, the mass flow rate is known. Therefore, the system of equations is,

$$\frac{\partial}{\partial x}(\rho u u) + \frac{\partial}{\partial y}(\rho v u) = -\frac{\partial \overline{p}}{\partial x} + \frac{\partial}{\partial y}\left(\mu \frac{\partial u}{\partial y}\right) \tag{9.9}$$

$$\frac{\partial}{\partial x}(\rho u) = -\frac{\partial}{\partial y}(\rho v) \tag{9.10}$$

$$\dot{M} = \int_A \rho u \, dA, \tag{9.11}$$

in which $\dot{M}$ is the mass flow in the duct. This problem is, again, of marching in the parabolic direction, and an algorithm should be devised for calculating the pressure gradient along the duct such that the global mass conservation, Eq. (9.11) is satisfied. The coupling among the velocity and the pressure gradient will be discussed in Sect. 9.5. It was assumed that the pressure variation in y is small, but if it necessary to calculate the pressure along y it is an easy matter using the y momentum equation with the velocities already determined.

9.3 Three-dimensional Parabolic Flows

9.3.1 External Three-Dimensional Parabolic Flows

The most common external flows that can be treated parabolically are jets and plumes, as shown in Fig. 9.2, where only one plane is shown to simplify the visualization. However, care should be exercised when considering the parabolic approach, as the

9.3 Three-dimensional Parabolic Flows

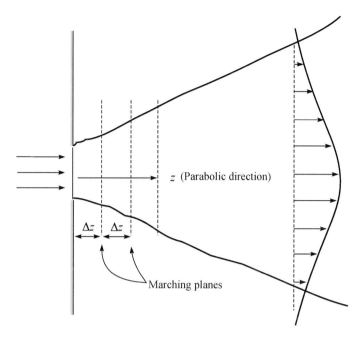

Fig. 9.2 Free-jet parabolic flow

jet will lose momentum by incorporation of the ambient fluid, ceasing to be parabolic at a certain distance from the discharge.

Pressure decoupling in the axial direction is done according to Eq. (9.1). A relevant detail now is to realize that the transversal equations, as done in the 2D case with the y momentum equation, cannot be discarded. The transversal equations in x and y directions must be kept in the system to find $\hat{p}$, since its gradient will determine the distribution of u and v velocities in the transversal plane.

It is easy to see that, even if having the velocity w calculated, it is not possible to extract u or v from the equation of conservation of mass, as was done in the two-dimensional problem, because there are now three velocities involved in the equation. The distribution of u and v in the plane (x, y) will be given by the solution of the two-dimensional elliptic problem in the plane, involving the momentum conservation equations in the x and y directions and the local mass conservation.

With the decoupling of the pressure field, the unknown p is replaced by two new variables, $\hat{p}$ and $\overline{p}$. The closure equation comes, as for the external two-dimensional flow, from the relationship with the undisturbed flow outside the solution region of the problem. For example, being a jet, the pressure along the axial direction is taken as constant and equal to the ambient pressure. This means that the unknown $\overline{p}$ will have its own relation with the free stream, which does not involve the variables of the problem and can, therefore be assumed to be known. This leaves the unknowns u, v, w and $\hat{p}$, whose system of equations to be solved is, already eliminating from the system the gradient of $\overline{p}$, in the case of constant free stream with a corresponding constant $\bar{p}$,

$$\frac{\partial}{\partial x}(\rho uu) + \frac{\partial}{\partial y}(\rho vu) + \frac{\partial}{\partial z}(\rho wu) = -\frac{\partial \hat{p}}{\partial x} + \frac{\partial}{\partial x}\left(\mu \frac{\partial u}{\partial x}\right) + \frac{\partial}{\partial y}\left(\mu \frac{\partial u}{\partial y}\right) \quad (9.12)$$

$$\frac{\partial}{\partial x}(\rho uv) + \frac{\partial}{\partial y}(\rho vv) + \frac{\partial}{\partial z}(\rho wv) = -\frac{\partial \hat{p}}{\partial y} + \frac{\partial}{\partial x}\left(\mu \frac{\partial v}{\partial x}\right) + \frac{\partial}{\partial y}\left(\mu \frac{\partial v}{\partial y}\right) \quad (9.13)$$

$$\frac{\partial}{\partial x}(\rho u) + \frac{\partial}{\partial y}(\rho v) = -\frac{\partial}{\partial z}(\rho w) \quad (9.14)$$

$$\frac{\partial}{\partial x}(\rho uw) + \frac{\partial}{\partial y}(\rho vw) + \frac{\partial}{\partial z}(\rho ww) = \frac{\partial}{\partial x}\left(\mu \frac{\partial w}{\partial x}\right) + \frac{\partial}{\partial y}\left(\mu \frac{\partial w}{\partial y}\right) \quad (9.15)$$

The way the previous equations have been arranged is intended to emphasize that Eqs. (9.12–9.14) constitutes the two-dimensional elliptic problem in the (x, y) plane, coupled with the momentum equation in z, Eq. (9.15). For the solution of the problem, Eqs. (9.12–9.15) must be discretized following what was seen in the previous chapters. All integrations in z, for being in the parabolic coordinate, are done between the calculation plane and the previous one, as shown in Fig. 9.3. For example, for the term $\partial(\rho u\phi)/\partial z$, it will be

$$\int_{z}^{z+\Delta z} \frac{\partial}{\partial z}(\rho u\phi)dz = \rho u\phi|_P - \rho u\phi|_B \quad (9.16)$$

in which all quantities evaluated at B are known and take part in the independent term of the linear system. In this way, after the discretization process is done, the algebraic equation becomes,

$$A_P\phi_P = A_e\phi_E + A_w\phi_W + A_n\phi_N + A_s\phi_S + B_P \quad (9.17)$$

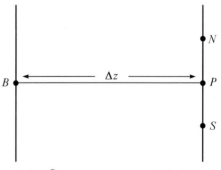

Fig. 9.3 Integration in the parabolic direction

9.3 Three-dimensional Parabolic Flows

in which it is recognized that it is an elliptic 2D problem in the transversal direction. Equations (9.12), (9.13) and (9.15) will have the form of Eq. (9.17) after discretization.

The techniques seen in Chap.4 for the solution of the linear system, the segregated and simultaneous solution seen in Chap. 7 and all comments about solutions strategies apply here. This mathematical model, however, has subtle differences which deserves few words. A full simultaneous solution for $u, v, w, \hat{p}$ is one choice, while a simultaneous solution for u, v and $\hat{p}$, followed by the solution of a linear system for w, is also possible. A segregated solution for all variables can also be done, but the convergence rate will be slower and computer time higher. Recall that in this case a SIMPLE-like method for handling the coupling, for finding u, v and $\hat{p}$, is required. A possible segregated solution procedure now follows, employing the SIMPLEC method.

1. At $z = 0$ the values of u, v and w are known;
2. Advance to the plane $z + \Delta z$ and estimate the values of u, v, w and $\hat{p}$. The values from the previous plane are adopted;
3. Calculate the coefficients for the momentum equation in z. Solve the linear system to obtain w;
4. Calculate the coefficients of the momentum equations for u and v;
5. With the estimated $\hat{p}^*$, solve the linear systems and determine $u*$ and v^*;
6. Obtain the equation for p' and solve for p'. Correct the velocities $u*$ and v^*, using the correcting equations of the SIMPLE or SIMPLEC methods;
7. Calculate $\hat{p} = \hat{p}^* + p'$;
8. Go back to item 4 or 5 and iterate until the two-dimensional problem converges. The solution is not correct, because the w velocity is not correct;
9. Go back to item 3 and iterate until convergence. At that point, the velocities (u, v, w) and pressure $\hat{p}$ calculated for that position z, are known;
10. Advance to a new station z and repeat the calculations advancing up to the required z position.

It is instructive realize that this marching problem in z is the same marching problem usually done in time. The transient term in the fluid mechanics and heat transfer equations is a parabolic term, that is, the transient term has a first order derivative. In space, neglecting diffusion, it will remain a first order derivative in space. If it is an explicit marching in time, there will be a maximum Δt allowable for stability. Analogous, if it is an explicit marching in space, there will a maximum Δz allowable. In Chap. 3 the explicit formulation was studied.

9.3.2 Internal Three-Dimensional Parabolic Flows

All three-dimensional flows inside straight ducts with any type of cross section without obstacles along the flow, can be treated with the parabolic procedure. The treatment is very similar to that already described for external flows. The fundamental difference is that it is now not possible to relate the pressure gradient in the axial direction to the flow outside the boundary layer, as already discussed for 2D flows. Therefore, $(\partial \overline{p}/\partial z)$ is not known, but rather is one more unknown to be determined. Fortunately, the physics offers one more equation to use, which is the global mass conservation.

Since the pressure gradient in the axial direction (z direction), when substituted into the axial momentum conservation equation must generate axial velocities that satisfy global mass conservation, there is a new coupling to be solved. Let us call it the pressure–velocity coupling in the axial direction. Then, there is a two-dimensional coupling in the (x, y) plane, as seen for 3D external flows, and a one-dimensional coupling in the z direction, both connected through the axial velocity that appears in both sets of equations (transversal and axial). The equation system to be solved is,

$$\frac{\partial}{\partial x}(\rho uu) + \frac{\partial}{\partial y}(\rho vu) + \frac{\partial}{\partial z}(\rho wu) = -\frac{\partial \hat{p}}{\partial x} + \frac{\partial}{\partial x}\left(\mu \frac{\partial u}{\partial x}\right) + \frac{\partial}{\partial y}\left(\mu \frac{\partial u}{\partial y}\right) \quad (9.18)$$

$$\frac{\partial}{\partial x}(\rho uv) + \frac{\partial}{\partial y}(\rho vv) + \frac{\partial}{\partial z}(\rho wv) = -\frac{\partial \hat{p}}{\partial y} + \frac{\partial}{\partial x}\left(\mu \frac{\partial v}{\partial x}\right) + \frac{\partial}{\partial y}\left(\mu \frac{\partial v}{\partial y}\right) \quad (9.19)$$

$$\frac{\partial}{\partial x}(\rho u) + \frac{\partial}{\partial y}(\rho v) = -\frac{\partial}{\partial z}(\rho w) \quad (9.20)$$

$$\frac{\partial}{\partial x}(\rho uw) + \frac{\partial}{\partial y}(\rho vw) + \frac{\partial}{\partial z}(\rho ww) = -\frac{d\overline{p}}{dz} + \frac{\partial}{\partial x}\left(\mu \frac{\partial w}{\partial x}\right) + \frac{\partial}{\partial y}\left(\mu \frac{\partial w}{\partial y}\right) \quad (9.21)$$

$$\dot{M} = \int_A \rho w \, dx \, dy \quad (9.22)$$

in which the Eqs. (9.18–9.20) form the mathematical model for the elliptic problem in (x, y) plane, same as for external flows, and Eqs. (9.21) and (9.22) the mathematical model for the axial direction. These two systems of equations are coupled, of course, but each one has its own coupling to be treated. The parabolic coupling is the novelty, since the elliptic one was already deeply discussed in Chap. 7.

9.3.2.1 $(\overline{p} - w)$ Coupling in the Parabolic Direction

The philosophy for creating a method that handles coupling in the parabolic direction is the same as that employed for two and three-dimensional elliptic couplings. That is, in the parabolic direction, one must find the pressure gradient in that direction that generates velocities that satisfy the global mass conservation. The process can be iterative, estimating a pressure gradient and correcting the velocities so that mass is conserved. Such an iterative process was proposed in Patankar and Spalding [1] and is now described.

Patankar and Spalding Method—$(\overline{p} - w)$ Coupling

9.3 Three-dimensional Parabolic Flows

The same procedure applied to determine the velocity correction equations shown in Chap. 7 will now be employed for the equation of motion in the z direction. Integrating Eq. (9.21) for the w velocity, it gives

$$A_P w_P = A_e w_E + A_w w_W + A_n w_N + A_s w_S + B - \left(\frac{d\overline{p}}{dz}\right)\Delta V \qquad (9.23)$$

For an estimated pressure gradient $(d\overline{p}/dz)^*$, Eq. (9.23) reads

$$A_P w_P^* = A_e w_E^* + A_w w_W^* + A_n w_N^* + A_s w_S^* + B - \left(\frac{d\overline{p}}{dz}\right)^* \Delta V \qquad (9.24)$$

Subtracting Eq. (9.24) from Eq. (9.23) and neglecting the differences $(w - w^*)$, one finds

$$w_P = w_P^* - \left(\frac{d\overline{p}}{dz}\right)' \frac{\Delta V}{A_P}, \qquad (9.25)$$

in which

$$\left(\frac{d\overline{p}}{dz}\right) = \left(\frac{d\overline{p}}{dz}\right)^* + \left(\frac{d\overline{p}}{dz}\right)' \qquad (9.26)$$

The numerical integration of Eq. (9.22) is

$$\dot{M} = \sum \rho w \Delta x \Delta y \qquad (9.27)$$

Substituting Eq. (9.25) into Eq. (9.27), results

$$\left(\frac{d\overline{p}}{dz}\right)' = -\frac{\dot{M} - \dot{M}^*}{\sum \rho \frac{\Delta V}{A_P} \Delta x \Delta y} \qquad (9.28)$$

in which $\dot{M}^*$ is the mass flow rate calculated with the w^* velocity that doesn't satisfy the global mass conservation, Eq. (9.27). The procedure is iterative inside the cycle treating the $(\overline{p} - w)$ coupling. The steps are the same as in the SIMPLE method, which creates a correction equation subtracting the momentum equations for $p*$ and p, and creating one equation for p' inserting the velocity correcting equation into mass conservation. In this problem the function is $(d\overline{p}/dz)'$.

Raithby and Schneider Method—$(\overline{p} - w)$ Coupling

The method proposed in Raithby and Schneide [2] aim at eliminating the iterative process in the correction of the pressure gradient in the z direction, invoking the linearity between the velocity w and the pressure gradient in Eq. (9.23), for a given set of coefficients. Exploring this characteristic, it was proposed the following new variables

$$Q = -\frac{d\overline{p}}{dz}; \quad f_P = \frac{\partial w_P}{\partial Q} \tag{9.29}$$

The variable f_P is obtained, according to its definition, by deriving Eq. (9.23) with respect to Q, with the following result,

$$A_P f_P = A_e f_E + A_w f_W + A_n f_N + A_s f_S + \Delta V \tag{9.30}$$

The numerical approximation of Eq. (9.29), gives.

$$\Delta Q = \frac{d\overline{p}}{dz} - \frac{d\overline{p}^*}{dz} \tag{9.31}$$

and

$$f_P \Delta Q = \Delta w_P = w_P - w_P^* \tag{9.32}$$

Integrating Eq. (9.32), one obtains

$$\Delta Q = \frac{\dot{M} - \dot{M}^*}{\sum \rho \Delta x \Delta y f_P} \tag{9.33}$$

The boundary conditions for Eq. (9.30) are obtained by inspecting Eq. (9.32). For prescribed velocity, there is no velocity correction, what forces f_P to be zero. For the prescribed velocity derivative, the equation requires the derivative of f_P to be zero. Figure 9.4 shows the strategy for correcting the pressure gradient in z without requiring iterations in the coupling in the parabolic direction, what is due to the linear relation among w and the pressure gradient.

The numerical procedure for the parabolic direction is very simple.

1. Estimate $\frac{d\overline{p}^*}{dz}$ and calculate w^* solving the linear system given by Eq. (9.24);
2. Find $\dot{M}^*$ using Eq. (9.22) with w^*;
3. Solve Eq. (9.30) for finding f_P;
4. Calculate ΔQ using Eq. (9.33);
5. Correct w_P^* using $w_P = w_P^* + f_P \Delta Q$. Global mass is then satisfied;
6. Using Eq. (9.31) calculate the new pressure gradient.

This procedure has no iteration in the parabolic coupling but requires the solution of an additional linear system for finding f_P. To conclude this chapter, it is presented a possible iterative procedure for solving a three-dimensional parabolic internal flow. Say, for example, it is the flow in the entrance region of a rectangular duct. To deal with the pressure–velocity coupling in the elliptic (x, y) plane, the PRIME method is used, and in the parabolic direction, the Raithby and Schneider's method. Let's remember that the solution marches along the duct solving a 2D elliptic problem in each cross-section. The duct cross-section can be of any shape, and even not constant,

9.3 Three-dimensional Parabolic Flows

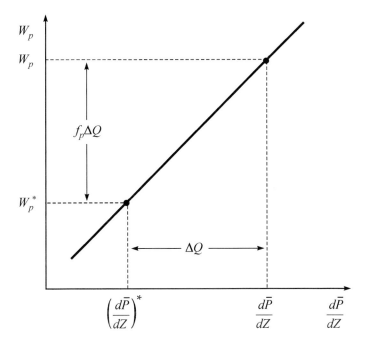

Fig. 9.4 Non-iterative correction of the pressure gradient

respecting the restrictions of the model, which don't allow flow recirculation to obey the parabolic approximation.

1. At duct entrance all variables are known;
2. Advance Δz and estimate all variables $(u, v, w, \hat{p}, \bar{p})$ in that plane. The values from the previous plane are used;
3. Calculate the coefficients for Eq. (9.23) and with $(d\bar{p}/dz)^*$ find w^* (solution of a linear system). Calculate $\dot{M}^*$;
4. Determine f (solution of a linear system). Calculate ΔQ;
5. Correct w^* and $(d\bar{p}/dz)*$ using Eqs. (9.32) and (9.31), respectively;

At this point, as the parabolic coupling is not iterative, one as the solution of the parabolic problem with a specified set of coefficients which involve u and v. Now, the elliptic problem in the plane should be solved;

6. Calculate the coefficients for the momentum equations for u and v;
7. Find $\hat{u}$, $\hat{v}$ and $\nabla \cdot \hat{\mathbf{V}}$ using the equations of Chap. 7 when the PRIME method was presented;
8. Solve for the pressure $\hat{p}$ using equation, Eq. (7.86), Chap. 7. Recall that $\hat{p}$ here plays the same role as p in that equation;
9. Correct u and v obtaining a velocity field which satisfy local mass conservation for the 2D elliptic problem;

10. The alternatives are going back to item 5 and iterate in the 2D elliptic problem until convergence, and then back to the problem in the parabolic direction, or back to item 2. There is no single rule that always works. The author's experience with these problems shows that it is possible to go back to item 2 without iterating the elliptic coupling [3]. It is, in fact, a compromise, since it should iterate enough to guarantee the 2D coupling stability, but avoiding excessive iterations because the velocity w is not correct. In CFD one is always faced with this compromise, since iterations are due to couplings and non-linearities, and it is not easy to adequately balance the iterations in each cycle;
11. Advance any other scalar, like temperature, concentration etc.;
12. Since the coefficients in item 2 were calculated with estimated velocities, go back to item 2 and iterate until convergence is achieved;
13. Move to a next z plane choosing Δz. This problem is parabolic and, therefore, the solution marches in z until it sweeps the entire domain of interest. The increments to be given in the z direction depend on the gradients of the variables with respect to z, and should be analyzed carefully. It is recommended that after advancing a few planes, and the singularity of the boundary conditions at the duct inlet has been taken into account with a finer grid, to increase the Δz to speed up the marching process, taking care, of course, of the accuracy of the solution in the various planes. If only the solution of the fully developed flow is of interest (recall that it is being considered the inlet region problem in a duct), it is possible to march with large Δz, since the fully developed solution will not be affected by the size of Δz during the marching. Think about the full similarities of this marching in z problem, with the marching in time in transient problems.

The advantage of the parabolic procedure is the economy in saving storage space for the variables, since only the variables in two adjacent planes must be stored, those in the calculation plane and those in the upstream plane. Furthermore, a two-dimensional problem is solved in each plane, instead of solving a full 3D problem. For example, a problem with a mesh of 400 volumes in the cross section and 100 steps in z would, if solved elliptically, represent a linear system of 40,000 points for each variable. With the parabolic procedure, the solution of 100 two-dimensional problems with 400 volumes will be required. The difference in the computational effort is considerably large.

Another possibility is to consider the problem partially elliptic, that is, to solve elliptically only for the pressure. The parabolic march presented in this section would be employed by sweeping the domain several times and updating the three-dimensional pressure field. This is an alternative that has not been frequently employed in the literature.

9.4 Conclusions

This chapter has shown a strategy for solving fluid flows with a predominant direction, reducing the efforts in computational time and variable storage. Two-dimensional boundary layer problems can be reduced as a series of one-dimensional problem, and a three-dimensional problem as a series of two-dimensional ones, with a very large economy in computational effort. For three-dimensional flows in straight pipelines, and even those with moderate curvature, the parabolic approximation should be considered, because it allows the complete problem to be solved, without prejudice in the investigation of the secondary flow, as in coupled natural convection, with an small computation time compared to the fully elliptic solution. Jet flows and plumes discharged in quiet environment can also be solved, covering a large class of environmental problems.

9.5 Exercises

9.1. Solve numerically the isothermal flow problem in the inlet region of two infinite parallel plates with distance h between them. The mass flow rate is known at the inlet. Calculate the product $f\text{Re}$ for the fully developed profile condition and compare it with the analytical result. Use Raithby and Schneider's method to treat the pressure–velocity coupling in the parabolic direction and recognize that there is no need to treat the coupling in the transverse direction, since the velocity v is determined from Eq. (9.10)

9.2. For Exercise 9.1, show that if advancing the solution along the duct axis explicitly, there is maximum step for advancing such that the coefficients do not result negative. Determine this maximum.

9.3. Obtain the coefficients of Eq. (9.17).

9.4. Estimate the variable storage requirement for the three-dimensional, laminar, incompressible flow problem inside a duct when the elliptic and parabolic solutions are employed.

References

1. Patankar SV, Spalding DB (1972) A calculation procedure for heat, mass and momentum transfer in three-dimensional parabolic flows. Int J Heat Mass Transf 15:1787–1806
2. Raithby GD, Schneider GE (1979) Numerical solution of problems in incompressible fluid flow: treatment of the velocity-pressure coupling. Numer Heat Transf 2:417–440
3. Maliska CR (1981) A solution method for 3D parabolic fluid flow problems in nonorthogonal coordinates. Ph.D. thesis, University of Waterloo, Canada

Chapter 10
General Recommendations for Conceiving and Testing Your Code

10.1 Introduction

This chapter is dedicated to beginners and students of numerical fluid mechanics and heat transfer when they are faced with the need to write their algorithms and codes for the assignment or dissertations. The idea is just to have guidelines and not rules that are used when full commercial software is developed and tested. For these, there are plenty of books in software development, and validation of engineering codes is a quite complex activity.

The task of developing a good computational code to numerically simulate any physical problem requires harmony between the use of available computational tools, such as languages, libraries for interface and visualization, strategies for the development and debugging of programming errors, and the correct interpretation of the results as a final step. It will be inevitable for any analyst developing programs to be confronted with numerous errors, both in programming logic and in the incorrect implementation of expressions. It is said that every code with more than 100,000 lines has on average 5% of lines with some kind of error. Apparently, this is an absurdly large percentage, especially because we are almost always tempted to believe that our codes are error-free.

Writing the program, testing it for logic and implementation errors, validating it numerically and physically are tasks that require knowledge of the physics of the problem and of the numerical methodology, and patience. Finding a difficult error requires some of a detective's skill, checking clues, analyzing them carefully, and eliminating areas of the code where the error will certainly not be etc.

There are, however, a few little rules that, although they may seem trivial, are often not followed by people beginning their studies in numerical fluid flow. These points are now described in a very general way, but they serve as a help and warning for the developers.

10.2 Writing Your Code

10.2.1 Generalities

In the academic environment, where a given computer code is implemented step-by-step, with the interference of several people introducing different features, its organization, reusability and maintenance are extremely important, but, at the same time, a difficult job. It is well known that one of the most frequent complaints from the students is the time lost reimplementing parts or even repeating the whole code, because the modules were not well documented or even without documentation. The university labs do not work as a software house, since the goal is to have students finishing their Master and Ph.D. works and not delivering final software products. However, it is mandatory to have a developing environment in which all the objects used in all developments are easily accessible by the developers. For example, dealing with unstructured grids all the geometric calculations are cumbersome to be done and should be ready to those implementing a new physics. Connections among elements, control volumes and nodes should be also available for a given grid. Modernly, there are resources and methodologies that help to build a computational abstraction suitable for reusability and maintenance of the code.

In the current developments, it should be tried, whenever possible, to provide the computational codes with friendly graphical interfaces (GUI), because these greatly facilitate and speed up the use, both by the developer and by those who will make new additions to the program. A good graphical user interface is compulsory when you want to develop a software. For academic and research programs it is not crucial, but even a basic GUI helps enormously, even for the developer to restart the use of the code after a certain time without use.

Object-oriented programming [1] as C++ and Phyton guide and provide the necessary subsidies to build a computational implementation that allows reusability and facilitates the construction of a GUI. The idea that the computational design of the numerical core of the program is independent of the GUI, and that it is enough to build a GUI that encapsulates and links the class and objects, is incorrect. The way of conceiving, modeling and programming the numerical core, that is, its computational abstraction, with its classes and objects well-defined in function of the data manipulated and functionalities available in the GUI, will be a decisive factor in the quality, performance, and ease of constructing the GUI.

As for visualization, the previous recommendations are even more important. The main concern is to avoid making the data transfer between the numerical kernel and the visualization application a laborious and time-consuming task each time a new simulation is run. This can easily occur if the input file format is inappropriate for the type of data being simulated, or when the type of visualization you want to obtain is not exactly the right visualizer used.

To solve these problems, there are several solutions. Before any decision is made, it is necessary to verify that the visualizer supports the types of data manipulated by the numerical kernel in question: type of mesh (structured, unstructured, hybrid), mesh

10.2 Writing Your Code

characteristics (mobile, adaptive, time-varying), types of stored properties (scalar, vector, tensor), position of variable storage (cell center, face center, edge center, vertex), features for presenting the results.

There are cases, however, when the visualization must be performed on-line with the simulation, synchronized with the real-time change of simulation parameters (parametric analysis, optimization processes, trial and error design). In these cases, saving a data file and later calling up the data manually becomes impractical, and it is extremely suitable and convenient to build the visualization directly integrated in the program. That is, the visualization windows become part of the computational abstraction. If this level of programming is desired, the developer will need to use software development tools that provide objects for creating the graphical interface (windows, buttons, edit boxes, sliders, etc.) and the scientific visualization.

The creation of a numerical code or a software should not be a coding-only activity. Coding is just one part of the many steps that exist in the creation of an application. Among the other steps, it can be mentioned, design, architectural project, detailed project, sizing, documentation, testing, debugging, validation, creation of the final package, and other steps specific to each development phase. Just as for the coding stage there is the integrated development environment and its compiler, part of the process known to all, for these other stages there are other tools of great importance. Below are some general comments on details of these tools.

10.2.2 Coding Languages

Traditionally, one of the first decisions that are made when developing a code is the choice of coding language. Making this choice as the first decision is the first mistake the programmer makes in the project. Several development steps that must be done before coding are independent of the coding language. In fact, the coding language should be seen merely to formalize, in a language interpretable by the compiler, all the design and modeling done in the previous steps, which include the computational abstraction. There are lots of languages available to the developer for writing his code.

Currently, on the top of the list are C++ and Phyton, each one with its own advantage according to the application to be created. Let's not forget the FORTRAN (FORmula TRANSlator) language, created to attend the scientific programming needs in the 50s. Believe or not it still used today for massive computation in astrophysics, molecular dynamics and weather predictions, among other applications which requires HPC (High Performance Computing). FORTRAN provides few programming resources, thus, coding a program in FORTRAN forces the programmer to work at a low level of abstraction, unable to bring into the "computational world" characteristics and, especially, concepts present in the "real world". Recently FORTRAN introduces some tools to have interoperability with C++, what gives extended life to it. In its way, C++, due to its more modern conception and recent major developments, allows the programmer to insert and use in his code the concepts on which he based his entire design and conception. It is because of these factors that a C++ program is said to be more readable, more organized, and easier to eliminate errors and to perform further maintenance and expansion. It is now in a level which is suitable for coding large scientific applications.

With this picture, one could say that for CFD programing C++ would be the choice nowadays, while Phyton, supported by its strong and large community, also advocates Phyton for developing large scientific applications. Experts says if rapid development of your project is needed, Phyton is recommended, while if computing speed is the requirement, use C++. Both are object-oriented languages and can help in developing codes with some degrees of reusability.

10.2.3 Tools to Aid the Development

Besides the compiler, as a coding tool, there are other extremely useful tools for development. It can be mentioned the computational design, which will help, together with the compiler, the programmer to conceive and visualize the various parts of his program. The design is responsible for thinking in all parts of the program its inter-relationships, such that a modification in one part doesn't affect the whole program. Design is important even in a simple application and absolutely necessary in large software. Designing is when the experience in physics and numerics of the designer appears, especially in the design of important classes and the interrelationship among them. Without a clear understanding, it is impossible to create an efficient design to be implemented, since it will be required frequent re-design of parts.

Online code documentation, besides being important for the programmer himself, is fundamental for the understanding of the code by people who will expand it. There are several on-line documentation tools on the market that, in addition to documenting what has been coded, allow high-level documents to be created from information inserted into the code. These documents can then be provided along with the program, as well as published on a website.

Another extremely important tool when the program development is done by a group of programmers, is the concurrent version control. One of the most traditional tools on the market for this type of solution is CVS (Concurrent Version System), which allows different developers in a group to simultaneously encode a program in one or more source files. The level of flexibility provided by this kind of tool is extraordinary, allowing several simultaneous modifications to the same source file to be made by different users. The system, in turn, on demand, can unite all the modifications of the various programmers in a single location, called a repository, maintaining the integrity of all the work done by the team. In addition, the system can inform the changes made by each user in the various versions of the software, assisting in the correction of bugs introduced during the coding of new features.

Once the coding stage is complete, the next step is to correct errors and optimize the code. Two tools are fundamental to help eliminate runtime errors, better known as bugs: a runtime debugging tool and a bug tracking and management tool.

The runtime debugging tools help the programmer to inspect the program's execution sequence, as well as the values of variables at any time during execution, in addition to identifying memory leaks due to non-deleted dynamic variables, the use of uninitialized variables, and other variable management errors, whether dynamic or static.

The bug tracking and management tools work as a database that concentrates all the information discovered by users or programmers regarding the improper functioning of the application. These tools usually work in a network environment (internet or intranet) and allow bugs to be added, viewed, associated with the person responsible for the problematic piece of code, commented on, and finally closed after deletion. In this way it is possible to follow the whole "life cycle" of a bug, from its discovery to its correction in the code.

The bug tracking and management tool, associated with the version control and online documentation tool, allows the documents of new features and deleted bugs to be easily created between any version of the tool.

Finally, after the execution error elimination stage, the optimization of the program begins. For this activity, it is recommended the use of a tool capable of measuring the total processing time consumed by each function or method of the program. With this data, this tool can inform where the processing bottlenecks are and indicate to the programmer the best places to perform the optimizations.

Once a good computational abstraction has been made and the benefits of the various development tools have been utilized, it is important for the beginner to implement his program in parts, making it grow slowly and always well tested, part by part. One should try to start by solving simple problems, preferably in Cartesian coordinates, without trying to incorporate all possible types of boundary conditions right from the start. The program should always be under absolute control, and as the various assignments are implemented and evaluated, others can be safely introduced.

It is advisable to check whether vector and parallel processors are available in the available computing infrastructure in order to take advantage of these architectures, if desired. The important thing is to write the program knowing what kind of processor it will run on. A program written without care for vectorization can be executed on a vector processor, but logically the beneficial effects will be less than if the program is written with the objective of vectorization. For parallel processing, this recommendation is even more important.

10.3 Running Your Application

The testing part of a computer program requires some important initial care. The compilation, the convergence criteria, the analysis of the results, the search for errors etc. are tasks that cannot be performed without strategies. The testing process must be conducted in a systematic way. In the following, these points are analyzed.

10.3.1 Compiling

There is nothing better than using an efficient compiler. It is very uncomfortable not being sure whether the program has undefined variables or not, loops that you don't know if they correctly cover the sweep you want, common areas of undefined variables etc. You should not compile the program in a non-rigorous compiler, trying to convince yourself that the program is correctly implemented. The chance that there are no errors is minimal. A rigorous compiler should always be employed in the initial phase. After the program has been compiled in "debug mode" (compiler option optimized to perform the most rigorous inspections possible) and "clean" both in terms of compilation errors and execution errors, you can compile it in "release mode" (option optimized for performance) to run it with maximum speed and minimum compiler overhead and memory consumption.

The main program, on the other hand, should not be a mixture of large module calls and small commands. The main program should uniquely manage the execution sequence, giving the programmer greater visibility of the flowchart to be followed.

10.3.2 Size of the Mesh

After choosing the problem to be solved (it will be comments on this choice shortly), it should be solved on a very coarse mesh. In a two-dimensional problem one should not be over a 10 × 10 mesh, because this mesh is generally able to capture the physics of the problem and will allow to verify if the results are qualitatively correct. Unfortunately, although it may seem trivial, the use of very refined meshes during testing is a very frequent mistake, and the consequences are many: high computational time, difficulty and expense of printing the results and coefficients etc. Moreover, if the results are examined on the screen, an excessively large mesh will cause difficulties to fit them into the available viewing space, and the analysis will be difficult.

The program should be run initially with very few iterations, just those necessary to check that the whole iterative cycles are correct. Only after everything seems coherent should the number of mesh volumes and the number of iterations be increased.

10.3.3 Convergence Criteria

The criterion to stop the program execution is not an easy decision. There are problems that have slow convergence, and if the execution is stopped by a poorly chosen criterion, the solution may still be far from the converged one. The other side of the coin is to use a to severe criterion, which keeps the program iterating unnecessarily. Therefore, it is advisable to initially leave the program without a criterion (there should be only one safety criterion in the total number of iterations to avoid a loop)

10.3 Running Your Application

and have two or three solutions printed, in separate iterations, and check how the problem behaves. From this, a convergence criterion can be chosen.

The choice of criterion is easy when the range of variation of the unknown is known, in which case even an absolute criterion can be efficient. When, however, the order of magnitude and the range of variation of the unknown to be determined is not known, the task is more difficult. A pressure field may vary, for example, between 10^{-3} and 10^4, and a relative convergence criterion, such as the one commonly used, can keep a program running when everything that matters from a physical point of view is already settled. This is the danger of using relative criteria in fields that have both small and large values.

To illustrate, imagine a variable with the minimum value 10^{-3} and the maximum value 10^3. Consider that in iteration k the value of ϕ at two points in the domain were 2×10^{-3} and 9×10^2, and in iteration $k+1$ were 1.8×10^{-3} and 1×10^3. Note that the change from 2×10^{-3} to 1.8×10^{-3}, compared to the magnitude of the variable in the domain, has no important physical meaning. However, using the criterion usually employed in the literature, given by

$$\left|\frac{\phi_P^{k+1} - \phi_P^k}{\phi_P^k}\right| \leq \epsilon \qquad (10.1)$$

the error found is 10^{-1}, while for the other point it is 10^{-3}. The result is that the error at the point in the domain that doesn't have physical influence will keep the program running, spending a large amount of computational time unnecessarily. The observation tells us that, for this problem, an error of the order of 10^{-3} should terminate the iterative procedure. A criterion that avoids this problem is to determine the range of variation of the ϕ in the domain, that is, the modulus of the difference between the maximum and minimum values, and use this as a reference, in the form

$$\left|\frac{\phi_P^{k+1} - \phi_P^k}{R}\right| \leq \epsilon \qquad (10.2)$$

in which

$$R = \phi_{max} - \phi_{min} \qquad (10.3)$$

Using this criterion, the errors become 2×10^{-6} and 10^{-3}, respectively, making the criterion of the unimportant point in the solution satisfied in advance. Notice that a relative criterion of the order of 10^{-3} now halts the calculations.

Attention should be paid to the fact that convergence criteria are needed at several levels within a simulation program, but mainly to stop calculations when variables have converged, or within linear system solvers. During the process of testing the program, the criteria used should be loose to avoid unnecessary computation time. Given the delicacy of this topic and the repercussion in computer time, the recommendation is to be very careful when choosing convergence criteria.

10.4 Choosing Test Problems—Finding Errors

To numerically validated a CFD code it would be necessary to have available analytical solutions of the problems. But, of course, they are not, and the validation should be made by parts. At the end the entire equation is not validated, but with well-chosen problems, good validation can be achieved. For example, initially it is always wise to solve a pure conduction problem that has an analytical solution. By doing so, a reasonable portion of the program is checked, such as the part of the coefficients (diffusive part), iterative structure, solution of the linear system etc. The same can be done with an advection problem. In the following section some simple test problems with analytical and numerical solutions will be presented.

10.4.1 Heat Conduction—2D Steady State

There are numerous analytical solutions of two-dimensional conduction problems that can be used [2]. In the following, it is suggested solutions with extremely simple implementations. The two-dimensional conduction problem on a plate with prescribed temperatures, as shown in Fig. 10.1, whose differential equation is

$$\frac{\partial^2 T}{\partial x^2} + \frac{\partial^2 T}{\partial y^2} = 0, \qquad (10.4)$$

has the following analytical solution

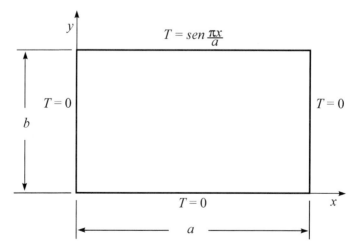

Fig. 10.1 Domain for the 2D heat conduction problem

10.4 Choosing Test Problems—Finding Errors

$$T(x, y) = \frac{\text{senh}\left(\frac{\pi y}{a}\right)}{\text{senh}\left(\frac{\pi b}{a}\right)} \text{sen}\left(\frac{\pi x}{a}\right) \tag{10.5}$$

With this problem the solution in two axes are checked. If the code is 3D, the same problem can be solved in the (x, z) and (y, z) planes. In doing so, the implementation of coefficients will be checked in all directions.

10.4.2 Transient Heat Conduction—One Dimensional

The solution in time, that is, the transient term can be also checked. Complex two and three-dimensional transient solutions with different boundary conditions exist in the literature [2]. For our goal, which is to verify if the implementation of the transient terms is correct, and if the update of the variables is being done correctly in the iterative cycle in time, it is sufficient to solve the transient one-dimensional heat conduction problem, whose differential equation is given by

$$\frac{\partial T}{\partial t} = \alpha \frac{\partial^2 T}{\partial x^2} \tag{10.6}$$

Two problems will be analyzed [3], differing only in the initial condition. Consider a flat plate, infinite in the y and z directions, with thickness L, initially with a given temperature distribution that, at time $t = 0$ undergoes a transient process in which the temperatures of the faces $x = 0$ and $x = L$ are kept at zero. For the case in which the initial temperature distribution is given by

$$T(x, 0) = T_o \text{sen}\left(\frac{\pi x}{L}\right) \tag{10.7}$$

the analytical solution is

$$T(x, t) = T_o e^{-\alpha \lambda_1^2 t} \text{sen}(\lambda_1 x) \tag{10.8}$$

in which $\lambda_1 = \pi/L$. For the case in which the initial temperature is T_i the analytical solution is

$$T(x, t) = \frac{4}{\pi} T_i \sum_{n=1,3,5,7}^{\infty} \frac{1}{n} e^{-\alpha \lambda_n^2 t} \text{sen}(\lambda_n x) \tag{10.9}$$

in which $\lambda_n = n\pi/L$

It is also possible to derive Eq. (10.9) and obtain the heat flux to be checked with the heat flux numerically calculated. This is a good test because it acts on the derivative of the function, which loses accuracy when evaluated. By checking the accuracy of the heat flux calculation, one will be evaluating an important numerical approximation at the interface. It will be possible to verify that to obtain good results for the heat flux, the numerical approximation cannot be of first order, unless very refined meshes are used.

It is also suggested to realize the marching in time with different time intervals, performing a grid resolution in time to check the accuracy of the transient calculation. It should be remembered that in a fully implicit formulation the time interval is limited by accuracy, not stability, at least for linear and simple problems. However, remember that in the solution of coupled, nonlinear equations, even with the fully implicit formulation, the time interval has limitations, not only by precision. In this case, the reasons are couplings and nonlinearities, which, by the mode they are treated, introduce explicit features to the procedure.

10.4.3   One Dimensional Advection/Diffusion

To begin the analysis of advection/diffusion problems, the unidimensional flow with a known constant velocity field is useful. This problem has already been discussed extensively in Chap. 5, when interpolation functions were considered. This problem serves to analyze the interpolation functions and their dependence on the Peclet number. The differential equation of the problem is

$$\frac{\partial}{\partial x}(\rho u \phi) = \frac{\partial}{\partial x}\left(\Gamma^\phi \frac{\partial \phi}{\partial x}\right) \tag{10.10}$$

and the analytical solution, in exponential form, can be found in Chap. 5. Remember that the numerical diffusion error now appears, and if one is using the exact interpolation function in the program, it should reproduce the analytical solution with any mesh. If WUDS is being used, the solution will differ very little from the exact one, even with coarse mesh, since the WUDS interpolation is nearly exact.

Most likely, the program being tested is two or three-dimensional, and the test suggested here is a one-dimensional problem. There are no difficulties, since it is enough to make a mesh in the y (or y and z) direction just large enough (in general 3 volumes suffice) and apply zero derivative conditions on the boundaries in the other directions. The results should be identical, and this constitutes in another test for the program. If the results don't repeat along the other coordinates there are something wrong with the implementation.

10.4.4   Two-Dimensional Advection/Diffusion

When the interest is to test interpolation functions to assess the numerical diffusion added by the scheme, the advection/diffusion of a pulse is recommended. The differential equation of this problem [4], is

10.4 Choosing Test Problems—Finding Errors

$$\frac{\partial}{\partial x}(\rho u \phi) + \frac{\partial}{\partial y}(\rho v \phi) = \frac{\partial}{\partial x}\left[\Gamma^\phi \frac{\partial \phi}{\partial x}\right] + \frac{\partial}{\partial y}\left[\Gamma^\phi \frac{\partial \phi}{\partial y}\right] \quad (10.11)$$

with boundary conditions given in Fig. 10.2 in which θ is the inclination of the velocity vector with respect to the x axis. Observe that above this line $\phi = 1$ and below, $\phi = 0$. The solution of Eq. (10.11) neglecting the diffusion along the flow direction is, for $\Delta x = \Delta y$, is

$$\phi = \frac{1}{2}\left\{1 + \mathrm{erf}\left[\frac{1}{2}\left(\frac{\rho}{\Gamma^\phi}\right)^{0.5}\left(\frac{(y-y_c)u - xv}{((y-y_c)v + xu)^{0.5}}\right)\right]\right\} \quad (10.12)$$

in which

$$y_c = \frac{L}{2}(1 - \tan\theta)$$

It is common, for avoiding interpolations when comparing the results, to plot the analytical solution along line AB, creating a mesh for the numerical solution that has velocities stored on the same line.

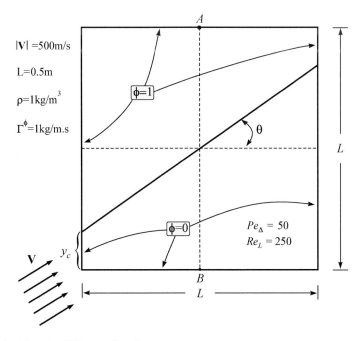

Fig. 10.2 Advection/diffusion of a pulse

10.4.5 Entrance Flow Between Parallel Plates

The isothermal fluid flow between two parallel plates can be used to test several details of the numerical solution. Figure 10.3 depicts the problem whose mathematical model is

$$\frac{\partial u}{\partial x} + \frac{\partial v}{\partial y} = 0 \tag{10.13}$$

$$\frac{\partial}{\partial x}(\rho u u) + \frac{\partial}{\partial y}(\rho v u) = -\frac{\partial P}{\partial x} + \mu \nabla^2 u \tag{10.14}$$

$$\frac{\partial}{\partial x}(\rho u v) + \frac{\partial}{\partial y}(\rho v v) = -\frac{\partial P}{\partial y} + \mu \nabla^2 v \tag{10.15}$$

The boundary conditions are uniform u, and $v = 0$ at the inlet, and $u = v = 0$ at the walls. At the outlet, the fully developed parabolic profile for u, given by Eq. (10.16), and $v = 0$ can be used, or zero derivative (locally parabolic) boundary conditions can also be used for both u and v. The results of the present problem and the characteristics of this flow will be used in the coming section.

10.5 Observing Details of the Solution

First, it is important to point out that it is not possible to check whether a given coefficient of the linear system is correct. The most that can be done is to check that the expression of the coefficient is apparently correctly implemented in the code. The absolute check of coefficients it is only possible with another code that uses the same methodology and the same numerical approximations. In general, such a code is not available. However, some extremely useful procedures in the search for errors involving the solution and the coefficients can be applied, and are commented on below.

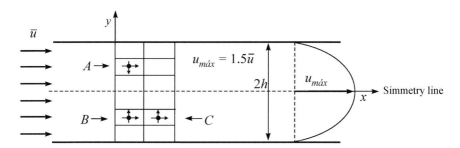

Fig. 10.3 Laminar flow between parallel plates

10.5 Observing Details of the Solution

10.5.1 Symmetry of the Solution

Always explore all possibilities to observe the symmetry of the solution. In the flow problem between parallel plates, Fig. 10.3, we know that $u_A = u_B$ and that $v_A = -v_B$. Pressure at A must be equal to that at B.

The conservation of mass is one of the most important requirements in the solution. Therefore, it is advisable to print the residue of this equation for all control volumes. Remember that in order to calculate the residuals, the velocities at the interfaces of the volume must be used. When the arrangement is staggered, the velocities are already at the interfaces, and when it is co-located, they are obtained as a function of the velocities at the center.

The residuals should be normalized to a reference mass flow representing the order of magnitude of the mass flows in the problem, usually the inlet mass flow. This normalized residual should be on the order of 10^{-6} to 10^{-7}. The residue field should also be symmetric.

In this particular problem of flow between parallel plates, using locally parabolic conditions at the outlet (which is recommended), the profile obtained should be that given by Eq. (10.16), with the velocity at the center equal to 1.5 times the mean inlet velocity. The u profile is,

$$\frac{u}{\bar{u}} = 1.5\left(1 - \left(\frac{y}{h}\right)^2\right) \qquad (10.16)$$

10.5.2 The Coefficients

Investigating the characteristics of the coefficients, relying on the physical analysis of the problem, is not a common practice among developers of computational codes for fluid flows. In general, we feel satisfied checking the expression implemented in the code, check if it is correctly written. But there is much more that can be extracted from the coefficients. And those resources need to be used when there are great difficulties in finding out why the program is not presenting the correct results, or why it is diverging. To start, it is always recommended to print (or have them at hand) the coefficients and the source term of the equation. It is natural to inspect them, since the coefficients and source term form the linear system to be solved.

It should be verified which is the weight (importance) of each coefficient in the system, remembering that the coefficients have a diffusive and an advective part. If the scheme has an interpolation function that cancels out the diffusive term when the velocity increases, this will cause the coefficient to tend to zero downstream of the flow. Recurring to Fig. 10.3, it means to say that the coefficient A_e for the velocity u stored in the center of the control volume B, should be close to zero for increasing

Table 10.1 Values of $u/\bar{u}$

X/Y	0.016	0.032	0.056	0.072	Fully Dev
0.90	0.761	0.528	0.380	0.348	0.285
0.70	1.066	1.033	0.923	0.874	0.765
0.50	1.077	1.155	1.180	1.170	1.125
0.30	1.055	1.154	1.256	1.288	1.365
0.00	1.040	1.130	1.262	1.322	1.500

velocities. On the other hand, the coefficient A_w for the same velocity should be the most important.

Still about the coefficients, one should have $(A_w)_A = (A_w)_B$, $(A_e)_A = (A_e)_B$ and $(A_s)_A = (A_n)_B$.

If the rules of Chap. 5 have been followed, the coefficients should be positive and the central coefficient, A_P, should be at least equal to the sum of the neighbors. Check the source term. Check its sign and magnitude. It is important in the stability of the solution, because it is always treated explicitly, that is, with fixed values from the last iteration.

When the difficulties in finding the errors are serious and a lot of time has already been spent on this, it may be necessary to use a very coarse mesh and to do some manual calculations, checking some coefficients, source terms and mass balances.

For the problem of flow between parallel plates, the values of $u/\bar{u}$ for four stations along the flow are shown below in Table 10.1 for comparison purposes. It is logical that the numbers given and those in the program under test may not match exactly, because they may have different numerical approximations, or the mesh used for calculation was not sufficiently refined.

In the mesh 9×27 of Fig. 10.4, the vectors denote the positions where the velocity $u/\bar{u}$ are reported in Table 10.1. The channel has length $x = 0.288$ m and the results were obtained with Re $= 20$, where the Reynolds number is based on half the channel height. The mesh in the x direction is equally spaced.

10.5.3 Testing the Solver of the Linear System

Nowadays, with the availability of several good libraries of solvers it is rare that a developer of a CFD code will write its own solver, unless it is a course assignment. Independently, if it is an in-house solver or from a library on the web, sometimes one wants to be sure that the solver we are using is giving the correct answer. The practice is to feed the solver with coefficients whose solution is known and check the output of the solver. Below is found a set of coefficients of a very simple problem, whose solution is independent of the size of the grid, what allows to check the results in the level of machine errors. The linear system is,

10.5 Observing Details of the Solution

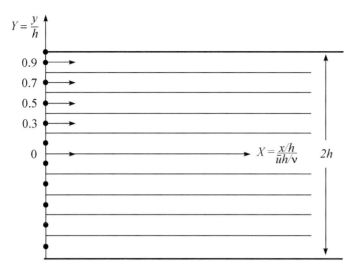

Fig. 10.4 Grid for the flow between parallel plates

$$
\overbrace{\begin{pmatrix} 2 & -1 & & & & & & & \\ -1 & 2 & -1 & & & & & & \\ & -1 & 2 & -1 & & & & & \\ & & -1 & 2 & -1 & & & & \\ & & & -1 & 2 & -1 & & & \\ & & & & -1 & 2 & -1 & & \\ & & & & & -1 & 2 & -1 & \\ & & & & & & -1 & 2 & -1 \\ & & & & & & & -1 & 2 \end{pmatrix}}^{\text{Matrix of coeficients}} \begin{pmatrix} \phi_1 \\ \phi_2 \\ \phi_3 \\ \phi_4 \\ \phi_5 \\ \phi_6 \\ \phi_7 \\ \phi_8 \\ \phi_9 \end{pmatrix} = \begin{pmatrix} 2 \\ 0 \\ 0 \\ 0 \\ 0 \\ 0 \\ 0 \\ 0 \\ 1 \end{pmatrix} \quad (10.17)
$$

The vector $S_1\{\phi_1, \ldots, \phi_9\}$ is the solution of this problem, while the vector $S_2\{\phi_1, \ldots, \phi_9\}$ is the solution if the two non zeros elements in the independent vector are changed from 2 to 1 and from 1 to 0.

$$
S_1 = \begin{pmatrix} 1.9 \\ 1.8 \\ 1.7 \\ 1.6 \\ 1.5 \\ 1.4 \\ 1.3 \\ 1.2 \\ 1.1 \end{pmatrix} \quad S_2 = \begin{pmatrix} 0.9 \\ 0.8 \\ 0.7 \\ 0.6 \\ 0.5 \\ 0.4 \\ 0.3 \\ 0.2 \\ 0.1 \end{pmatrix}
$$

10.6 Conclusions

As already pointed out in other chapters, unfortunately, there are no theories that guarantee that a system of nonlinear partial differential equations solved in a segregated manner, or even simultaneously, has stable iterative processes. The couplings, the ways of advancing the nonlinearities, the size of the time interval, the number of iterations in each cycle etc., are all factors that can cause the solution to diverge.

For this reason, avoid interacting with the computer on the "changing-and-run" rule, hoping the changes you have done will work. Instead, make a careful and detailed analysis of the problem, the influence of the variables in the process, the weight of the coefficients etc., and the progress in finding the error will be faster than the trial-and-error method. Finding the errors and making the program converge is not an easy task when complex systems of coupled nonlinear equations are being solved. It is in this difficulty that the numerical knowledge and the expertise of the analyst is indispensable.

About the user of a commercial CFD code, one could say that the knowledge embedded in a software can be divided in two main groups: numerical and physical knowledge. Numerical knowledge refers to the user's ability to judge the quality of a mesh, to infer the behavior of a solver for a given linear system, to choose a suitable interpolation function, to judge which parameters should be changed when the solution diverges etc. The physical knowledge concerns the user's ability to create his mathematical model, specify boundary conditions, interpret the simulation results, and accept the solution according to what is required for the engineering application.

By now both, numerical and physical expertise are required by the users. The trend is that the numerical requirements will become, in some time, less dependent of the user. That is, there will be efficient algorithms to calculate the truncation error volume by volume and in time, refining the mesh spatially and temporally until the error set by the user is reached. The algorithms and solvers will be robust such that convergence will be almost always achieved. Perhaps computer storage and capacity will be so high that allow to use simple and stable numerical algorithms. The developing research in the numerical techniques goes in this direction. It does not exist at the level of commercial software yet, and so, even today, knowledge of the numerical methodologies built into commercial software is a requirement that must be met by the user.

Physical requirements, on the other hand, will never be replaced by a "virtual engineer". The user of a numerical method should always know the physics of what he intends to simulate. Numerical simulation is equivalent to performing an experiment in a laboratory. Can anyone conceive the idea of a person walking into a laboratory and simply taking measurements without having any notion of what he is measuring and for what purpose? The physical knowledge to perform a numerical experiment (simulation) is the same as for laboratory experiments. Otherwise, the numbers from the simulation or laboratory experimentation will have no application whatsoever.

Machine learning, artificial intelligence and massive data management may train a computer code "to be an engineer" for a class of engineering problems, but never will substitute a human brain trained in physics to propose the ideal mathematical model for that specific problem in consideration, especially when it is a new one. This is my belief, perhaps I am wrong.

Besides physics and numerics, it is recommended to always use the simplest and most general criterion, also used in any human activity: common sense. Without it, numerical simulation also becomes difficult.

References

1. Buzzi-Ferraris G (1993) Scientific C++: building libraries the oriented-object way. Addison-Wesley
2. Carslaw HS, Jaeger JC (1959) Conduction of heat in solids. Oxford University Press, London
3. Ozisik MN (1985) Heat transfer—basic text. Editora Guanabara Dois
4. Raithby GD (1976) Skew upstream differencing schemes for problems involving fluid flow. Comp Meth Appl Mech Eng 9:153–164

Chapter 11
Introducing General Grids Discretization

11.1 Introduction

The first 10 chapters of this book had the task of presenting the fundamentals of the finite volume method. The concepts discussed are general and valid for any finite volume method written for any coordinate system, respecting the peculiarities of each system. The Cartesian coordinate system was used for simplicity, but this system is very limited if the interest is to solve real engineering problems, where, almost always, the geometry is irregular. The growth of the finite volume methods can be divided in three major stages of discretization: orthogonal coordinate systems, with the Cartesian coordinates being the important representative, general curvilinear coordinate systems and unstructured discretization. Figure 11.1 shows these three levels using a 2D domain with a hole to demonstrate the difficulties of the Cartesian discretization. In Fig. 11.1a, it is seen that this discretization is not suitable for the inner boundary, with difficulties in applying boundary conditions. This type of discretization survived up to the 80 s, besides having works on unstructured grids by this time. Figure 11.1b and c follows the boundary and, therefore, are possible discretization for this geometry.

The discretization shown in Fig. 11.1b, which follows a generalized coordinate system, avoids the stair-shape control volumes at the boundary and was in the beginning of the 80 s the great revolution on finite volume methods for arbitrary geometries. In this option, a global curvilinear coordinate system exists, and it is possible to write the differential equations and integrate them into this new curvilinear system.

The discretization shown in Fig. 11.1c, which is unstructured, is preferable, since it adapts to very complex geometries, it is easy to generate and it is, nowadays, the standard practice in computational fluid dynamics.

This chapter intends to make a bridge between the Cartesian grids and the curvilinear and unstructured grids, presenting the concept of element and the possibilities of creating the control volumes for the finite volume integration based on the elements. Curvilinear and unstructured discretization are the topics of Chaps. 12 and 13.

© The Author(s), under exclusive license to Springer Nature Switzerland AG 2023
C. R. Maliska, *Fundamentals of Computational Fluid Dynamics*, Fluid Mechanics and Its Applications 135, https://doi.org/10.1007/978-3-031-18235-8_11

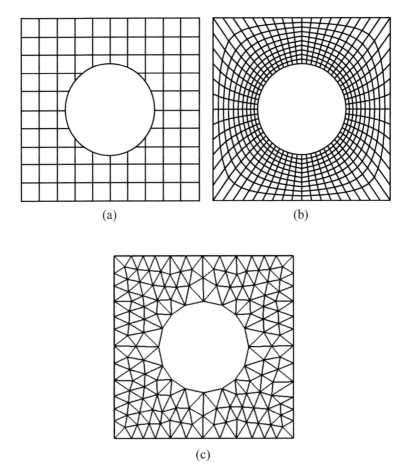

Fig. 11.1 Cartesian, curvilinear and unstructured discretization

11.2 Structured and Non-structured Grids

When control volumes are obtained with a discretization that follows a global coordinate system, as in Figs. 11.1b and 11.2a, the discretization or the resulting mesh is structured, since each internal volume always has the same number of neighbors. This definition of structured mesh, therefore, considers only geometric aspects of the mesh. It says that, if the volumes of the mesh have a certain construction law, always with the same number of neighbors, the mesh is said to be structured. In fact, what is most important for the numerical method is the structure of the matrix of coefficient resulting from the integration process. A structured mesh with its volumes ordered according to a global coordinate system, as shown in Fig. 11.2a, and employing an

11.2 Structured and Non-structured Grids

interpolation scheme that uses only the neighboring volumes, will give rise to a diagonal type coefficient matrix with 3, 5, and 7 diagonals for 1D, 2D, and 3D problems, respectively.

A structured mesh can also originate matrices that do not have defined diagonals by just numbering the volumes randomly, or by using different interpolation schemes for different volumes. But it is logical that this shouldn't be done, because whenever there is available a geometrically structured mesh, the numbering of the volumes should be done in a way as to get the most out of the structure. Structured meshes have, therefore, the advantages of allowing easy ordering and, consequently, obtaining diagonal matrices that allow to use simpler and more efficient solvers.

Unfortunately, real problems, due to the complexity of the geometry, do not allow structured meshes to always be employed. Unstructured meshes then come into play. They are more versatile, easier to adapt, and extremely well suited to discretize irregular geometries with corners and bumps. In many problems, only unstructured meshes can adequately discretize the domain.

They present, however, the difficulty of ordering, which, in turn, will give rise to non-diagonal matrices. Figure 11.2b shows a simple unstructured mesh, but one can see that it is difficult to choose a numbering path. The result of this ordering establishes the size of the bands in the matrix. For example, for the chosen ordering, volume 3 is connected to volumes 2, 4, 9 and 10, with a band size therefore of 2 to 10 in that row, while volume 9 is connected to volumes 3, 5, 8, 14, and 15, with a band size of 3 to 15. This variation in the band size of the matrix makes it difficult to apply many methods for solving linear systems. In addition to the varying bandwidth, the number of neighbors varies from volume to volume, as was the case with volume 3, which has four neighbors, and volume 9, which has five.

There are, therefore, advantages and disadvantages to each of the discretization, and the choice depends on the nature of the problem. In the coming chapters numerical models applicable to both types of grids, curvilinear and structured will be presented,

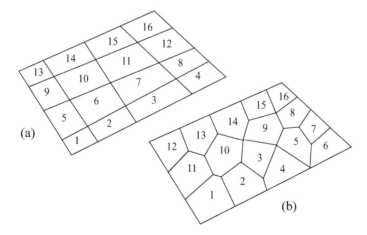

Fig. 11.2 Discretization: structured (**a**) unstructured (**b**)

and the characteristics of each discretization will be highlighted in due course. In Chap. 12 it will discussed generalized curvilinear structured meshes, presenting the mathematical support required, as base vectors, vector representation in the curvilinear system, concepts of physical and transformed planes, the transformation of the equations from the physical to the transformed domain and, finally, the integration of these equations into the transformed domain. In Chap. 13, methods for unstructured meshes will be presented. The element-based finite volume method (EbFVM) will be presented in detail, as well as its particularization for Voronoi diagrams, in which case the elements form a Delaunay triangulation and exhibit local orthogonality. For the finite volume method, it does not matter how and in what way the control volumes were created. The basic characteristic of this class of methods is the integration of equations in the conservative form over an elemental control volume. To introduce the reader to the geometric aspects of these methods is the purpose of this Chapter.

11.3 The Concept of Element

The element is not traditionally used in the finite volume method, because in this method it is enough to define the control volumes for integration purposes. However, initially defining the elements, as done in finite elements, and then relating them to the control volumes, allows a number of generalizations, including the creation of a more general computational abstraction, based on the elements, which results in an algorithm that can be used for any type of mesh, structured or not.

The elements are always the product of the mesh generator. In fact, the mesh generator furnishes the elements, defined by the coordinates of its nodes and, later, in the proper grid generator or in the numerical code, the control volumes are created. It depends on the dedication of the grid generator, it can supply the elements, nodes and connectivity. Consider Fig. 11.3, in which a curvilinear structured mesh is shown with the definition of the element 1234. In this mesh, a local coordinate system can be defined for the element, but it is common to define a global coordinate system, (ξ, η, γ), due to its structured nature, in which the conservation equation can be transformed and solved in this new system. In this case, the new global coordinate system is related to the original (x, y, z) coordinate system through a coordinate transformation, as will be seen in Chap. 12.

For an unstructured mesh of triangles and quadrilaterals, as shown in Fig. 11.4, it can be identified the elements 1234 (quadrangular) and 235 (triangular). Note that the definition of the elements precedes the creation of the control volumes, which is done based on the elements. The element is a geometric entity that covers the whole computational domain without overlapping and without broken parts at the boundaries. In the element, it is defined a local coordinate system, since a new global coordinate system can't be defined. The local coordinate system, (ξ, η, γ) is related to the (x, y, z) through a coordinate transformation.

The element, therefore, is an entity defined by the coordinate of its vertices. The grid generator can offer elements of different shapes. The most common employed

11.3 The Concept of Element

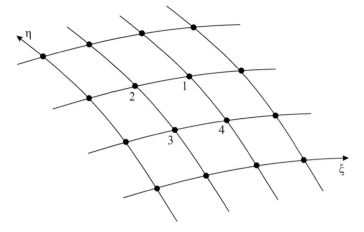

Fig. 11.3 Curvilinear grid and the element

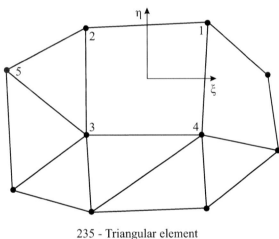

Fig. 11.4 Unstructured grid with quadrangular and triangular elements

235 - Triangular element
1234 - Quadrangular element

are triangle and quadrangular in 2D, and tetrahedron, pyramid, hexahedron and prims, in 3D, as shown in Fig. 13.1. Figure 11.5 shows the types of elements which comes from the grid generator and how they can be used in the numerical discretization. For a 1D element we may have a linear interpolation among the two nodes, called a first order element. If a third node is introduced, the interpolation function will be quadratic, and the element is called a 2nd order element. The same applies for a triangular in 2D and a hexahedron for 3D.

Each new node that is introduced, it is an additional unknown to the linear system and a corresponding increase in CPU time, therefore, this is a question which should be judiciously decided. Higher order elements also increase the complexity in the

Fig. 11.5 1D, 2D and 3D elements of 1st and 2nd order

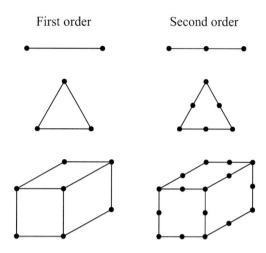

implementation and the size of the matrix band. Wouldn't it be wise to refine the grid in locals of higher gradients and use first order elements, rather than to increase the order of the element for the whole domain? In finite volume calculations this is the approach commonly used.

Hence, considering a 3D hexahedral element coming from the grid generator with its eight vertices identified by the (x, y, z) coordinates, the developer of the numerical method will decide about the shape functions, or interpolation functions, for this element, which means to decide about the order of the element.

11.4 Construction of the Control Volume

The creation of the control volumes, as already pointed out, is done based on the elements. There are two basic classes of methods based on the relative geometrical position between the control volume and the element. The formulations in which the control volume is chosen as the element itself, and the variables to be determined are stored in the center of the element, are called cell-center, because the unknown variable is the center of the element. It is the classical, or traditional, finite volume method and, in Cartesian coordinates the integration points (ip) are symbolized by (e, w, n, s, f, b). For a curvilinear coordinate system, as in Fig. 11.6, if a cell-center approach is used, nothing changes in terms of notation, since one still has a control volume with four faces (in 2D), and integration points are on the middle of the faces.

What changes, related to Cartesian system, with strong influence on the implementation of the method is the need of having all Cartesian components of the velocity vector at these points to obtain the normal velocities for mass flow calculations.

If an unstructured mesh of triangles is employed and the elements are used as control volumes, as shown in Fig. 11.7, one has, again, the classical formulation of

11.4 Construction of the Control Volume

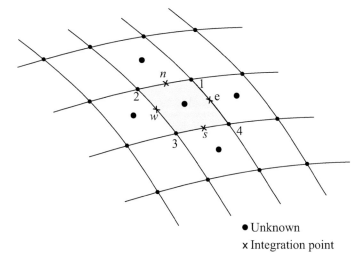

Fig. 11.6 Curvilinear grid. Cell-center

the finite volume method. For the calculation of the diffusion fluxes, for example, in both, Figs. 11.6 and 11.7 it will need to use the neighbor control volumes. In other words, fluxes in general will not be possible to be calculated on an element-based approach, precluding the use of a cleaner programming.

The other class of methods, called cell vertex, with the center of the control volumes at the vertex of the elements (location of the unknown variable), as shown in Fig. 11.8 for curvilinear grids, constructs the control volume with parts of the elements which shares the same grid node. The control volume is constructed by connecting the centroids of the elements to the midpoint of their faces. Two alternatives are possible. In the first case, shown in Fig. 11.8a, the integration point is in the middle of the face. For an element-base construction it is not convenient, since

Fig. 11.7 Unstructured grids. Cell-center

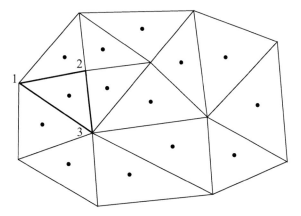

the integration point at the south face, for example, has no clear definition to what element it belongs to, if to element 1234 or element 1456. The clear definition is mandatory because in an element-base method all fluxes must be calculated inside a single element.

In the second case, shown in Fig. 11.8b, there are two integration points on each face, a construction that leaves no ambiguities, since every two integration points of a control volume are located within the same element. In addition, the accuracy of the numerical scheme is increased, as the flow across the face is now calculated using a better discretization of the integration surface. This construction allows the use of the element as the geometric entity on which all calculations are performed. In this way, the computational implementation is facilitated, resulting in algorithms that are applicable to both structured and unstructured meshes.

For unstructured meshes, the construction of the control volumes is identical, that is, the centroids of each element are connected to the midpoint of each face. The face of the control volume between the centroid and the midpoint of the element face houses an integration point. Figure 11.9 shows an unstructured mesh with triangular and quadrangular elements, where control volumes centered at nodes 3 and 4 are shown. Note that a triangular element will always have three integration points, and two of them are always used together for the fluxes balance for a given control volume.

A quadrangular element will always have four integration points, and, again, two of them are used for fluxes balance of given control volume. For example, in the balance for the control volume centered at node 3, two integration points from the triangular element 253 and two from the quadrangular element 1234 take part, in addition to the other integration points belonging to the other elements that also contribute to the volume centered at 3. These integration points are on the faces of the so-called sub-control (SCV), or, sub-elements, i.e. the parts of the element that will form the control volume. Figure 11.8b identifies the SVC1 of element 1234 that

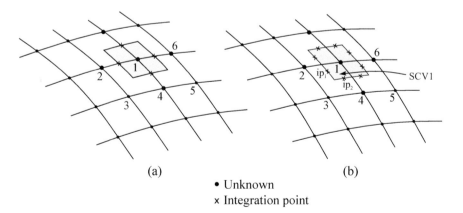

Fig. 11.8 Curvilinear grids. Cell-vertex

Fig. 11.9 Unstructured grid. Cell-vertex (EbFVM)

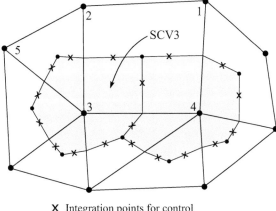

X Integration points for control volumes centered in 3 and 4

takes part in the balance of the control volume centered on 1, and Fig. 11.9 shows the SCV3 of element 1234 that takes part in the control volume centered on 3.

This construction allows, therefore, the balance over the control volume to be done through a summation of flows calculated at the integration points of the elements. All calculations, therefore, can be done for the elements, which gives rise to the name of the element-based finite volume method (EbFVM), that is, a finite volume methodology where the basis is the element. This method will be discussed in detail in Chap. 13.

Nowadays, the cell-center unstructured grid, as shown in Fig. 11.7, and the element-based approach, as in Fig. 11.9, are the two numerical technology mainly used in commercial software for the simulation of fluid flows.

11.5 Conclusions

The main goal of this chapter was to introduce the reader to what comes next. The elements in curvilinear coordinate systems and in unstructured grids with the corresponding creation of control volume were shown. The alternatives of using the element to build a cell-center method or a cell-vertex method was presented. The location of the integration points allowing to create element-based methods was outlined. By its turn, the cell-vertex method, whose control volumes are built with parts of the elements, allows the determination of the fluxes such that the construction of the global matrix can be in an element-basis.

Chapter 12
Coordinate Transformation—General Curvilinear Coordinate Systems

12.1 Introduction

Having completed the basic presentation of the finite volume method and its fundamental details using Cartesian coordinates, and the introduction of curvilinear and unstructured grids, our next chapters seek to extend the knowledge to these two types of non-Cartesian discretization. Curvilinear grids are types of meshes that can be represented in a new global coordinate system using a transformation of coordinates from the Cartesian system (x, y, z) to the new system (ξ, η, γ). This means that the differential equations to be solved can be transformed to this new coordinate system, whose dimensions are arbitrarily chosen. The algorithms could also be developed using local coordinate systems. This chapter is devoted for the solution on the global coordinate system.

In order to have mathematical support when working with the transformed equations, and the required basis for interpreting the terms and elements in the new system, let us start with a brief description of the relationships in a coordinate transformation. Since the finite-volume method is based on the development of physically based algorithms, and such basic derivations are usually made for the Cartesian system, as done in the previous chapters, only with the geometric interpretation of the transformation it is possible to transfer this reasoning to more complex shapes.

We have observed that the lack of these interpretations has cost to new users of general coordinate systems and unstructured grids, considerable learning time and has inhibited their ability to search for errors in computer codes. Furthermore, it becomes difficult to generalize and create new algorithms if the coordinate transformation and its influence on the calculation of physical quantities are not well understood. In this chapter, therefore, it is aimed to understand geometrical aspects of the irregular domains and its effects on the creation of a numerical algorithm. No mathematical deepening will be exercised, but just the essential for the geometrical interpretation. Readers interested should consult classical texts on tensor calculus [1, 2]. Those interested in expressions for the various operators in generalized curvilinear coordinates can consult [2, 3].

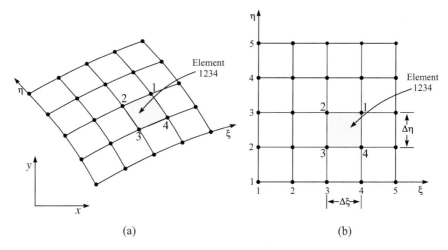

Fig. 12.1 Curvilinear coordinate system. Mapping

When it is possible to discretize the computational domain with a structured mesh, as in Fig. 12.1a, for example, the lines of the mesh form a new coordinate system (ξ, η), called curvilinear or generalized. For example, if one decides to map an irregular physical 2D domain (Fig. 12.1a into a rectangle (into a parallelepiped in 3D), as in Fig. 12.1b, it may be adopted unitary increments of the coordinates in the new system, i.e., $\Delta\xi = \Delta\eta = 1$, and the computational domain loses is physical dimensions, being necessary to know in which parameters the information about the real shape and size of the domain are embedded. The metrics of the transformation of coordinates tell us that.

12.2 Global Coordinate Transformation

12.2.1 General

In this section it is considered the situation where all elements are arranged in a structured way and are formed by the lines of the generalized curvilinear coordinate system as shown in Fig. 12.1a with the respective mapping, or transformed, or computational plane, as in Fig. 12.1b. Note that element 1234 is part of a global coordinate system and is defined by lines $\xi = 3$ and $\xi = 4$, and lines $\eta = 2$ and $\eta = 3$, arbitrarily chosen. The size of $\Delta\xi$ and $\Delta\eta$ could be of any value, but usually chosen as unity.

Figure 12.2 shows a curvilinear coordinate system (ξ, η, γ) referred to the Cartesian system (x, y, z) with the identification of point A on the η axis. The curvilinear coordinates of a point are related to the Cartesian system by a transformation equation

12.2 Global Coordinate Transformation

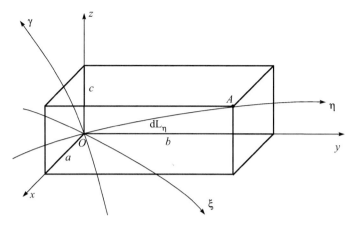

Fig. 12.2 Three-dimensional curvilinear coordinate system (ξ, η, γ)

of the type,

$$\xi = \xi(x, y, z) \tag{12.1}$$

$$\eta = \eta(x, y, z) \tag{12.2}$$

$$\gamma = \gamma(x, y, z) \tag{12.3}$$

There is a possibility that the new coordinate system may change with time, which would alter the functional form of Eqs. (12.1–12.3) to involve the time variable. This will be seen shortly in this chapter.

The differentials in each coordinate axis are,

$$d\xi = \xi_x dx + \xi_y dy + \xi_z dz \tag{12.4}$$

$$d\eta = \eta_x dx + \eta_y dy + \eta_z dz \tag{12.5}$$

$$d\gamma = \gamma_x dx + \gamma_y dy + \gamma_z dz, \tag{12.6}$$

or in the matrix form,

$$\begin{bmatrix} d\xi \\ d\eta \\ d\gamma \end{bmatrix} = \begin{bmatrix} \xi_x & \xi_y & \xi_z \\ \eta_x & \eta_y & \eta_z \\ \gamma_x & \gamma_y & \gamma_z \end{bmatrix} \begin{bmatrix} dx \\ dy \\ dz \end{bmatrix}, \tag{12.7}$$

or, again, as

$$[\mathbf{d}^T] = [\mathbf{A}][\mathbf{d}^F] \tag{12.8}$$

in which $\mathbf{d}^T$ and $\mathbf{d}^F$ are the differentials in the transformed and physical domains, respectively. By their turn, the differentials in the physical domain are given by

$$\begin{bmatrix} dx \\ dy \\ dz \end{bmatrix} = \begin{bmatrix} x_\xi & x_\eta & x_\gamma \\ y_\xi & y_\eta & y_\gamma \\ z_\xi & z_\eta & z_\gamma \end{bmatrix} \begin{bmatrix} d\xi \\ d\eta \\ d\gamma \end{bmatrix} \tag{12.9}$$

or

$$[\mathbf{d}^F] = [\mathbf{B}][\mathbf{d}^T] \tag{12.10}$$

Using Eqs. (12.8) and (12.10), one finds

$$\mathbf{A} = \mathbf{B}^{-1} = 1/J \begin{bmatrix} y_\eta z_\gamma - y_\gamma z_\eta & -(x_\eta z_\gamma - x_\gamma z_\eta) & x_\eta y_\gamma - x_\gamma y_\eta \\ -(y_\xi z_\gamma - y_\gamma z_\xi) & x_\xi z_\gamma - x_\gamma z_\xi & -(x_\xi y_\gamma - x_\gamma y_\xi) \\ y_\xi z_\eta - y_\eta z_\xi & -(x_\xi z_\eta - x_\eta z_\xi) & x_\xi y_\eta - x_\eta y_\xi \end{bmatrix} \tag{12.11}$$

Comparing $\mathbf{A}$ with $\mathbf{B}^{-1}$, element by element of the matrices, the metrics of the transformation are found, as

$$\xi_x = \frac{1}{J}(y_\eta z_\gamma - y_\gamma z_\eta)$$

$$\xi_y = -\frac{1}{J}(x_\eta z_\gamma - x_\gamma z_\eta)$$

$$\xi_z = \frac{1}{J}(x_\eta y_\gamma - x_\gamma y_\eta) \tag{12.12}$$

$$\eta_x = -\frac{1}{J}(y_\xi z_\gamma - y_\gamma z_\xi)$$

$$\eta_y = \frac{1}{J}(x_\xi z_\gamma - x_\gamma z_\xi)$$

$$\eta_z = -\frac{1}{J}(x_\xi y_\gamma - x_\gamma y_\xi) \tag{12.13}$$

$$\gamma_x = \frac{1}{J}(y_\xi z_\eta - y_\eta z_\xi)$$

$$\gamma_y = -\frac{1}{J}(x_\xi z_\eta - x_\eta z_\xi)$$

$$\gamma_z = \frac{1}{J}(x_\xi y_\eta - x_\eta y_\xi) \tag{12.14}$$

in which

12.2 Global Coordinate Transformation

$$J = \det[\mathbf{B}] = \frac{1}{\det[\mathbf{A}]} \tag{12.15}$$

or

$$J = x_\xi(y_\eta z_\gamma - y_\gamma z_\eta) - x_\eta(y_\xi z_\gamma - y_\gamma z_\xi) + x_\gamma(y_\xi z_\eta - y_\eta z_\xi), \tag{12.16}$$

is the Jacobian of the transformation, which has an important geometric interpretation.

Equations (12.1–12.3) represent the transformation from the system (x, y, z) to the system (ξ, η, γ). The inverse function theorem, which allowed obtaining the relations given by Eq. (12.12–12.14) admits the existence of the inverse of the transformation given by

$$x = x(\xi, \eta, \gamma)$$
$$y = y(\xi, \eta, \gamma)$$
$$z = z(\xi, \eta, \gamma), \tag{12.17}$$

in which the metrics of the inverse function are,

$$x_\xi = J(\eta_y \gamma_z - \gamma_y \eta_z)$$
$$x_\eta = -J(\xi_y \gamma_z - \xi_z \gamma_y)$$
$$x_\gamma = J(\eta_z \xi_y - \xi_z \eta_y) \tag{12.18}$$

$$y_\xi = -J(\eta_x \gamma_z - \eta_z \gamma_x)$$
$$y_\eta = J(\xi_x \gamma_z - \gamma_x \xi_z)$$
$$y_\gamma = -J(\xi_x \eta_z - \xi_z \eta_x) \tag{12.19}$$

$$z_\xi = J(\eta_x \gamma_y - \eta_y \gamma_x)$$
$$z_\eta = -J(\xi_x \gamma_y - \gamma_x \xi_y)$$
$$z_\gamma = J(\xi_x \eta_y - \xi_y \eta_x) \tag{12.20}$$

To exemplify, consider a transformation from the Cartesian to the polar coordinate system, given by

$$r = \sqrt{x^2 + y^2}$$
$$\theta = \tan^{-1}\left(\frac{y}{x}\right) \tag{12.21}$$

in which (r, θ, z) represents (ξ, η, γ). It is easy to show that the inverse are the well-known relations,

$$x = r\cos\theta$$
$$y = r\,\text{sen}\,\theta \tag{12.22}$$

12.2.2 Length Along a Coordinate Axis

The relations among the components of the metric tensor and lengths in a curvilinear system are helpful interpretations when performing balances inherent to the finite volume method. Considering again Fig. 12.2, let's call the length of the segment along the η axis measured from the origin to the point A by dL_η. The coordinates (a, b, c) of the point A are given by

$$a = \frac{\partial x}{\partial \eta} \Delta\eta \tag{12.23}$$

$$b = \frac{\partial y}{\partial \eta} \Delta\eta \tag{12.24}$$

$$c = \frac{\partial z}{\partial \eta} \Delta\eta \tag{12.25}$$

since, along $\overline{OA}$, $\Delta\xi$ and $\Delta\gamma$ are equal to zero. Using the Pythagorean theorem, one finds

$$dL_\eta = \sqrt{\left(\frac{\partial x}{\partial \eta}\right)^2 + \left(\frac{\partial y}{\partial \eta}\right)^2 + \left(\frac{\partial z}{\partial \eta}\right)^2}\, \Delta\eta \tag{12.26}$$

In a similar way, the lengths along the ξ and γ coordinates are,

$$dL_\xi = \sqrt{\left(\frac{\partial x}{\partial \xi}\right)^2 + \left(\frac{\partial y}{\partial \xi}\right)^2 + \left(\frac{\partial z}{\partial \xi}\right)^2}\, \Delta\xi \tag{12.27}$$

$$dL_\gamma = \sqrt{\left(\frac{\partial x}{\partial \gamma}\right)^2 + \left(\frac{\partial y}{\partial \gamma}\right)^2 + \left(\frac{\partial z}{\partial \gamma}\right)^2}\, \Delta\gamma \tag{12.28}$$

According to the definition of the metric tensor given by

$$g_{ij} = \frac{\partial x}{\partial x^i}\frac{\partial x}{\partial x^j} + \frac{\partial y}{\partial x^i}\frac{\partial y}{\partial x^j} + \frac{\partial z}{\partial x^i}\frac{\partial z}{\partial x^j}, \tag{12.29}$$

one can see that the lengths dL_ξ, dL_η and dL_γ are, respectively,

12.2 Global Coordinate Transformation

$$dL_\xi = \sqrt{g_{11}}\Delta\xi \tag{12.30}$$

$$dL_\eta = \sqrt{g_{22}}\Delta\eta \tag{12.31}$$

$$dL_\gamma = \sqrt{g_{33}}\Delta\gamma, \tag{12.32}$$

that is, a length along one of the coordinate axes is related to only one component of the metric tensor. A generic length ds is given by

$$ds^2 = dx^2 + dy^2 + dz^2 \tag{12.33}$$

Recalling the expressions for the differentials dx, dy and dz given by Eq. (12.9) and the definition of the metric tensor, $\mathbf{g}_{ij}$, one finds

$$ds^2 = dx^2 + dy^2 + dz^2 = \sum_{i=1}^{3}\sum_{k=1}^{3} g_{ik} dx^i dx^k \tag{12.34}$$

Since $\mathbf{g}_{ij}$ has nine components, a generic length will involve all components. In a matrix form, it reads

$$\mathbf{g}_{ik} = \begin{bmatrix} g_{11} & g_{12} & g_{13} \\ g_{21} & g_{22} & g_{23} \\ g_{31} & g_{32} & g_{33} \end{bmatrix} \tag{12.35}$$

In Eq. (12.34), x^i and x^k represent the curvilinear coordinates or, for $i = 1$, $x^1 = \xi$, for $i = 2$, $x^2 = \eta$, and for $i = 3$, $x^3 = \gamma$.

For orthogonal coordinate systems all cross components ($i \neq k$) of the metric tensor are zero, and, for example, for the cylindrical coordinate system the diagonal components are

$$\begin{aligned} g_{11} &= 1 \\ g_{22} &= r^2 \\ g_{33} &= 1 \end{aligned} \tag{12.36}$$

12.2.3 Areas (or Volumes) in the Curvilinear System

Considering a two-dimensional situation, for simplicity, as shown in Fig. 12.3, and using the expressions obtained for the 3D case, one can write,

Fig. 12.3 Area in the computational domain

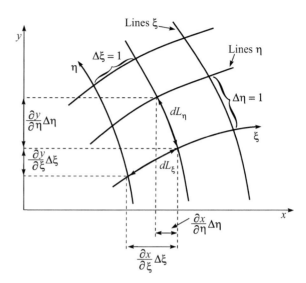

$$dL_\xi = \sqrt{(y_\xi^2 + x_\xi^2)}\Delta\xi = \sqrt{g_{11}}\Delta\xi \qquad (12.37)$$

$$dL_\eta = \sqrt{(y_\eta^2 + x_\eta^2)}\Delta\eta = \sqrt{g_{22}}\Delta\eta \qquad (12.38)$$

According to Fig. 12.3, these lengths can be represented by vectors, as

$$d\mathbf{L}_\xi = x_\xi \Delta\xi \mathbf{i} + y_\xi \Delta\xi \mathbf{j}$$
$$d\mathbf{L}_\eta = x_\eta \Delta\eta \mathbf{i} + y_\eta \Delta\eta \mathbf{j} \qquad (12.39)$$

The area of the parallelogram formed by these two vectors is given by the absolute value of the resulting vector from the cross product of them. The cross product is

$$\mathbf{dS} = \begin{bmatrix} \mathbf{i} & \mathbf{j} & \mathbf{k} \\ x_\xi \Delta\xi & y_\xi \Delta\xi & 0 \\ x_\eta \Delta\eta & y_\eta \Delta\eta & 0 \end{bmatrix} = \mathbf{k}(x_\xi y_\eta - x_\eta y_\xi)\Delta\xi \Delta\eta, \qquad (12.40)$$

and the area dS is given by

$$dS = |\mathbf{dS}| = (x_\xi y_\eta - x_\eta y_\xi)\Delta\xi\Delta\eta = J\Delta\xi\Delta\eta \qquad (12.41)$$

An important geometric interpretation arises from this equation recognizing that the term inside the parentheses in Eq. (12.41) is the Jacobian of the transformation, given by

$$J = (x_\xi y_\eta - x_\eta y_x), \qquad (12.42)$$

12.2 Global Coordinate Transformation

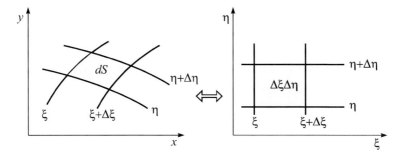

Fig. 12.4 Physical and computational control volume

Therefore,

$$\frac{dS}{\Delta\xi\,\Delta\eta} = J, \qquad (12.43)$$

that is, the Jacobian represents the ratio among the physical and the transformed area. If $\Delta\xi = \Delta\eta = 1$ the value of the Jacobian is exactly the area in the physical domain, which can be see that it is the determinant of matrix **B**, Eq. (12.9), for two dimensions.

Numerically approximating Eq. (12.42) and substituting in Eq. (12.43), it is observed that the product $\Delta\xi\,\Delta\eta$ cancel out, what means that they are of free choice. For a 3D transformation one gets,

$$\frac{dV}{\Delta\xi\,\Delta\eta\,\Delta\gamma} = J \qquad (12.44)$$

It is left to the reader to demonstrate that

$$J = \sqrt{g}, \qquad (12.45)$$

in which $g = \det[g_{ik}]$. Figure 12.4 depicts the physical and the transformed domain.

12.2.4 Basis Vectors

When the Cartesian coordinate system is used, a vector varying with space is described in terms of the Cartesian components referenced to a local basis of vectors **i**, **j** and **k**. The magnitude and direction of each basis vector are the same for any point in space. The vectors **i**, **j** and **k** are unit vectors and that the magnitude of a component of a given vector represents a proportionality to the basis vector on that axis.

Fig. 12.5 Contravariant, co-variant, Cartesian and unitary basis vectors

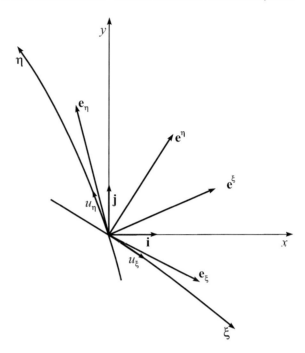

When generalized curvilinear coordinates are employed and we wish to describe the same vector in these coordinates, it is convenient to employ local basis vectors that are aligned with or normal to the coordinate lines. Since the coordinates are nonorthogonal, and to avoid ambiguities, there are two systems of basis vectors that can be used: the covariant and the contravariant. In addition to these, a unitary system is often employed along the covariant basis. Figure 12.5 shows these basis vectors together with the rectangular Cartesian basis vector system.

12.2.4.1 Covariant Basis Vector

Covariant basis vectors, by definition, are tangent to the coordinate lines, as shown in Fig. 12.6 for the case of the covariant vector tangent to the axis ξ. In this case it is given by

$$\mathbf{e}_\xi = \lim_{\Delta\xi \to 0} \frac{\mathbf{r}(\xi + \Delta\xi) - \mathbf{r}(\xi)}{\Delta\xi} = \frac{\partial \mathbf{r}}{\partial \xi} \tag{12.46}$$

Knowing that the vector $\mathbf{r}$ is given by

$$\mathbf{r} = x\mathbf{i} + y\mathbf{j} + z\mathbf{k}, \tag{12.47}$$

12.2 Global Coordinate Transformation

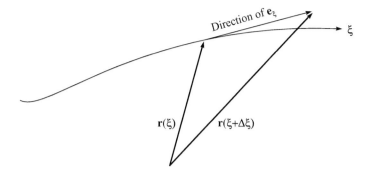

Fig. 12.6 Covariant basis vectors definition

the covariant basis vector, e_ξ results,

$$\mathbf{e}_\xi = \frac{\partial x}{\partial \xi}\mathbf{i} + \frac{\partial y}{\partial \xi}\mathbf{j} + \frac{\partial z}{\partial \xi}\mathbf{k} \quad (12.48)$$

With an analogous procedure for the other coordinate directions, one finds

$$\mathbf{e}_\eta = \frac{\partial x}{\partial \eta}\mathbf{i} + \frac{\partial y}{\partial \eta}\mathbf{j} + \frac{\partial z}{\partial \eta}\mathbf{k} \quad (12.49)$$

$$\mathbf{e}_\gamma = \frac{\partial x}{\partial \gamma}\mathbf{i} + \frac{\partial y}{\partial \gamma}\mathbf{j} + \frac{\partial z}{\partial \gamma}\mathbf{k}, \quad (12.50)$$

or, in matrix form, as

$$\begin{bmatrix} \mathbf{e}_\xi \\ \mathbf{e}_\eta \\ \mathbf{e}_\gamma \end{bmatrix} = \begin{bmatrix} x_\xi & y_\xi & z_\xi \\ x_\eta & y_\eta & z_\eta \\ x_\gamma & y_\gamma & z_\gamma \end{bmatrix} \begin{bmatrix} \mathbf{i} \\ \mathbf{j} \\ \mathbf{k} \end{bmatrix} \quad (12.51)$$

or, using the coordinate x^i to denote the ξ, η and γ coordinates, as

$$\mathbf{e}_i = \frac{\partial x}{\partial x^i}\mathbf{i} + \frac{\partial y}{\partial x^i}\mathbf{j} + \frac{\partial z}{\partial x^i}\mathbf{k} \quad (12.52)$$

The recognition that covariant basis vector is tangent to generalized coordinate lines greatly assists the numerical analyst, since, when analyzing results of physical problems, it is common to calculate entities tangent to coordinate lines. It is therefore enough to have the basis vectors of the transformation, given by Eqs. (12.48–12.50) and determine the vector in the desired tangential direction.

To exercise, consider the polar coordinate system, Eq. (12.22). The interest is in calculating the covariant basis vector in the θ direction. It is a vector tangent to the θ coordinate. Using Eqs. (12.48) and (12.49), the basis vector for these directions are

Fig. 12.7 Covariant basis vectors for the polar system

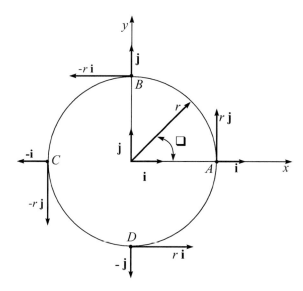

$$\mathbf{e}_\theta = r(-\text{sen}\theta\, \mathbf{i} + \cos\theta\, \mathbf{j}), \tag{12.53}$$

while, for the r direction, it is

$$\mathbf{e}_r = \cos\theta\, \mathbf{i} + \text{sen}\theta\, \mathbf{j} \tag{12.54}$$

Figure 12.7 shows for the point A, B, C and D, the basis vectors for the polar system obtained using Eqs. (12.53) and (12.54). Notice that $\mathbf{e}_\theta$ has module r, and $\mathbf{e}_r$ has unitary module.

12.2.4.2 Contravariant Basis Vector

The contravariant basis has its vectors normal to the coordinate surfaces, as illustrated in Fig. 12.8. Therefore, we must seek, for its definition, the mathematical entity that represents this condition. It is known that the gradient has this property. Therefore, the contravariant basis vectors are defined by

$$\mathbf{e}^\xi = \nabla\xi = \frac{\partial \xi}{\partial x}\mathbf{i} + \frac{\partial \xi}{\partial y}\mathbf{j} + \frac{\partial \xi}{\partial z}\mathbf{k} \tag{12.55}$$

$$\mathbf{e}^\eta = \nabla\eta = \frac{\partial \eta}{\partial x}\mathbf{i} + \frac{\partial \eta}{\partial y}\mathbf{j} + \frac{\partial \eta}{\partial z}\mathbf{k} \tag{12.56}$$

$$\mathbf{e}^\gamma = \nabla\gamma = \frac{\partial \gamma}{\partial x}\mathbf{i} + \frac{\partial \gamma}{\partial y}\mathbf{j} + \frac{\partial \gamma}{\partial z}\mathbf{k} \tag{12.57}$$

12.2 Global Coordinate Transformation

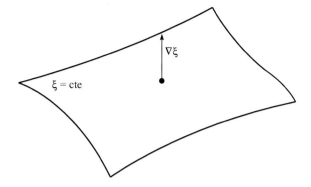

Fig. 12.8 Contravariant basis vector definition

or, in a matrix form, as before,

$$\begin{bmatrix} \mathbf{e}^\xi \\ \mathbf{e}^\eta \\ \mathbf{e}^\gamma \end{bmatrix} = \begin{bmatrix} \xi_x & \xi_y & \xi_z \\ \eta_x & \eta_y & \eta_z \\ \gamma_x & \gamma_y & \gamma_z \end{bmatrix} \begin{bmatrix} \mathbf{i} \\ \mathbf{j} \\ \mathbf{k} \end{bmatrix} \qquad (12.58)$$

It is left to the reader, as an exercise, to obtain the contravariant basis vectors for the polar coordinate system, as was done for the covariant basis.

The covariant and contravariant bases vectors have been written in terms of the unit vectors $\mathbf{i}$, $\mathbf{j}$ and $\mathbf{k}$. These can, of course, be expressed in the covariant and contravariant bases. Following, these and other important relationships are collected [2],

$$\mathbf{e}_i = \frac{\partial x}{\partial x^i}\mathbf{i} + \frac{\partial y}{\partial x^i}\mathbf{j} + \frac{\partial z}{\partial x^i}\mathbf{k} \qquad (12.59)$$

$$\mathbf{e}^i = \frac{\partial x^i}{\partial x}\mathbf{i} + \frac{\partial x^i}{\partial y}\mathbf{j} + \frac{\partial x^i}{\partial z}\mathbf{k} \qquad (12.60)$$

$$\mathbf{i} = \sum_{i=1}^{3} \frac{\partial x^i}{\partial x}\mathbf{e}_i = \sum_{i=1}^{3} \frac{\partial x}{\partial x^i}\mathbf{e}^i \qquad (12.61)$$

$$\mathbf{j} = \sum_{i=1}^{3} \frac{\partial x^i}{\partial y}\mathbf{e}_i = \sum_{i=1}^{3} \frac{\partial y}{\partial x^i}\mathbf{e}^i \qquad (12.62)$$

$$\mathbf{k} = \sum_{i=1}^{3} \frac{\partial x^i}{\partial z}\mathbf{e}_i = \sum_{i=1}^{3} \frac{\partial z}{\partial x^i}\mathbf{e}^i \qquad (12.63)$$

$$\mathbf{u}_i = \frac{1}{\sqrt{g_{ii}}}\mathbf{e}_i \qquad (12.64)$$

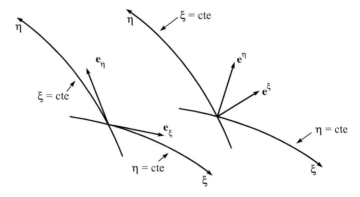

Fig. 12.9 Covariant and contravariant basis vectors

$$g_{ij} = \mathbf{e}_i \cdot \mathbf{e}_j \tag{12.65}$$

$$g^{ij} = \mathbf{e}^i \cdot \mathbf{e}^j \tag{12.66}$$

In Eq. (12.64), the vector $\mathbf{u}_i$ is unitary along the covariant basis and is obtained by dividing the vector $\mathbf{e}_i$ by its modulus. Again, taking the polar system as an example, the vector $\mathbf{e}_r$ is already unitary and $\mathbf{e}_\theta$ has modulus r, as already seen. Therefore, the unitary vector $\mathbf{u}_\theta$ in the polar system is

$$\mathbf{u}_\theta = -\mathrm{sen}\theta\, \mathbf{i} + \cos\theta\, \mathbf{j} \tag{12.67}$$

From Eq. (12.64) it is concluded that

$$|\mathbf{e}_i| = \sqrt{g_{ii}}, \tag{12.68}$$

and, therefore, it should be kept in mind that when a vector is described by its covariant or contravariant components, the basis vector to this representation is not unitary. Comparing Eqs. (12.30–12.32) with Eq. (12.68), one at a time, this fact can be proved. Note that in the previous equations it is always possible to recover the three components by making $x^1 = \xi$, $x^2 = \eta$ and $x^3 = \gamma$.

To conclude this section, Fig. 12.9 shows, in the plane, a nonorthogonal curvilinear coordinate system and the corresponding covariant and contravariant base vectors.

12.2.5 Vector Representation in the Curvilinear System

There are two possible bases to represent a vector in the curvilinear ordinate system, the covariant and the contravariant. The third, auxiliary basis, is unitary and derived from the covariant.

12.2 Global Coordinate Transformation

When representing vectors in coordinate systems whose basis vectors are not unitary, precaution should be taken in interpreting the meaning of a component of a vector. A component of a vector is a scalar that is multiplied by the basis vector to obtain a length. The important point is that the length is an invariant, that is, it has the same value for any coordinate system. For generalized coordinate systems, with base vectors of non-unitary modulus, this is crucial, because the scalar represented by the component is not physically interpretable if it is not associated with a base vector. The product of the value of the component by the modulus of the base vector results in a length which is, therefore, invariant and physically representable.

Let $\mathbf{V}$ be a vector variable in space, that is $\mathbf{V} = \mathbf{V}(x, y, z)$. This vector can be represented by

$$\mathbf{V} = \hat{V}_1 \mathbf{u}_1 + \hat{V}_2 \mathbf{u}_2 + \hat{V}_3 \mathbf{u}_3 \tag{12.69}$$

$$\mathbf{V} = V^1 \mathbf{e}_1 + V^2 \mathbf{e}_2 + V^3 \mathbf{e}_3 \tag{12.70}$$

$$\mathbf{V} = V_1 \mathbf{e}^1 + V_2 \mathbf{e}^2 + V_3 \mathbf{e}^3 \tag{12.71}$$

which give rise to the physical, contravariant, and covariant components, as shown in Fig. 12.10. There is no possibility to draw on a graph with scales the magnitudes of the contravariant and covariant components of a vector, since those magnitudes have no meaning. However, if they are multiplied by their corresponding bases vectors the results are invariants, with physical meaning. These can be drawn and measured, since they represent lengths, as shown in Fig. 12.10. In fact, what is being represented, then, are physical components, invariants.

Fig. 12.10 Contravariant, covariant and physical components

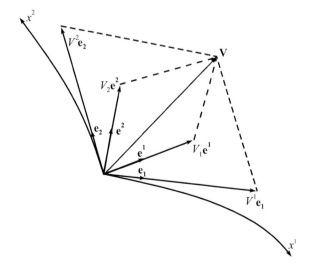

Notice that the representation of the vector in the covariant basis gives the contravariant components of the vector. This becomes clear when it is realized that the contravariant components are obtained by the scalar product of the vector in question by the corresponding contravariant basis vector. It is this operation that gives its name. Furthermore, it is important to note that the products $V^1 e_1$ and $V_1 e^1$ have physical and geometrical sense because they originate lengths, while the products $V^1 e^1$ and $V_1 e_1$ have no interpretation.

The expressions for each of the components represented in a possible basis vectors can be obtained with the help of Eqs. (12.59) and (12.60). Thus,

$$\hat{V}_i = \sqrt{g_{ii}} V^i \tag{12.72}$$

$$V^i = \frac{\partial x^i}{\partial x} V_x + \frac{\partial x^i}{\partial y} V_y + \frac{\partial x^i}{\partial z} V_z \tag{12.73}$$

$$V_i = \frac{\partial x}{\partial x^i} V_x + \frac{\partial y}{\partial x^i} V_y + \frac{\partial z}{\partial x^i} V_z, \tag{12.74}$$

and, for the Cartesian components,

$$V_x = \sum_{i=1}^{3} \frac{\partial x}{\partial x^i} \frac{\hat{V}_i}{\sqrt{g_{ii}}} = \sum_{i=1}^{3} \frac{\partial x}{\partial x^i} V^i = \sum_{i=1}^{3} \frac{\partial x^i}{\partial x} V_i \tag{12.75}$$

$$V_y = \sum_{i=1}^{3} \frac{\partial y}{\partial x^i} \frac{\hat{V}_i}{\sqrt{g_{ii}}} = \sum_{i=1}^{3} \frac{\partial y}{\partial x^i} V^i = \sum_{i=1}^{3} \frac{\partial x^i}{\partial y} V_i \tag{12.76}$$

$$V_z = \sum_{i=1}^{3} \frac{\partial z}{\partial x^i} \frac{\hat{V}_i}{\sqrt{g_{ii}}} = \sum_{i=1}^{3} \frac{\partial z}{\partial x^i} V^i = \sum_{i=1}^{3} \frac{\partial x^i}{\partial z} V_i \tag{12.77}$$

As our goal is the simulation of fluid flows, at this point it is didactic to emphasize the importance of choosing which components of the velocity vector will take part in the numerical model. In the finite volume method, the balances of properties (mass, momentum, energy, etc.) in the elementary volume give rise to the approximate equations. These balances involve the quantification of the advective flows of these properties on the faces of the elementary volumes. The evaluation of these fluxes, regardless of the property in question, depends on the mass flow, since any property is transported by advection by the mass flow. Therefore, the evaluation of the mass flow at the interfaces of the elementary volumes is one of the most important operations performed in the algorithm, since its influence appears in all the equations to be solved. And this knowledge is not only applied to curvilinear grids, but in any control volume which is not aligned with the Cartesian system, case in which the mass flow is easily computed using just one component of the velocity vector.

Consider Fig. 12.11 in which the vector **V** is the velocity vector. In order to calculate the flux of any given property by unit mass ϕ, it is necessary to find **V.n**,

12.2 Global Coordinate Transformation

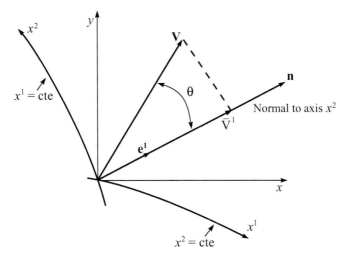

Fig. 12.11 Projection of **V** in the normal to the axis x^2

such that $\rho(\mathbf{V}\cdot\mathbf{n})\phi$ is the flux of ϕ per unit area and time. In this case it is considered **n** normal to the x^2 axis, that is, normal to constant x^1 lines. One wants to determine the relationship between the physical component of the velocity vector normal to the x^2 axis and the contravariant and covariant components.

The contravariant components V^i of the velocity vector are, by definition, given by

$$V^i = \mathbf{V} \cdot \mathbf{e}^i \tag{12.78}$$

Using a 2D framework and taking the $i = 1$ as example, one gets

$$\mathbf{V} \cdot \mathbf{e}^1 = (V^1 \mathbf{e}_1 + V^2 \mathbf{e}_2) \cdot \mathbf{e}^1 = V^1 \tag{12.79}$$

Expanding the scalar product in the right-hand side of Eq. (12.78), and according to the nomenclature used in Fig. 12.11,

$$\mathbf{V} \cdot \mathbf{e}^1 = |\mathbf{V}||\mathbf{e}^1|\cos\theta = \overline{V}^1 |\mathbf{e}^1| = V^1 \tag{12.80}$$

From Eq. (12.80) and from expressions of $|\mathbf{e}^1|$ and $\sqrt{g}$, given by Eqs. (12.45) and (12.60), respectively, one finds

$$\overline{V}^1 = \frac{V^1}{|\mathbf{e}^1|} = \frac{V^1 \sqrt{g}}{\sqrt{g_{22}}} = \frac{J V^1}{\sqrt{g_{22}}} = \frac{U}{\sqrt{g_{22}}}, \tag{12.81}$$

in which, in Eqs. (12.80) and (12.81), $\overline{V}^1$ is the projection of the vector **V** normal to the axis x^2, required for the calculation of the advective flux of ϕ. In Eq. (12.81), U is called the contravariant component without metric normalization of the vector **V**. Paying attention to Eq. (12.31), in which the geometric meaning of the length $\sqrt{g_{22}}$ can be obtained, one sees that the relation among U and $\overline{V}^1$ is just a length. Since $\overline{V}^1$ is the physical component of **V** in the direction normal to the x^2 axis, it is found that U, when multiplied by Δx^2 ($\Delta \eta$ in Eq. (12.31)), represents the volumetric flow rate across the face.

In two dimensions, the contravariant components, without metric normalization, of the vector **V** are represented by U and V. Later in this section we will make use of these components, and will show that these quantities arise directly when the conservation equations are transformed from the physical domain (x, y, z) to the transformed domain (ξ, η, γ).

It is important to recognize in Fig. 12.11 that $\overline{V}^1$, the portion of the velocity vector responsible for the mass flow through the face, is related to only one of the contravariant components of the velocity vector. That is, when it is required to calculate a flux normal to the x^2 axis (which is a constant x^1 line), it is needed only V^1, and when it is needed normal to the x^1 axis (which is a constant x^2 line), it is needed only V^2. A similar situation happens in the Cartesian system, in which, for calculating the advective flux normal to the constant y and x lines, one needs only v and u, respectively. In a generalized coordinate system, if the covariant components were employed, the portion responsible for the mass flow, $\overline{V}^1$, would depend on the two covariant components, since

$$\mathbf{V} \cdot \mathbf{e}^1 = (V_1 \mathbf{e}^1 + V_2 \mathbf{e}^2) \cdot \mathbf{e}^1 = \overline{V}^1 |\mathbf{e}^1| \quad (12.82)$$

In this case both products $\mathbf{e}^1 . \mathbf{e}^1$ and $\mathbf{e}^1 . \mathbf{e}^2$ are non zeros and, therefore,

$$\overline{V}^1 = f(V_1, V_2) \quad (12.83)$$

12.2.6 Mass Flow Calculation

It is important to bear in mind that one is talking about the velocity entity which enters in the mass flow computation, the so-called advecting velocity. In a moving control volume, it is the relative velocity which carries the mass flow in and out of the control volume. The advected velocity, even in a general curvilinear system, will continue to be the Cartesian components, by choice, since it would complicate severely the numerical method if one attempts to use the contravariant components also as advected velocity, or better saying, as the dependent variable of the problem. There were some initiatives of doing this in the past, but the idea didn't prosper.

12.2 Global Coordinate Transformation

From the point of view of constructing the numerical method, since the Cartesian components are always kept as dependent variables, all components (u, v, w) are required at the faces of the control volume for mass calculation.

Digging a little deeper into the physical interpretation of each component of a vector when curvilinear coordinate systems are employed, consider Fig. 12.12, in which, again the velocity vector **V** is shown, and the coordinates x^1 and x^2 have been replaced by ξ and η, respectively. The interest now is to correlate the contravariant velocity components with the Cartesian ones, considering the physical interpretation. The goal is to calculate the mass flow rate through the segment $\overline{AB}$. Bringing back to mind the knowledge from the basic calculus and fluid mechanics courses, this mass flow rate is given by

$$\dot{m} = \rho(\mathbf{V} \cdot \mathbf{n})_p \overline{AB}, \tag{12.84}$$

or,

$$\dot{m} = \rho|\mathbf{V}|\cos\theta \sqrt{g_{22}}\Delta\eta, \tag{12.85}$$

in which the representation of $\overline{AB}$ came from Eq. (12.31) which calculates the length of a segment over a coordinate axis, η in this case. Knowing that $|\mathbf{V}|\cos\theta = \overline{V}^1$, according to Fig. 12.11,

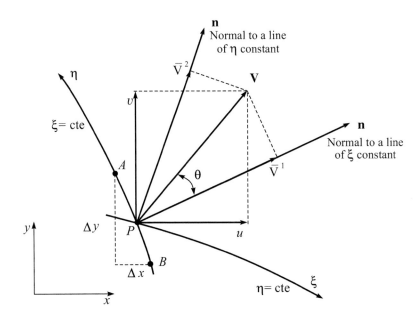

Fig. 12.12 Projection of **V** normal to ξ and η constant lines

$$\dot{m} = \rho \overline{V}^{1} \sqrt{g_{22}} \Delta \eta, \tag{12.86}$$

or, with help of Eq. (12.81),

$$\dot{m} = \rho J V^{1} \Delta \eta = \rho U \Delta \eta \tag{12.87}$$

The beauty of this interpretation is that starting from the basic knowledge of physics and vector calculus, Eq. (12.84), one reaches Eq. (12.87), which involves in the mass flow calculation the entities of a curvilinear coordinate system, as Jacobian, contravariant velocity components and dimensions in the transformed plane. The term $U \Delta \eta$ in Eq. (12.87), with U being the contravariant velocity component without metric normalization, is the volumetric flow rate crossing the segment. It is shown again that for calculating the mass flow rate in a segment along an axis it is required only one component of the contravariant velocity vector.

The segment $\overline{AB}$ could be a surface of any irregular control volume in unstructured grids. Therefore, it is emphasized that the learning of this section is helpful for creating and understanding finite volume methods in curvilinear coordinate systems, as well as in unstructured grids, which uses a local coordinate transformation. The mass flow calculation uses the same entities in control volumes of any shape.

Still with the help of Fig. 12.12, and using the Cartesian components of the velocity vector and the projection of the $\overline{AB}$ segment in the x and y axis, it is easy to show that

$$\dot{m} = \rho \left(u \frac{\Delta y}{\Delta \eta} - v \frac{\Delta x}{\Delta \eta} \right) \Delta \eta \tag{12.88}$$

Using Eqs. (12.81) and (12.87) one finds,

$$U = u \frac{\Delta y}{\Delta \eta} - v \frac{\Delta x}{\Delta \eta} \tag{12.89}$$

In unstructured grids, in which there are no global coordinate transformation, the volumetric flow rate is calculated by

$$Q = u \Delta y - v \Delta x, \tag{12.90}$$

and the mass flow rate by

$$\dot{m} = \rho Q = \rho (u \Delta y - v \Delta x) \tag{12.91}$$

Equations (12.89) and (12.90) calculates the same quantity. But, in Eq. (12.90) Δx and Δy should carry a sign, which is given by a convention on how to sweep the boundary of the control volume (counterclockwise) while in Eq. (12.89) the sign is given automatically by the metrics of the transformation, $\Delta x / \Delta \eta$ and $\Delta y / \Delta \eta$. Figure 12.13 explains this, depicting a situation in which $\Delta x / \Delta \eta$ is negative,

12.2 Global Coordinate Transformation

resulting in a positive sign in Eq. (12.89), which is the correct one, because both the component u and v contribute positively to the mass flow through segment $\overline{AB}$ (see that the projections of u and v on the normal add up, giving rise to $\overline{\mathbf{V}}^1$). Figure 12.14 shows a situation in which $\Delta x/\Delta \eta$ is positive (x grows with increasing η), resulting in the negative sign in the second term on the right-hand side of Eq. (12.89). This shows that the v component contributes negatively to the mass flow at $\overline{AB}$, which can also be seen by the projections of u and v on the normal.

Manipulating the equations, it is also possible to show that

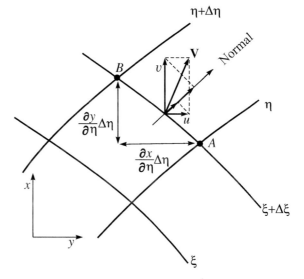

Fig. 12.13 Projection of **V** on the normal. $\Delta x/\Delta \eta$ negative

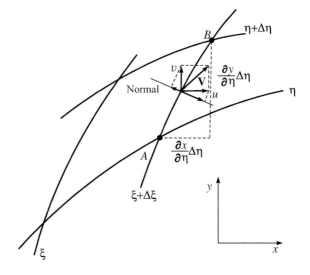

Fig. 12.14 Projection of **V** over the normal. $\Delta x/\Delta \eta$ positive

$$\rho \overline{V}^1 \sqrt{g_{22}} \Delta \eta = \rho \sqrt{g} \left(u \frac{\Delta \xi}{\Delta x} + v \frac{\Delta \xi}{\Delta y} \right) \Delta \eta, \tag{12.92}$$

The expression in the parentheses of Eq. (12.92) is identified as the contravariant component V^1 (see Eq. (12.73)). The relation among the velocity projected in the normal do the face (responsible for the mass flow) and the contravariant velocity component is only related to geometry, as

$$\frac{\overline{V}^1}{V^1} = \frac{\sqrt{g}}{\sqrt{g_{22}}} \tag{12.93}$$

To calculate the mass flow rate through a segment along ξ, by analogy, it is employed the other contravariant component, which is related to the projection of **V** onto the normal to the axis ξ, by

$$\frac{\overline{V}^2}{V^2} = \frac{\sqrt{g}}{\sqrt{g_{11}}} \tag{12.94}$$

The mass flow rate and the corresponding contravariant velocity component without metric normalization, are respectively,

$$\dot{m} = \rho \left(v \frac{\Delta x}{\Delta \xi} - u \frac{\Delta y}{\Delta \xi} \right) \Delta \xi$$

$$V = v \frac{\Delta x}{\Delta \xi} - u \frac{\Delta y}{\Delta \xi} \tag{12.95}$$

in which the negative sign should be interpreted in the same way as previously done for the U component, considering now the sign of $\Delta y / \Delta \xi$.

12.2.7 Example of a Nonorthogonal Transformation

Learning numerical methods using generalized coordinates, or local coordinates in unstructured grids, becomes considerably easier when the student is familiar with all the parameters and metrics of the transformation. This allows how to interpret the terms geometrically and understanding the respective consequences on the conception and writing of the code. Finding errors in the balances of quantities is only possible if this ability is developed. This is what this section is about.

For extract the maximum benefit from the following exercise, use a graph paper (or a software suitable for drawing and reading vector sizes) to draw the mesh and record all the basis vectors and components of the vector that will be represented in this nonorthogonal system. In a real complex geometry, the transformation is always discrete, represented by a tabular data. For simplicity, and without prejudice of the learning

12.2 Global Coordinate Transformation

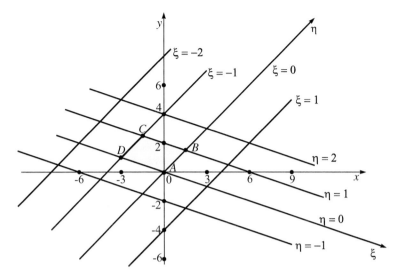

Fig. 12.15 Example of a non-orthogonal transformation

outcome, a two-dimensional analytic transformation will be employed. Figure 12.15 shows the mesh, given by the following transformation,

$$\xi = -0.25y + 0.25x \tag{12.96}$$

$$\eta = 0.5y + \frac{x}{6} \tag{12.97}$$

The inverse of the transformation can be easily found, as

$$x = 3\xi + 1.5\eta \tag{12.98}$$

$$y = -\xi + 1.5\eta \tag{12.99}$$

Note that if the transformation were not analytic, Eqs. (12.96–12.99) wouldn't exist. Hence, the metrics given by Eq. (12.101) would be obtained numerically, by central differences, from the known points (x, y) of the mesh, as will be seen in Sect. 12.2.8.

The metrics of the transformation are given by

$$\xi_x = 0.25$$
$$\xi_y = -0.25$$
$$\eta_x = \frac{1}{6}$$

$$\eta_y = 0.5, \tag{12.100}$$

while the metrics of the inverse are,

$$x_\xi = 3$$
$$y_\xi = -1$$
$$x_\eta = 1.5$$
$$y_\eta = 1.5 \tag{12.101}$$

Using Eq. (12.16) one can find, for two dimensions, the Jacobian, given by

$$J = \det \begin{bmatrix} x_\xi & y_\xi \\ x_\eta & y_\eta \end{bmatrix} = 6 \tag{12.102}$$

It can be also checking the values given by Eq. (12.100) using the relation given by the theorem of the inverse function,

$$\eta_x = -\frac{1}{J} y_\xi = -\frac{1}{6}(-1) = \frac{1}{6}$$
$$\eta_y = \frac{1}{J} x_\xi = \frac{1}{6}(3) = 0.5$$
$$\xi_x = \frac{1}{J} y_\eta = \frac{1}{6}(1.5) = 0.25$$
$$\xi_y = -\frac{1}{J} x_\eta = -\frac{1}{6}(1.5) = -0.25 \tag{12.103}$$

The area of the parallelogram $ABCDA$, shown in Fig. 12.15 can be calculated by

$$\frac{dS}{\Delta\xi\,\Delta\eta} = J = 6, \tag{12.104}$$

which can be checked its dimension geometrically in Fig. 12.15 drawn in a graph paper. The components of the metric tensor are

$$g_{11} = x_\xi^2 + y_\xi^2 = \gamma = 10$$
$$g_{22} = x_\eta^2 + y_\eta^2 = \alpha = 4.5$$
$$g_{12} = g_{21} = x_\xi x_\eta + y_\xi y_\eta = \beta = 3, \tag{12.105}$$

in which one can see that the component g_{12} is not zero, since the system is nonorthogonal. All orthogonal systems have the non-diagonal entries of the metric tensor matrix equal to zero. In Eq. (12.105) it is named the tensor metric components by α, β and γ for the 2D transformation, since this nomenclature is widely used in the literature

12.2 Global Coordinate Transformation

and will be used in this book. The γ variable cannot be confused with the coordinate axis of the curvilinear system. The lengths $\overline{AB}$ and $\overline{CB}$, as already seen, are lengths along the η and ξ axis, respectively, and their dimension are given by,

$$\overline{AB} = dL_\eta = \sqrt{g_{22}}\Delta\eta = \sqrt{4.5}$$
$$\overline{CB} = dL_\xi = \sqrt{g_{11}}\Delta\xi = \sqrt{10} \qquad (12.106)$$

The vectors of the covariant and contravariant base don't change in space because the transformation is linear among (x, y) and (ξ, η). The vectors of the covariant basis are,

$$e_\xi = 3\mathbf{i} - \mathbf{j}$$
$$e_\eta = 1.5\mathbf{i} + 1.5\mathbf{j}, \qquad (12.107)$$

while the vectors of the contravariant basis are,

$$e^\xi = 0.25\mathbf{i} - 0.25\mathbf{j}$$
$$e^\eta = \frac{1}{6}\mathbf{i} + 0.5\mathbf{j} \qquad (12.108)$$

Figure 12.16 shows the vector $\mathbf{V} = 6\mathbf{i} + 3\mathbf{j}$, its components in the Cartesian and curvilinear coordinate systems and the corresponding contravariant and covariant base. The contravariant components, using Eq. (12.73), are

$$V^\xi = 0.25 \times 6 - 0.25 \times 3 = 0.75$$
$$V^\eta = \frac{1}{6} \times 6 + 0.5 \times 3 = 2.5, \qquad (12.109)$$

resulting, for the velocity vector $\mathbf{V}$ represented by the contravariant velocity components,

$$\mathbf{V} = 0.75\, e_\xi + 2.5\, e_\eta \qquad (12.110)$$

Using Eq. (12.74), one finds the expressions for the covariant components

$$V_\xi = 3 \times 6 - 1 \times 3 = 15$$
$$V_\eta = 1.5 \times 6 + 1.5 \times 3 = 13.5 \qquad (12.111)$$

The expression for $\mathbf{V}$ represented by the covariant components is, therefore,

$$\mathbf{V} = 15\, e^\xi + 13.5\, e^\eta \qquad (12.112)$$

Fig. 12.16 Vector **V** and its components in a nonorthogonal system

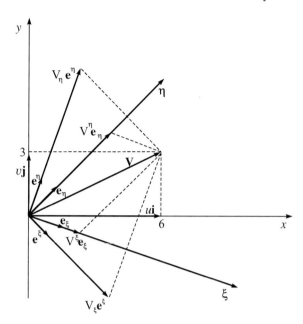

In this example, it was suggested to use a graph paper, so that all the lengths and the geometric meanings of the metrics could be interpreted. Influenced by the use of the Cartesian coordinate system, we are tempted to measure, on graph paper, the size of the contravariant and covariant components of the vector **V**. Remember that the length of the components which are possible to measure in a graph paper are its numerical value time the basis vector, as can be seen in Fig. 12.16, where it is shown the products $V^\xi \mathbf{e}_\xi$, $V^\eta \mathbf{e}_\eta$, $V_\xi \mathbf{e}^\xi$ and $V_\eta \mathbf{e}^\eta$. These products are lengths that can be measured in a graph paper. There are many other relationships involving the components of the metric tensor that can be shown for the type of transformation used in this exercise. This can be practiced solving the Exercise 12.8.

12.2.8 Calculation of the Metrics of a Transformation

When generalized curvilinear meshes are employed, with the conservation equations written in the domain (ξ, η, γ), it will be necessary to know the metrics of the transformation which relates the physical to the transformed domain, that is, the quantities $\partial x/\partial \xi$, $\partial x/\partial \eta$, $\partial y/\partial \xi$, $\partial y/\partial \xi$, in two dimensions. In three dimensions there will be nine metrics.

With these metrics, one can calculate the Jacobian, via Eq. (12.16), and all the expressions given by Eqs. (12.12–12.14). The transformation is almost never analytic, as in the example shown in Sect. 12.2.7 and, therefore the metrics must be calculated numerically. Figure 12.17 shows a curvilinear grid coming from the grid generator.

12.2 Global Coordinate Transformation

As discussed in previous chapter, the grid generator furnishes, at least, the coordinates of the vertices of the element, shown shaded in Fig. 12.17. With those coordinates, the metrics can be calculated in any position of the element. For example, the four metrics at point e are,

$$\left.\frac{\partial x}{\partial \eta}\right|_e = \frac{x_A - x_B}{\Delta \eta}$$

$$\left.\frac{\partial y}{\partial \eta}\right|_e = \frac{y_A - y_B}{\Delta \eta}$$

$$\left.\frac{\partial x}{\partial \xi}\right|_e = \frac{x_E - x_P}{\Delta \xi}$$

$$\left.\frac{\partial y}{\partial \xi}\right|_e = \frac{y_E - y_P}{\Delta \xi}, \qquad (12.113)$$

in which the values of x and y at points E and P are determined by interpolating the four neighboring x and y points. Other metrics, at different points in the domain, are calculated in an analogous way. An important recommendation is to always interpolate x and y where they are need for the metric calculation, and never average the metrics, since the metric are connected directly with the real size of the domain, and balances of the property can carry large errors in the computational domain. Figure 12.18 exemplifies the error that can be committed when the metrics are interpolated. For example, if the interest is to calculate $\partial x / \partial \xi$ at the point s the correct way to find it is through

$$\left.\frac{\partial x}{\partial \xi}\right|_s = \frac{x_D - x_C}{\Delta \xi} \qquad (12.114)$$

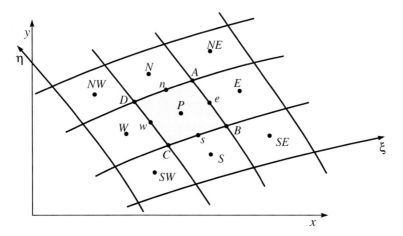

Fig. 12.17 Global coordinate system. Metrics of the transformation

Fig. 12.18 Metrics calculation

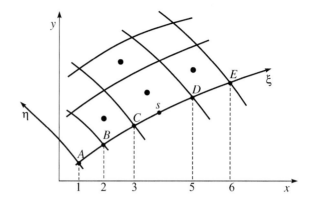

If $\Delta\xi = 1$, the final value is 2, using the dimensions given in Fig. 12.18. If the metric at point s is calculated using the values of the neighbor metrics, the value will be 1.5, and the length of $\overline{CD}$, which is calculated using this metric and $\partial y/\partial \xi$, will be wrong, affecting mass conservation and all other conserved properties. This warning is point out herein because it is customary to find the metrics at the vertices of the element (vertices of the control volume if it is a cell-center method) and obtain others required metrics by interpolating the existing ones.

12.3 Nature of the Discrete Transformation

12.3.1 Preliminaries

Generalized curvilinear coordinates began to be employed in the early 1970s among researchers of the finite difference method and were one of the most important steps taken by this community for the treatment of fluid flow problems in complex geometries.

Probably with the primary concern of finding efficient ways to deal with the pressure–velocity coupling and the nonlinearities of the Navier–Stokes equations, little effort had been devoted by finite difference researchers to solve the issue of irregular geometries. Therefore, developments in the area of finite differences on irregular geometries were in slow pace until that decade. The use of finite differences, for example, became so associated with orthogonal coordinate systems that there are, even today, those who believe that the method only applies to these systems. It is often encountered in the literature that the alternative for the treatment of irregular geometries in numerical simulation is only through the finite element method.

In the late 70s, with the demand from engineering for the simulation of flows in complex geometries, and with the experience in the treatment of fluid flows accumulated by researchers in the area of finite differences, an intense research activity

12.3 Nature of the Discrete Transformation

was initiated in the development of methods seeking to extend to complex problems the classical methodologies of conformal mapping, technique widely used in fluid mechanics [4], but limited to simple operators. The possibility of finding the transformation, even numerical, between the original coordinate system, usually Cartesian, and the generalized coordinate system coincident with the irregular geometry, allows the mapping of the irregular geometry, written in the system (x, y, z), into a regular geometry in the system (ξ, η, γ), as shown in Fig. 12.19 for two dimensions.

Pioneering efforts to extend the use of mappings to more general situations were made by Winslow [5] and Chu [6]. Winslow solved magnetostatics problems using a transformation that maps a grid of irregular triangles onto a grid of regular triangles. He called the regular plane the logical plane.

Chu [6] followed the same strategy, solving magnetohydrodynamics problems, presenting a numerical transformation, which he called machine transformation. Although Chu's work was truly the initial milestone in the use of generalized coordinates, the title given to his work probably did not reflect the importance and advances it contained. Barfield [7], Amsden and Hirt [8], Godunov and Prokopov [9]

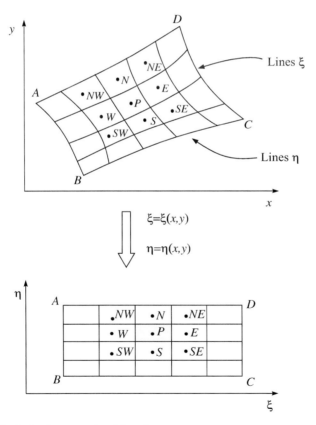

Fig. 12.19 Physical and computational domains

contributed in the same direction. These pioneering works, also, did not have enough resonance in the scientific environment at the time.

The beginning of the stupendous development experienced by numerical methods using generalized coordinates are due to the works of Thompson et al. [10–12] who, with interesting titles in their papers, called the attention of the community for the great potential of the methodology. After these works, the use of generalized coordinates spread in an amazing way. Thompson and his colleagues applied the methodology, using finite differences, to two-dimensional aerodynamics problems.

Back to Fig. 12.19, notice that the irregular geometry, defined in the physical plane, can change its shape without changing its representation in the transformed plane, which is fixed. Thus, by writing the conservation equations also in the transformed plane, the computational code will be written for a fixed geometry in this plane. Changing the geometry in the physical plane does not require changes in the computational code. Information about the physical geometry is provided to the computational code through the metrics of the transformation $\xi = \xi(x, y, z)$, $\eta = \eta(x, y, z)$ and, $\gamma = \gamma(x, y, z)$ and their inverses, which appear in the transformed conservation equations.

Generalized coordinate systems played a very important role in the numerical simulation for two decades. Nowadays the leading technique is the use of unstructured grids with its great flexibility and easiness in discretizing the domain. Curvilinear grids can be used in the same framework of unstructured grids, therefore, there are no reason for developing a specific code for this type of discretization. Structured meshes is required in certain situations, for example, in turbulent boundary layer problems. Good practices recommend always using structured meshes close to the wall, because good refinement for the most exact application of the law of the wall are only achieved with structured meshes in these regions. The use of few layers of structured meshes and then fill in the rest of the domain with unstructured meshes, is a common approach. Whenever the interest is in the calculation of fluxes at the wall, whether they are momentum quantities (stresses) or heat fluxes, mesh refinement on the wall is critical, and structured meshes are well suited for this.

In the following sections, the use of generalized curvilinear meshes will be presented, starting with numerical nature of the transformation, a brief description of the types of mappings, the transformation of the differential to the (ξ, η, γ) system for a general variable ϕ, ending with the integration of the equation in the transformed domain.

12.3.2 The Nature of the Transformation

The example given in Sect. 12.2.7, was of an analytical transformation. In real problem this is rare, and the transformation is discrete, what means that the points in space, represented by its (x, y, z) coordinates, are related in tabular form to the (ξ, η, γ) points in the computational or transformed domain. Consider now the geometry shown in Fig. 12.20, and the corresponding computational domain in Fig. 12.21.

12.3 Nature of the Discrete Transformation

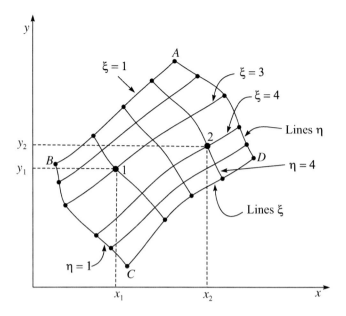

Fig. 12.20 Functional correspondence of points 1 and 2

The arbitrary shape of the geometry precludes having an analytic transformation, but rather a discrete correspondence between the points on the physical and transformed planes. The correspondence of the point 1 and 2 are

$$(\xi, \eta) = (3, 2)$$
$$(x, y) = (x_1, y_1)$$
$$(\xi, \eta) = (4, 4)$$
$$(x, y) = (x_2, y_2) \tag{12.115}$$

In curvilinear discretization it is possible to have different physical coordinate system mapping onto the same computational domain. To have a glance on these mappings is the goal now. We are not concerned how the curvilinear systems are generated, by some means the transformation in discrete form exists.

12.3.2.1 Simply Connected Geometries

The mapping of simply connected geometries has already been indirectly seen through Figs. 12.20 and 12.21. Exercising a bit more, let's collapse points B and C in Fig. 12.20 to obtain a geometry shown in Fig. 12.22 (left), with the same transformed plane (right) as shown in Fig. 12.21. Along $\overline{BC}$ $\Delta \xi$ is unitary, and an

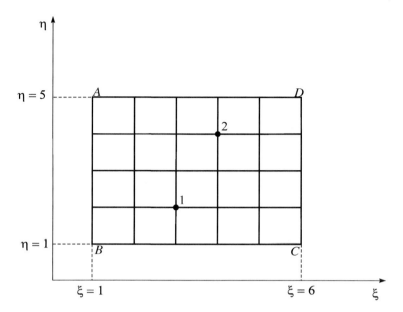

Fig. 12.21 Normalized computational domain of Fig. 12.20

information from the coordinate transformation will calculate the physical dimension of $\overline{BC}$ as zero, as it should be.

The same geometry can be discretized in different ways but having the same mapping. Figures 12.22 and 12.23 show a triangle being discretized with two different generalized coordinate systems and with the same transformed plane. The choice of which system is best suited for a given geometry depends on the physical problem that will be solved on it. This shows the versatility of creating grids with the same computational domain.

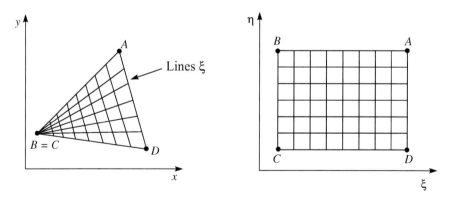

Fig. 12.22 Grid for a triangle and the computational domain

12.3 Nature of the Discrete Transformation

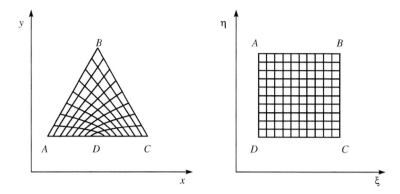

Fig. 12.23 Grid for a triangle. Same computational domain as Fig. 12.22

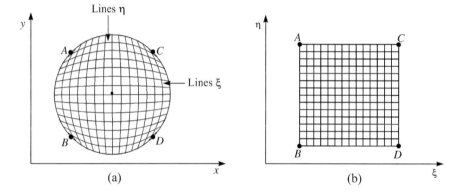

Fig. 12.24 Grid for the circle (**a**) and its computational domain (**b**)

To complete the examples of mappings of simply connected geometries, Fig. 12.24a shows a circle with a generalized discretization and, in Fig. 12.24b, its transformed plane. This is an example in which the mesh is very nonorthogonal in the vicinity of the four points A, B, C and D, chosen to divide the circle into four segments. At these locations, the nonorthogonality is maximal, because the lines ξ and η at the boundary are collinear.

12.3.2.2 Mapping by Blocks

In all the mappings shown so far, situations have been considered in which it is possible to generate a mesh whose mapping is a single, rectangular block. In such cases, the boundary of the physical domain is divided into four segments, two of which will be ξ constant lines and two η constant lines.

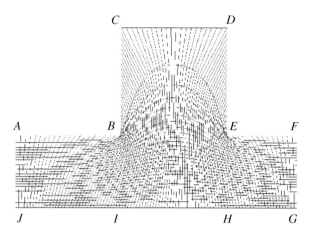

Fig. 12.25 Geometry requiring multiblock discretization

Certain geometries, with sharp protuberances, for example, are not suitable for generating a mesh with a single-block mapping, because, as can be seen in Fig. 12.25, it is difficult to have the coordinate lines to discretize the protuberance efficiently, since the lines collapsed over each other, as can be seen by the difficult in visualizing the grid. To concentrate coordinate lines near the wall of the protuberance, it would be created a very nonorthogonal mesh with the coordinate lines almost collinear in the protuberance region. The solution in these cases, without the use of unstructured meshes, is the use of multi-blocks, solving the problem, in the case of Fig. 12.25 for example, in two separate blocks, blocks $ABEFGHIJA$ and $IBCDEHI$, as in Fig. 12.26, iterating between blocks until convergence of the process is achieved. For each of the blocks, the entire methodology developed for single block mapping can be applied. The solution of one block serves as boundary condition for the other. Boundary conditions at the BI and EH are internal surfaces of one block, while the BE is internal for the other block. The procedure iterates between blocks. The number of blocks can be any, however, too much blocks resembles an unstructured grid, and it is preferable to solve the problem in a full unstructured grid.

To conclude the discussion of simply connected domain mapping, it is important to note that whether the mapping is single block or multiblock, the boundaries of the transformed domain are exactly the boundaries of the physical domain. Thus, the boundary conditions for the transformed plane are those that exist in the physical plane.

12.3.2.3 Double and Multiple Connected Geometries

Doubly connected geometries can be mapped into a single block or into more blocks, depending on their complexity. For example, Fig. 12.27a shows a doubly connected

12.3 Nature of the Discrete Transformation

Fig. 12.26 Suitable grid for multiblock discretization of Fig. 12.25

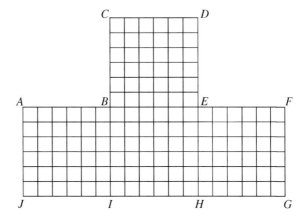

geometry being mapped into a single block, as shown in Fig. 12.27b. The mapping was made possible by "cutting" along a ξ line, connecting the inner with the outer boundary, "opening" the geometry at this location, and representing it in the plane (ξ, η). A basic rule for mapping doubly and multiply connected geometries into a single block is to start in a point, sweep all boundaries, and be back to it without lifting the pencil from the paper.

This process determines the required cuts. For instance, in Fig. 12.27a, it was started at point A, and followed the entire inner boundary, passing through B and arriving at C, which is point A itself. To reach the outer border without lifting the pencil from the paper it is necessary to cut along CD, follow the entire outer surface, passing through E, arriving at F and returning through the cut, until reaching point A, completing the process. By passing twice through the same cut, the two boundaries in the transformed plane are generated, as shown in Fig. 12.27b. These segments, CD and AF, are coincident and, in the physical plane, they are not boundaries, but an internal line of the domain. It is easy to see that in the transformed plane there

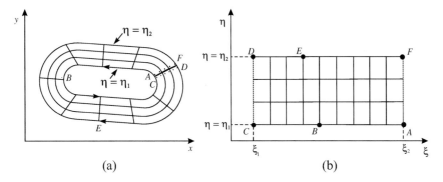

Fig. 12.27 Double connected geometry and the computational domain

will be no boundary conditions on these boundaries. The way to treat these boundary conditions is just apply the repetitive boundary condition.

The geometry of Fig. 12.27a can be mapped, if desired, by means of blocks, using a discretization of the type shown in Fig. 12.28a. In this case, the outer and inner boundaries are divided into four segments, the same technique applied for mapping simply connected geometries, and mapped as shown in Fig. 12.28b. The transformed domain is no longer a single block. Since there are no cuts, the boundary conditions existing on the boundaries of the physical domain transfer exactly to the boundaries of the transformed domain.

For multiply connected geometries, the procedure is the same as for doubly connected geometries. Here, again, one can make use of cuts, which generate a single rectangular computational domain, or the option of creating a computational plane composed of rectangular blocks. The mapping of the geometry shown in Fig. 12.29a in the physical plane is shown in Fig. 12.29b.

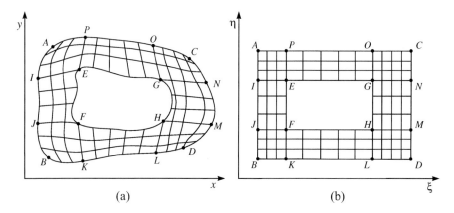

Fig. 12.28 Mapping keeping the hole in the computational domain

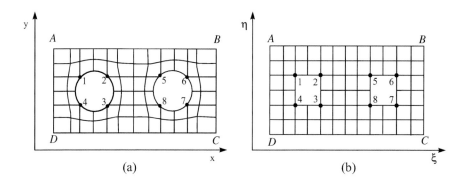

Fig. 12.29 Multiple connected domain (**a**). Transformed domain (**b**)

12.3.2.4 Three-dimensional Geometries

All the concepts just seen for obtaining the transformed plane for two-dimensional geometries can also be applied for three-dimensional regions. The cuts, previously made along coordinate lines, are now made along coordinate surfaces.

The mapping of a simply connected three-dimensional region is shown in Fig. 12.30. When moving from two-dimensional to three-dimensional regions, the difficulty of mesh generation grows considerably. When such a meshes are required, it means that a commercial software is behind the simulation, and those packages provide grid generators of quality. For academic purposes, developing and implementing small codes, training in numerical methods, there are free grid generators available in the literature. In general, the involvement of the analyst with the grid is related to the quality of the grid for the problem under consideration, like refinement in regions of large gradients, distortion, homogeneity and nonorthogonality of the control volumes. Three-dimensional grid generation involves several tools of the CAE environment. These tools are reunited in commercial packages prepared to feed the numerical simulator kernel with full integration in a standalone application. The satellite tools of a numerical simulator are, nowadays, extremely sophisticated and complete.

If the code accepts only curvilinear structured grids, the use of multiblocks will be compulsory, since real engineering geometries rarely can be mapped onto a unique parallelepiped or in a small number of blocks.

Figure 12.31 presents an example in which a three-dimensional mesh for a solid of revolution is obtained with a simple methodology. In this case with the rotation of a two-dimensional mesh, generated algebraically, establishing as external boundary, for example, a hyperbola, and as internal boundary the geometry of the body. The

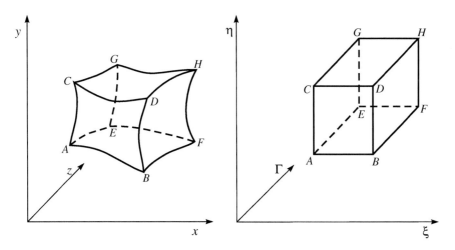

Fig. 12.30 Three-dimensional region and its mapping

two-dimensional mesh is generated for the geometry defined by $\overline{ABCDA}$. This mesh is rotated around the axis of the body, generating the 3D control volumes.

To better understand this mapping, Fig. 12.32(left) shows the 2D generating mesh and, in Fig. 12.32(right), the transformed domain. At each rotation increment in the two-dimensional mesh, the planes with the 2D meshes in the transformed domain are generated. The $\overline{DA}$ segment rotates on itself, generating in the transformed domain always the repetition of this segment, while the $\overline{CB}$ segment will coincide with the $\overline{FE}$ segment after rotating 180° (when the problem does not present any flow symmetry, the mesh must be rotated 360°). In the transformed domain there is the surface $\overline{CDDFC}$, which represents the external surface of the mesh, and the surface $\overline{BAAEB}$, representing the surface of the body. On these two surfaces there are the free flow boundary conditions and the existing boundary conditions on the body, respectively.

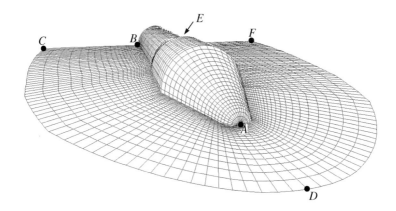

Fig. 12.31 Three-dimensional grid for a body of revolution

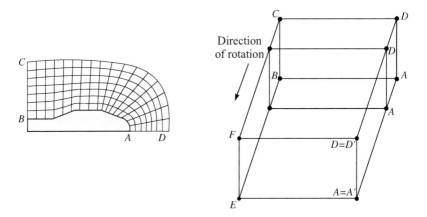

Fig. 12.32 Computational domain for the grid over a blunt body

For an external flow over the body with a vertical inclination (angle of attack), the flow is symmetric about the vertical plane. Therefore, the surfaces $\overline{CDABC}$ and $\overline{FDAEF}$ in the transformed domain are symmetric. The surface $\overline{DAA'D'D}$ is an internal line in the physical domain that is transformed into a plane in the transformed domain and, because it is internal, does not require boundary conditions.

12.4 Equations Written in the Curvilinear System

The modern general purposes fluid flow simulators have a numerical structure which accepts all kinds of grids, among them, curvilinear structured, of course. This means that creating fluid simulators that use curvilinear coordinates with the conservation equations transformed onto the computational domain applies only in very dedicated niches of applications. Among them problems in which the physical geometry in analysis changes such that the computational domain is always the same.

Even though, we are dedicating this section to register the important moves made in the area of numerical simulation, from Cartesian to unstructured, chronologically, and, at the same time, describing a tool that can be of interested for researchers in curvilinear coordinates for further studies. Besides that, this topic enriches the understandings of the mathematics behind a numerical discretization.

Once obtained the new discrete coordinate system, the next step is to derive the approximate equations. This section presents the transformation of the partial differential equations from the (x, y, z) coordinate system to the curvilinear system (ξ, η, γ). Keep in mind that all conservation equations will be written in the new curvilinear system, but the dependent variables will be those of the Cartesian system.

The transformation of the conservation equation starts from the equation in the Cartesian coordinate system in conservative form. The goal is to have them written in the curvilinear coordinate system keeping its conservative form. In vectorial form the conservation equations for a general variable ϕ are

$$\frac{\partial Q}{\partial t} + \nabla \cdot \mathbf{F} = S^\phi, \tag{12.116}$$

or, in the form,

$$\frac{\partial Q}{\partial t} + \frac{\partial E}{\partial x} + \frac{\partial F}{\partial y} + \frac{\partial G}{\partial z} = S^\phi, \tag{12.117}$$

in which

$$\mathbf{F} = E\mathbf{i} + F\mathbf{j} + G\mathbf{k} \tag{12.118}$$

$$Q = \rho\phi \tag{12.119}$$

$$E = \rho u \phi - \Gamma^\phi \frac{\partial \phi}{\partial x} \qquad (12.120)$$

$$F = \rho v \phi - \Gamma^\phi \frac{\partial \phi}{\partial y} \qquad (12.121)$$

$$G = \rho w \phi - \Gamma^\phi \frac{\partial \phi}{\partial z} \qquad (12.122)$$

in which ϕ is a general scalar which represents the conserved properties, like mass, momentum, energy, among others, and Γ^ϕ represents the transport coefficient. Recall that the letter F without bold face is a component of the vector $\mathbf{F}$ and not its modulus. In this case the components E, F and G are the Cartesian components of the advective + diffusive fluxes of ϕ.

The coordinate transformation required for the solution of 3D transient problems is,

$$\begin{aligned} \xi &= \xi(x, y, z, t) \\ \eta &= \eta(x, y, z, t) \\ \gamma &= \gamma(x, y, z, t) \\ \tau &= t \end{aligned} \qquad (12.123)$$

in which the Jacobian and the spatial metrics of the transformation are the same given in the beginning of this Chapter, when the transformation was done without considering the time coordinate. Adding the following relations which involves the time coordinate,

$$\xi_t = -\xi_x x_\tau - \xi_y y_\tau - \xi_z z_\tau \qquad (12.124)$$

$$\eta_t = -\eta_x x_\tau - \eta_y y_\tau - \eta_z z_\tau \qquad (12.125)$$

$$\gamma_t = -\gamma_x x_\tau - \gamma_y y_\tau - \gamma_z z_\tau, \qquad (12.126)$$

Our goal is to transform Eq. (12.117) to the $(\xi, \eta, \gamma, \tau)$ coordinate system. Using the chain rule, the derivatives of the components of the fluxes, read,

$$\frac{\partial E}{\partial x} = \frac{\partial E}{\partial \xi} \xi_x + \frac{\partial E}{\partial \eta} \eta_x + \frac{\partial E}{\partial \gamma} \gamma_x + \frac{\partial E}{\partial \tau} \tau_x \qquad (12.127)$$

$$\frac{\partial F}{\partial y} = \frac{\partial F}{\partial \xi} \xi_y + \frac{\partial F}{\partial \eta} \eta_y + \frac{\partial F}{\partial \gamma} \gamma_y + \frac{\partial F}{\partial \tau} \tau_y \qquad (12.128)$$

$$\frac{\partial G}{\partial z} = \frac{\partial G}{\partial \xi} \xi_z + \frac{\partial G}{\partial \eta} \eta_z + \frac{\partial G}{\partial \gamma} \gamma_z + \frac{\partial G}{\partial \tau} \tau_z \qquad (12.129)$$

12.4 Equations Written in the Curvilinear System

$$\frac{\partial Q}{\partial t} = \frac{\partial Q}{\partial \xi}\xi_t + \frac{\partial Q}{\partial \eta}\eta_t + \frac{\partial Q}{\partial \gamma}\gamma_t + \frac{\partial Q}{\partial \tau}\tau_t \qquad (12.130)$$

The last term in Eqs. (12.127–12.129) are zero, since τ is not function of x, y and z. Introducing these equations in Eq. (12.117) and multiplying by the Jacobian, it will appear terms of the type,

$$\frac{\partial E}{\partial \xi}(J\xi_x), \quad \frac{\partial F}{\partial \xi}(J\xi_y), \quad \frac{\partial Q}{\partial \tau}(J), \text{ etc}$$

To have the Jacobian and the metrics inside the derivative, searching for a conservative form, addition and subtraction of terms of the type

$$E\frac{\partial}{\partial \xi}(J\xi_x), \quad F\frac{\partial}{\partial \xi}(J\xi_y), \quad G\frac{\partial}{\partial \xi}(J\xi_z), \quad Q\frac{\partial}{\partial \xi}(J\xi_t), \text{ etc}$$

are introduced in the equations. These terms are part of the derivative of the product $\frac{\partial}{\partial \xi}(E(J\xi_x))$. With some algebraic manipulation, the conservative form of the equations appears,

$$\frac{\partial}{\partial \tau}(QJ)$$
$$+ \frac{\partial}{\partial \xi}\left(J(\xi_t Q + \xi_x E + \xi_y F + \xi_z G)\right)$$
$$+ \frac{\partial}{\partial \eta}\left(J(\eta_t Q + \eta_x E + \eta_y F + \eta_z G)\right)$$
$$+ \frac{\partial}{\partial \gamma}\left(J(\gamma_t Q + \gamma_x E + \gamma_y F + \gamma_z G)\right)$$
$$- Q\left[\frac{\partial}{\partial \xi}(J\xi_t) + \frac{\partial}{\partial \eta}(J\eta_t) + \frac{\partial}{\partial \gamma}(J\gamma_t) + \frac{\partial}{\partial \tau}(J)\right]+$$
$$- E\left[\frac{\partial}{\partial \xi}(J\xi_x) + \frac{\partial}{\partial \eta}(J\eta_x) + \frac{\partial}{\partial \gamma}(J\gamma_x)\right]$$
$$- F\left[\frac{\partial}{\partial \xi}(J\xi_y) + \frac{\partial}{\partial \eta}(J\eta_y) + \frac{\partial}{\partial \gamma}(J\gamma_y)\right]$$
$$- G\left[\frac{\partial}{\partial \xi}(J\xi_z) + \frac{\partial}{\partial \eta}(J\eta_z) + \frac{\partial}{\partial \gamma}(J\gamma_z)\right] = JS^\phi \qquad (12.131)$$

Using the expression for the Jacobian, Eq. (12.16) and Eqs. (12.124–12.126), it is easy to show that all terms inside brackets are zero, with the final form of transformed equation, as

$$\frac{\partial}{\partial \tau}(QJ)$$
$$+ \frac{\partial}{\partial \xi}\left(J(\xi_t Q + \xi_x E + \xi_y F + \xi_z G)\right)$$
$$+ \frac{\partial}{\partial \eta}\left(J(\eta_t Q + \eta_x E + \eta_y F + \eta_z G)\right)$$
$$+ \frac{\partial}{\partial \gamma}\left(J(\gamma_t Q + \gamma_x E + \gamma_y F + \gamma_z G)\right) = J S^\phi \qquad (12.132)$$

Recalling that the components E, F and G contains the advective and diffusive fluxes, Eq. (12.132) is in the divergence (conservative) form in the new coordinate system. Defining,

$$\hat{Q} = Q J \qquad (12.133)$$

$$\hat{E} = J(\xi_t Q + \xi_x E + \xi_y F + \xi_z G) \qquad (12.134)$$

$$\hat{F} = J(\eta_t Q + \eta_x E + \eta_y F + \eta_z G) \qquad (12.135)$$

$$\hat{G} = J(\gamma_t Q + \gamma_x E + \gamma_y F + \gamma_z G) \qquad (12.136)$$

$$\hat{S}^\phi = J S^\phi \qquad (12.137)$$

one finds the transformed equation in compact form, as

$$\frac{\partial \hat{Q}}{\partial \tau} + \frac{\partial \hat{E}}{\partial \eta} + \frac{\partial \hat{F}}{\partial \gamma} + \frac{\partial \hat{G}}{\partial \tau} = \hat{S}^\phi \qquad (12.138)$$

Substituting into Eq. (12.132) the relations for the components E, F and G, one finds the equation with the advection and diffusion terms separately, as

$$\frac{\partial}{\partial \tau}(J\rho\phi)$$
$$+ \frac{\partial}{\partial \xi}\left(J((\xi_t + \xi_x u + \xi_y v + \xi_z w)\rho\phi\right)$$
$$+ \frac{\partial}{\partial \eta}\left(J(\eta_t + \eta_x u + \eta_y v + \eta_z w)\rho\phi\right)$$
$$+ \frac{\partial}{\partial \gamma}\left(J(\gamma_t + \gamma_x u + \gamma_y v + \gamma_z w)\rho\phi\right)$$
$$= \frac{\partial}{\partial \xi}\left\{\left(\xi_x \frac{\partial \phi}{\partial x} + \xi_y \frac{\partial \phi}{\partial y} + \xi_z \frac{\partial \phi}{\partial z}\right) J \Gamma^\phi\right\}$$

12.4 Equations Written in the Curvilinear System

$$+ \frac{\partial}{\partial \eta}\left\{\left(\eta_x \frac{\partial \phi}{\partial x} + \eta_y \frac{\partial \phi}{\partial y} + \eta_z \frac{\partial \phi}{\partial z}\right)J\Gamma^\phi\right\}$$
$$+ \frac{\partial}{\partial \gamma}\left\{\left(\gamma_x \frac{\partial \phi}{\partial x} + \gamma_y \frac{\partial \phi}{\partial y} + \gamma_z \frac{\partial \phi}{\partial z}\right)J\Gamma^\phi\right\} + JS^\phi \tag{12.139}$$

Inspecting the second term in the LHS of Eq. (12.139), and based on the knowledge from the Cartesian system, the term which multiplies $\rho\phi$ must be a velocity which enters in the mass flow calculation and which transports ϕ, the property per unit of mass.

Defining the following entities,

$$\tilde{U} = J(\xi_t + \xi_x u + \xi_y v + \xi_z w) \tag{12.140}$$

$$\tilde{V} = J(\eta_t + \eta_x u + \eta_y v + \eta_z w) \tag{12.141}$$

$$\tilde{W} = J(\gamma_t + \gamma_x u + \gamma_y v + \gamma_z w), \tag{12.142}$$

and using the chain rule to expand the derivatives of ϕ related to x, y and z, such that all derivatives are written in terms of the new independent variables ξ, η and γ, it is found,

$$\frac{\partial}{\partial \tau}(J\rho\phi) + \frac{\partial}{\partial \xi}\left(\rho\tilde{U}\phi\right) + \frac{\partial}{\partial \eta}\left(\rho\tilde{V}\phi\right) + \frac{\partial}{\partial \gamma}\left(\rho\tilde{W}\phi\right)$$
$$= \frac{\partial}{\partial \xi}\left\{\left(a\frac{\partial \phi}{\partial \xi} + d\frac{\partial \phi}{\partial \eta} + e\frac{\partial \phi}{\partial \gamma}\right)J\Gamma^\phi\right\}$$
$$+ \frac{\partial}{\partial \eta}\left\{\left(d\frac{\partial \phi}{\partial \xi} + b\frac{\partial \phi}{\partial \eta} + f\frac{\partial \phi}{\partial \gamma}\right)J\Gamma^\phi\right\}$$
$$+ \frac{\partial}{\partial \gamma}\left\{\left(e\frac{\partial \phi}{\partial \xi} + f\frac{\partial \phi}{\partial \eta} + c\frac{\partial \phi}{\partial \gamma}\right)J\Gamma^\phi\right\} + JS^\phi, \tag{12.143}$$

in which,

$$a = \xi_x^2 + \xi_y^2 + \xi_z^2 \tag{12.144}$$

$$b = \eta_x^2 + \eta_y^2 + \eta_z^2 \tag{12.145}$$

$$c = \gamma_x^2 + \gamma_y^2 + \gamma_z^2 \tag{12.146}$$

$$d = \xi_x \eta_x + \xi_y \eta_y + \xi_z \eta_z \tag{12.147}$$

$$e = \xi_x \gamma_x + \xi_y \gamma_y + \xi_z \gamma_z \tag{12.148}$$

$$f = \eta_x \gamma_x + \eta_y \gamma_y + \eta_z \gamma_z \tag{12.149}$$

are components of the metric tensor and $\tilde{U}$, $\tilde{V}$ and $\tilde{W}$ are the contravariant velocity components which considers the movement of the grid with time. The relative velocity (the one which transport mass into the control volume) appeared automatically by considering the time coordinate in the transformation.

Compare, for example, the advective term, $\rho \tilde{U}$, with ρu for the Cartesian coordinate system. Both represent the mass flow by unit of area, recognizing that in the later it is enough to use the component u for the mass flow calculation, since u is normal to the flow area in the Cartesian system. In the curvilinear system the mass flow requires the contravariant velocity component ($\tilde{U}$) for the mass flow calculation, which involves u, v and w (see for instance Eq. (12.140)).

To exercise, use Eq. (12.143) to reproduce the mass conservation equation in the Cartesian system, in which $\phi = 1$, $J = 1$, $\tilde{U} = u$, $\tilde{V} = v$ and $\tilde{W} = w$.

Another observation is that the diffusive term in the right-hand side of Eq. (12.143) requires the three Cartesian components for its calculation normal to the face of the control volume. This should not be a surprise, since the diffusive flux is a vector. For orthogonal curvilinear systems, d, e and f are zero, the cross-derivatives are all zero and the diffusive terms reduces to the three second order derivative terms in the direct directions, as in the Cartesian system. Therefore, Eq. (12.143) is the general equation for a scalar ϕ written in the curvilinear coordinate system (ξ, η, γ), since it encompasses all orthogonal and non-orthogonal systems.

This transformation has a fixed computational domain. Thus, according to the transformation given by Eqs. (12.123), (12.143) can solve problems with moving mesh, keeping a fixed mesh in the transformed plane. Of course, the relative velocity between the flow velocity and the mesh velocity must be accounted in some term of the transformed equation. It is exactly in the $\tilde{U}$, $\tilde{V}$ and $\tilde{W}$ contravariant components that this effect is embedded, precisely in the first term in the parentheses of Eqs. (12.140–12.142).

To facilitate the physical interpretation, consider the two-dimensional problem, in which the velocities involved are,

$$\tilde{U} = J(\xi_t + \xi_x u + \xi_y v) \tag{12.150}$$

$$\tilde{V} = J(\eta_t + \eta_x u + \eta_y v) \tag{12.151}$$

Using the metrics relation, these velocities assume the following form,

$$\tilde{U} = y_\eta (u - x_\tau) - x_\eta (v - y_\tau) \tag{12.152}$$

$$\tilde{V} = x_\xi (v - y_\tau) - y_\xi (u - x_\tau) \tag{12.153}$$

12.4 Equations Written in the Curvilinear System

Recalling the definition of the contravariant velocity components without metric normalization for the case of fixed grids, given by

$$U = y_\eta u - x_\eta v \tag{12.154}$$

$$V = x_\xi v - y_\xi u, \tag{12.155}$$

one can find the relation among $\tilde{U}$ and U, and $\tilde{V}$ and V, as

$$\tilde{U} = U - U_M \tag{12.156}$$

$$\tilde{V} = V - V_M, \tag{12.157}$$

in which U_M and V_M are the contravariant velocity components without metric normalization of the grid velocity, given by

$$U_M = y_\eta x_\tau - x_\eta y_\tau \tag{12.158}$$

$$V_M = x_\xi y_\tau - y_\xi x_\tau, \tag{12.159}$$

in which the quantities x_τ and y_τ are the Cartesian velocity components of the grid velocity in the x and y directions, respectively. In Eqs. (12.156) and (12.157), $\tilde{U}$ and $\tilde{V}$ are the contravariant velocity components already considering the movement of the grid in the conservation balances, that is, they are the components of the relative velocity, the velocity which carries mass in or out of a control volume.

Equation (12.143) can be employed to solve problems in which the mesh varies with time while keeping the computational plane fixed and with the dimensions of the elementary volumes in this plane also fixed and unitary, if desired.

Figure 12.33 shows a moving mesh, where the Cartesian x and y components of the grid velocity vector for points A and B can be identified. If the mesh in the physical plane is moving with the same velocity of the flow, there will be no mass flow entering the control volumes from faces that have the same velocity as the flow, and therefore $\tilde{U} = \tilde{V} = 0$. There are numerous problems, with free or specified boundary movement, requiring grid motion in the physical plane that can be attacked with this transformation.

Figure 12.34 shows a mesh in two different time levels and its respective fixed transformed domain. The transformation involving the time coordinate will take into account the change in lengths, for example from $\overline{EA}$ to $\overline{E'A'}$.

The Geometric Conservation Law [13, 14], a name that catches our attention and is popular among finite difference analysts, is nothing more than a recommendation that algorithms should always be conservative, especially when moving meshes are employed. From a physical point of view, it is logical that the movement of the mesh should consistently increase or decrease the volume (m^3) of the control volume.

308   12 Coordinate Transformation—General Curvilinear Coordinate Systems

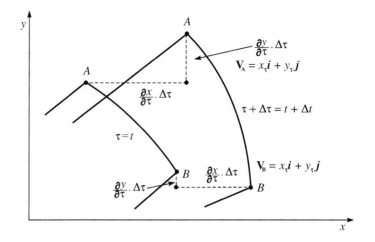

Fig. 12.33 Moving grid. Physical domain

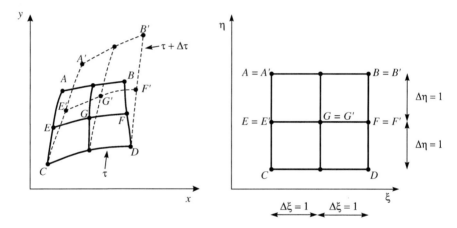

Fig. 12.34 Moving mesh in the physical and computational domain

For example, in a one-dimensional problem, for the same Δt a displacement of the left side of the control volume must be the same of the right side without the volume changing. The volume (of the control volume) must change and respect the conservation of volume. When finite differences are used, in which the concept of control volume does not exist, care should be exercised for avoiding the creation what could be called "numerical volumes", or "false volumes".

Still exercising the interpretation of terms of Eq. (12.143), it helps simplify it for a two-dimensional case by setting the parameters c, e and f to zero, and neglecting the advective term in the y direction, resulting in

12.4 Equations Written in the Curvilinear System

$$\frac{\partial}{\partial \tau}(\rho J \phi) + \frac{\partial}{\partial \xi}\left(\rho \tilde{U} \phi\right) + \frac{\partial}{\partial \eta}\left(\rho \tilde{V} \phi\right) = \frac{\partial}{\partial \xi}\left(\Gamma^\phi \frac{\alpha}{J} \frac{\partial \phi}{\partial \xi} - \Gamma^\phi \frac{\beta}{J} \frac{\partial \phi}{\partial \eta}\right)$$
$$+ \frac{\partial}{\partial \eta}\left(\Gamma^\phi \frac{\gamma}{J} \frac{\partial \phi}{\partial \eta} - \Gamma^\phi \frac{\beta}{J} \frac{\partial \phi}{\partial \xi}\right) + \hat{S}^\phi, \tag{12.160}$$

in which α, β and γ are the components g_{22}, g_{12} and g_{11} of the metric tensor, respectively, from the relations for a, b and d of Eq. (12.143) for the 2D case. The first term in the left-hand side takes care of the transient, while the second and third terms represents the advection of ϕ through the ξ and η faces, respectively. In the right hand side the diffusion of ϕ through ξ and η faces, and the source term.

It is the purpose of this book, whenever possible, to link the mathematics of the equations with physics. Here, in which Eq. (12.160) is written in a general curvilinear coordinate system, whose appearance is not familiar, it is an excellent opportunity for exercising that. Consider ϕ as the scalar temperature (T), and let's perform a heat balance in a 2D control volume centered in P shown in Fig. 12.35. The control volume has unity dimension in the third direction.

The heat balance gives,

$$\dot{Q}_w + \dot{Q}_s - \dot{Q}_e - \dot{Q}_n + Q_g \Delta V = \left.\frac{\Delta(\rho c_p T \Delta V)}{\Delta t}\right|_{CV} \tag{12.161}$$

in which $\dot{Q}$ represents the heat flux by advection and diffusion and Q_g represents a possible heat generation in W/m^3. To exemplify the calculation of this flux, the west boundary is used,

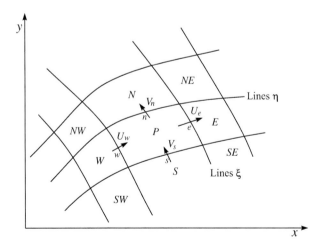

Fig. 12.35 Example. Energy balance in 2D

$$\dot{Q}_w = (\rho V_n A)_w (c_p T)_w + \left(-k \frac{\partial T}{\partial \mathbf{n}} A\right)_w$$

Recall that the velocity normal to the face is the physical contravariant velocity component, $\overline{V}^1$, as in Eq. 12.86. The area A is given by $dL_n = \sqrt{g_{22}} \Delta \eta = \sqrt{\alpha} \Delta \eta$, which is a length along the η axis. Introducing this heat flux and the similar ones for the other three faces in Eq. (12.161), it is obtained

$$\rho \overline{V}^1 \sqrt{\alpha} c_p T \Delta \eta \Big|_w - \rho \overline{V}^1 \sqrt{\alpha} c_p T \Delta \eta \Big|_e + \rho \overline{V}^2 \sqrt{\gamma} c_p T \Delta \xi \Big|_s - \rho \overline{V}^2 \sqrt{\gamma} c_p T \Delta \xi \Big|_n$$
$$- k \sqrt{\alpha} \frac{\partial T}{\partial \mathbf{n}} \Delta \eta \Big|_w + k \sqrt{\alpha} \frac{\partial T}{\partial \mathbf{n}} \Delta \eta \Big|_e - k \sqrt{\gamma} \frac{\partial T}{\partial \mathbf{n}} \Delta \xi \Big|_s + k \sqrt{\gamma} \frac{\partial T}{\partial \mathbf{n}} \Delta \xi \Big|_n$$
$$+ Q_g J \Delta \xi \Delta \eta = \frac{\partial}{\partial t} (\rho c_p J \Delta \xi \Delta \eta T) \qquad (12.162)$$

Expressing the normal derivatives of T as function of ξ and η, and recalling the relation among $\overline{V}^1$ and U and $\overline{V}^2$ and V, being U and V the contravariant velocity components without metric normalization, and for a fixed grid in space, the final equation is

$$J \frac{\partial}{\partial t}(\rho c_p T) + \frac{\partial}{\partial \xi}(\rho U c_p T) + \frac{\partial}{\partial \eta}(\rho V c_p T) = \frac{\partial}{\partial \xi}\left(\frac{\alpha}{J} k \frac{\partial T}{\partial \xi} - \frac{\beta}{J} k \frac{\partial T}{\partial \eta}\right)$$
$$+ \frac{\partial}{\partial \eta}\left(\frac{\gamma}{J} k \frac{\partial T}{\partial \eta} - \frac{\beta}{J} k \frac{\partial T}{\partial \xi}\right) + J Q_g, \qquad (12.163)$$

in which α, β and γ are the components of the metric tensor in 2D, repeated here for convenience,

$$\alpha = x_\eta^2 + y_\eta^2 \qquad (12.164)$$

$$\beta = x_\xi x_\eta + y_\xi y_\eta \qquad (12.165)$$

$$\gamma = x_\xi^2 + y_\xi^2 \qquad (12.166)$$

Equation (12.163) is exactly Eq. (12.160) for $\phi = T$. The other conservation equations, such as conservation of mass, momentum etc., can also be obtained by making the respective balances or by substituting the variable under consideration in the general equation. Note that the form of Eq. (12.163) is conservative, and therefore can be integrated over elementary volumes, generating a finite volume discretization.

12.5 Discretization of the Transformed Equations

In the previous section, the conservation equations were transformed from the physical domain, in this case the Cartesian domain (x, y, z), to the computational domain (ξ, η, γ). In this domain the control volumes are cubic regions and fixed, even if the grid in the physical domain moves. The transformation was performed by considering a general variable ϕ that retrieves the conservation equations for mass, momentum in the x, y and z directions, and energy, with ϕ assuming the values of 1, u, v, w and T, respectively.

In this section, the discretization of the 3D equation for the general variable ϕ will be performed. The general discretized equation will be particularized for the momentum equation, forming the system of u, v, w, and p. This system can be solved simultaneously, with the remaining scalars solved segregated, or all equations can be solved segregated. It will be considered a cell-center method, that is, the control volume is the element, what is recognized as the conventional finite volume method. Recall that when solving the $p - \mathbf{V}$ system simultaneously, if the grid is co-located, one needs to treat the co-located pressure–velocity coupling, while if it is solved segregated, both, the co-located and segregated couplings should be treated.

These coupling methods were already presented and discussed in Chapter 7 and can be applied here with no difficulties. Just guidelines on the details of those methods in curvilinear coordinates will be reported.

Re-writing Eq. (12.143) leaving explicitly the pressure term, one gets,

$$\frac{\partial}{\partial \tau}(J\rho\phi) + \frac{\partial}{\partial \xi}\left(\rho\tilde{U}\phi\right) + \frac{\partial}{\partial \eta}\left(\rho\tilde{V}\phi\right) + \frac{\partial}{\partial \gamma}\left(\rho\tilde{W}\phi\right)$$
$$= \frac{\partial}{\partial \xi}\left(\alpha_{11}\frac{1}{J}\Gamma^\phi\frac{\partial\phi}{\partial\xi} + \alpha_{12}\frac{1}{J}\Gamma^\phi\frac{\partial\phi}{\partial\eta} + \alpha_{13}\frac{1}{J}\Gamma^\phi\frac{\partial\phi}{\partial\gamma}\right)$$
$$+ \frac{\partial}{\partial \eta}\left(\alpha_{21}\frac{1}{J}\Gamma^\phi\frac{\partial\phi}{\partial\xi} + \alpha_{22}\frac{1}{J}\Gamma^\phi\frac{\partial\phi}{\partial\eta} + \alpha_{23}\frac{1}{J}\Gamma^\phi\frac{\partial\phi}{\partial\gamma}\right)$$
$$+ \frac{\partial}{\partial \gamma}\left(\alpha_{31}\frac{1}{J}\Gamma^\phi\frac{\partial\phi}{\partial\xi} + \alpha_{32}\frac{1}{J}\Gamma^\phi\frac{\partial\phi}{\partial\eta} + \alpha_{33}\frac{1}{J}\Gamma^\phi\frac{\partial\phi}{\partial\gamma}\right) - \hat{P}^\phi + \hat{S}^\phi \quad (12.167)$$

in which,

$$\alpha_{11} = J^2(\xi_x^2 + \xi_y^2 + \xi_z^2) \quad (12.168)$$

$$\alpha_{22} = J^2(\eta_x^2 + \eta_y^2 + \eta_z^2) \quad (12.169)$$

$$\alpha_{33} = J^2(\gamma_x^2 + \gamma_y^2 + \gamma_z^2) \quad (12.170)$$

$$\alpha_{31} = \alpha_{13} = J^2(\gamma_x\xi_x + \gamma_y\xi_y + \gamma_z\xi_z) \quad (12.171)$$

$$\alpha_{32} = \alpha_{23} = J^2(\eta_x \gamma_x + \eta_y \gamma_y + \eta_z \gamma_z) \qquad (12.172)$$

$$\alpha_{12} = \alpha_{21} = J^2(\xi_x \eta_x + \xi_y \eta_y + \xi_z \eta_z) \qquad (12.173)$$

All terms which contain α_{ij} are derived from the diffusion terms, and when $i \neq j$ is due to the nonorthogonality of the grid. These terms increase considerably the stencil of points involved in the discretization, increasing the band of the linear system. For example, if all α_{ij} with $i \neq j$ are zero the stencil will involve 7 control volumes, the central, and the 6 neighboring ones. When the grid is fully nonorthogonal, the stencil will involve 19 points.

Adopting a fully implicit formulation with two time levels for evaluation the transient term, the integration in time and in the ξ, η, γ directions in the control volume shown in Fig. 12.36, gives,

$$\frac{M_P \phi_P - M_P^o \phi_P^o}{\Delta t} + \dot{M}_e \phi_e - \dot{M}_w \phi_w$$
$$+ \dot{M}_n \phi_n - \dot{M}_s \phi_s + \dot{M}_f \phi_f - \dot{M}_b \phi_b$$
$$= \left[D_{11} \frac{\partial \phi}{\partial \xi} + D_{12} \frac{\partial \phi}{\partial \eta} + D_{13} \frac{\partial \phi}{\partial \gamma} \right]_e - \left[D_{11} \frac{\partial \phi}{\partial \xi} + D_{12} \frac{\partial \phi}{\partial \eta} + D_{13} \frac{\partial \phi}{\partial \gamma} \right]_w$$
$$+ \left[D_{21} \frac{\partial \phi}{\partial \xi} + D_{22} \frac{\partial \phi}{\partial \eta} + D_{23} \frac{\partial \phi}{\partial \gamma} \right]_n - \left[D_{21} \frac{\partial \phi}{\partial \xi} + D_{22} \frac{\partial \phi}{\partial \eta} + D_{23} \frac{\partial \phi}{\partial \gamma} \right]_s$$

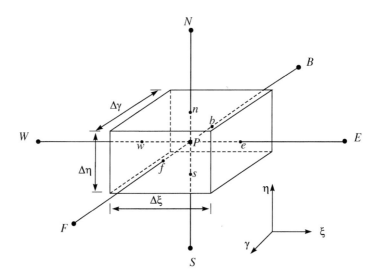

Fig. 12.36 3D control volume in the (ξ, η, γ) domain

12.5 Discretization of the Transformed Equations

$$+ \left[D_{31}\frac{\partial \phi}{\partial \xi} + D_{32}\frac{\partial \phi}{\partial \eta} + D_{33}\frac{\partial \phi}{\partial \gamma} \right]_f - \left[D_{31}\frac{\partial \phi}{\partial \xi} + D_{32}\frac{\partial \phi}{\partial \eta} + D_{33}\frac{\partial \phi}{\partial \gamma} \right]_b$$
$$- L\left[\hat{P}^\phi\right]_P \Delta V + L\left[\hat{S}^\phi\right]_P \Delta V \qquad (12.174)$$

in which the operator $L[\bullet]$ represents the numerical approximation of the term inside the brackets. The integration procedure applied to Eq. (12.174) is the same as seen in Chap. 6, when the 3D advection/diffusion equation was integrated. The terms in Eq. (12.174) are,

$$M_P = \rho_P J_P \Delta V, \qquad M_P^o = \rho_P^o J_P^o \Delta V$$
$$\dot{M}_e = \left(\rho \tilde{U}\right)_e \Delta \eta \Delta \gamma, \quad \dot{M}_w = \left(\rho \tilde{U}\right)_w \Delta \eta \Delta \gamma$$
$$\dot{M}_n = \left(\rho \tilde{V}\right)_n \Delta \xi \Delta \gamma, \quad \dot{M}_s = \left(\rho \tilde{V}\right)_s \Delta \xi \Delta \gamma$$
$$\dot{M}_f = \left(\rho \tilde{W}\right)_f \Delta \xi \Delta \eta, \quad \dot{M}_b = \left(\rho \tilde{W}\right)_b \Delta \xi \Delta \eta \qquad (12.175)$$

and they represent, respectively, the mass inside the control volume at time levels $t + \Delta t$ and t, and the mass flow rates through the six faces of the control volume. Figure 12.37 shows the connections between point P, which can't be seen is this figure, and its 18 neighbors control volumes, creating a matrix of the linear system with 19 diagonals.

The coefficients in the diffusive terms are,

$$D_{11} = \Gamma^\phi \frac{\alpha_{11}}{J} \Delta \eta \Delta \gamma, \quad D_{12} = \Gamma^\phi \frac{\alpha_{12}}{J} \Delta \eta \Delta \gamma, \quad D_{13} = \Gamma^\phi \frac{\alpha_{13}}{J} \Delta \eta \Delta \gamma$$

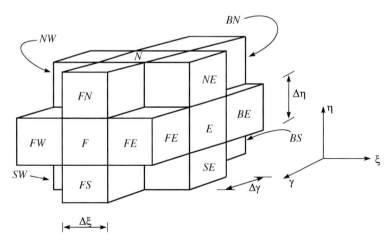

Fig. 12.37 Central control volume (hidden) and its 18 neighbors

$$D_{21} = \Gamma^\phi \frac{\alpha_{21}}{J} \Delta\xi \Delta\gamma, \quad D_{22} = \Gamma^\phi \frac{\alpha_{22}}{J} \Delta\xi \Delta\gamma, \quad D_{23} = \Gamma^\phi \frac{\alpha_{23}}{J} \Delta\xi \Delta\gamma$$

$$D_{31} = \Gamma^\phi \frac{\alpha_{31}}{J} \Delta\xi \Delta\eta, \quad D_{32} = \Gamma^\phi \frac{\alpha_{32}}{J} \Delta\xi \Delta\eta, \quad D_{33} = \Gamma^\phi \frac{\alpha_{33}}{J} \Delta\xi \Delta\eta \quad (12.176)$$

As a characteristic of the finite volume method the values of ϕ and $(\partial\phi/\partial x^i)$ are required at the integration points (e, w, n, s, f, b) for calculating the advective and diffusive terms at the interfaces of the control volume. Those evaluations demand the choice of a spatial interpolation function, topic already covered in Chap. 5, whose learnings apply directly for curvilinear coordinate systems. The interpolation scheme chosen is WUDS—Weighted Upstream Differencing Scheme, As seen in Chap. 5, the WUDS scheme uses the analytical solution of a 1D advection/diffusion problem written along a coordinate axis.

When the coordinate system is orthogonal, the diffusive flux can be represented just by one derivative, the direct derivative, while for nonorthogonal coordinate systems, the determination of the flux requires the derivative in all coordinate axis, which includes the cross-derivatives. Therefore, the 1D advection/diffusion problem used for determining the interpolation function uses the advective flux in full, but just the part of the diffusive flux in that direction [15].

In other words, all derivatives appearing in Eq. (12.174), multiplied by D_{ij}, with $i \neq j$, are evaluated by central differences, while those multiplied by D_{ii} will have the weighting factor depending on the advective and diffusive importance in the process. Of course, the entire diffusive term could be approximated by central differences without any weighting. This is in fact what happens, as the weighting factor for the diffusive terms quickly tends to unity with increasing velocity. As discussed previously, the diffusive terms pose no difficulties for the solution, neither introduces numerical diffusion, and can be treated with central differencing schemes.

To obtain the weighting factors for the WUDS scheme the following 1D problem is solved,

$$\frac{\partial}{\partial \xi}\left(\rho \tilde{U} \phi\right) = \frac{\partial}{\partial \xi}\left(a_{11} \frac{\Gamma^\phi}{J} \frac{\partial \phi}{\partial \xi}\right) \quad (12.177)$$

Following the procedures described in Chap. 5, the weighting factors $\overline{\alpha}$ and $\overline{\beta}$ for the advective and diffusive fluxes are, respectively,

$$\overline{\alpha} = \frac{r^2}{(10 + 2r^2)} \quad (12.178)$$

$$\overline{\beta} = \frac{(1 + 0.005r^2)}{(1 + 0.05r^2)}, \quad (12.179)$$

in which is the ratio $r = \dot{M}/D$ is the ratio of the advective and diffusive fluxes in the coordinate direction.

12.5 Discretization of the Transformed Equations

Knowing the weighting factors, the values and the derivatives of ϕ at the integration points become,

$$\phi_e = \left(\frac{1}{2}+\overline{\alpha}_e\right)\phi_P + \left(\frac{1}{2}-\overline{\alpha}_e\right)\phi_E, \quad \phi_w = \left(\frac{1}{2}+\overline{\alpha}_w\right)\phi_W + \left(\frac{1}{2}-\overline{\alpha}_w\right)\phi_P$$

$$\phi_n = \left(\frac{1}{2}+\overline{\alpha}_n\right)\phi_P + \left(\frac{1}{2}-\overline{\alpha}_n\right)\phi_N, \quad \phi_s = \left(\frac{1}{2}+\overline{\alpha}_s\right)\phi_S + \left(\frac{1}{2}-\overline{\alpha}_s\right)\phi_P$$

$$\phi_f = \left(\frac{1}{2}+\overline{\alpha}_f\right)\phi_P + \left(\frac{1}{2}-\overline{\alpha}_f\right)\phi_F, \quad \phi_b = \left(\frac{1}{2}+\overline{\alpha}_b\right)\phi_B + \left(\frac{1}{2}-\overline{\alpha}_b\right)\phi_P$$

(12.180)

The direct derivatives, which are part of the diffusive fluxes, are

$$\left.\frac{\partial\phi}{\partial\xi}\right|_e = \overline{\beta}_e\frac{(\phi_E-\phi_P)}{\Delta\xi}, \quad \left.\frac{\partial\phi}{\partial\xi}\right|_w = \overline{\beta}_w\frac{(\phi_P-\phi_W)}{\Delta\xi}$$

$$\left.\frac{\partial\phi}{\partial\eta}\right|_n = \overline{\beta}_n\frac{(\phi_N-\phi_P)}{\Delta\eta}, \quad \left.\frac{\partial\phi}{\partial\eta}\right|_s = \overline{\beta}_s\frac{(\phi_P-\phi_S)}{\Delta\eta}$$

$$\left.\frac{\partial\phi}{\partial\gamma}\right|_f = \overline{\beta}_f\frac{(\phi_F-\phi_P)}{\Delta\gamma}, \quad \left.\frac{\partial\phi}{\partial\gamma}\right|_b = \overline{\beta}_b\frac{(\phi_P-\phi_B)}{\Delta\gamma}, \quad (12.181)$$

while the cross derivatives approximated by central differences, are

$$\left.\frac{\partial\phi}{\partial\eta}\right|_e = \frac{\phi_N+\phi_{NE}-\phi_S-\phi_{SE}}{4\Delta\eta}, \quad \left.\frac{\partial\phi}{\partial\gamma}\right|_e = \frac{\phi_F+\phi_{FE}-\phi_B-\phi_{BE}}{4\Delta\gamma}$$

$$\left.\frac{\partial\phi}{\partial\eta}\right|_w = \frac{\phi_N+\phi_{NW}-\phi_S-\phi_{SW}}{4\Delta\eta}, \quad \left.\frac{\partial\phi}{\partial\gamma}\right|_w = \frac{\phi_F+\phi_{FW}-\phi_B-\phi_{BW}}{4\Delta\gamma}$$

$$\left.\frac{\partial\phi}{\partial\xi}\right|_n = \frac{\phi_E+\phi_{NE}-\phi_W-\phi_{NW}}{4\Delta\xi}, \quad \left.\frac{\partial\phi}{\partial\gamma}\right|_n = \frac{\phi_F+\phi_{FN}-\phi_B-\phi_{BN}}{4\Delta\gamma}$$

$$\left.\frac{\partial\phi}{\partial\xi}\right|_s = \frac{\phi_E+\phi_{SE}-\phi_W-\phi_{SW}}{4\Delta\xi}, \quad \left.\frac{\partial\phi}{\partial\gamma}\right|_s = \frac{\phi_F+\phi_{FS}-\phi_B-\phi_{BS}}{4\Delta\gamma}$$

$$\left.\frac{\partial\phi}{\partial\xi}\right|_f = \frac{\phi_E+\phi_{FE}-\phi_W-\phi_{FW}}{4\Delta\xi}, \quad \left.\frac{\partial\phi}{\partial\eta}\right|_f = \frac{\phi_N+\phi_{FN}-\phi_S-\phi_{FS}}{4\Delta\eta}$$

$$\left.\frac{\partial\phi}{\partial\xi}\right|_b = \frac{\phi_E+\phi_{BE}-\phi_W-\phi_{BW}}{4\Delta\xi}, \quad \left.\frac{\partial\phi}{\partial\eta}\right|_b = \frac{\phi_N+\phi_{BN}-\phi_B-\phi_{BS}}{4\Delta\eta} \quad (12.182)$$

Introducing the expressions for ϕ and its derivatives into Eq. (12.174), it results

$$A_P\phi_P = A_e\phi_E + A_w\phi_W + A_n\phi_N + A_s\phi_S + A_f\phi_F + A_b\phi_B$$
$$+ A_{ne}\phi_{NE} + A_{nw}\phi_{NW} + A_{se}\phi_{SE} + A_{sw}\phi_{SW}$$
$$+ A_{fe}\phi_{FE} + A_{fw}\phi_{FW} + A_{fn}\phi_{FN} + A_{fs}\phi_{FS}$$

$$+ A_{be}\phi_{BE} + A_{bw}\phi_{BW} + A_{bn}\phi_{BN} + A_{bs}\phi_{BS} + B \qquad (12.183)$$

or, in matrix form, as

$$[A][\phi] = [B] \qquad (12.184)$$

The coefficients of Eq. (12.183) are,

$$A_P = \sum (A_{nb}) + \frac{M_P^o}{\Delta t} \qquad (12.185)$$

$$A_e = -\dot{M}_e \left(\frac{1}{2} - \overline{\alpha}_e\right) + \frac{D_{11e}\overline{\beta}_e}{\Delta \xi} + \frac{(D_{21n} - D_{21s} + D_{31f} - D_{31b})}{4\Delta \xi} \qquad (12.186)$$

$$A_w = \dot{M}_w \left(\frac{1}{2} + \overline{\alpha}_w\right) + \frac{D_{11w}\overline{\beta}_w}{\Delta \xi} + \frac{(D_{21s} - D_{21n} + D_{31b} - D_{31f})}{4\Delta \xi} \qquad (12.187)$$

$$A_n = -\dot{M}_n \left(\frac{1}{2} - \overline{\alpha}_n\right) + \frac{D_{22n}\overline{\beta}_n}{\Delta \eta} + \frac{(D_{12e} - D_{12w} + D_{32f} - D_{32b})}{4\Delta \eta} \qquad (12.188)$$

$$A_s = \dot{M}_s \left(\frac{1}{2} + \overline{\alpha}_s\right) + \frac{D_{22s}\overline{\beta}_s}{\Delta \eta} + \frac{(D_{12w} - D_{12e} + D_{32b} - D_{32f})}{4\Delta \eta} \qquad (12.189)$$

$$A_f = -\dot{M}_f \left(\frac{1}{2} - \overline{\alpha}_f\right) + \frac{D_{33f}\overline{\beta}_f}{\Delta \gamma} + \frac{(D_{13e} - D_{13w} + D_{23n} - D_{23s})}{4\Delta \gamma} \qquad (12.190)$$

$$A_b = \dot{M}_b \left(\frac{1}{2} + \overline{\alpha}_b\right) + \frac{D_{33b}\overline{\beta}_b}{\Delta \gamma} + \frac{(D_{13w} - D_{13e} + D_{23s} - D_{23n})}{4\Delta \gamma}, \qquad (12.191)$$

The mass conservation should be introduced in the central coefficient A_P, as done in Chap. 5, to obtain the coefficients shown above. The coefficients due to the nonorthogonality of the grid are given by,

$$A_{ne} = \frac{D_{12e}}{4\Delta \eta} + \frac{D_{21n}}{4\Delta \xi}, \quad A_{nw} = -\frac{D_{12w}}{4\Delta \eta} - \frac{D_{21n}}{4\Delta \xi}, \quad A_{se} = -\frac{D_{12e}}{4\Delta \eta} - \frac{D_{21s}}{4\Delta \xi}$$

$$A_{sw} = \frac{D_{12w}}{4\Delta \eta} + \frac{D_{21s}}{4\Delta \xi}, \quad A_{fe} = \frac{D_{13e}}{4\Delta \gamma} + \frac{D_{31f}}{4\Delta \xi}, \quad A_{fw} = -\frac{D_{13w}}{4\Delta \gamma} - \frac{D_{31f}}{4\Delta \xi}$$

$$A_{fn} = \frac{D_{23n}}{4\Delta \gamma} + \frac{D_{32f}}{4\Delta \eta}, \quad A_{fs} = -\frac{D_{23s}}{4\Delta \gamma} - \frac{D_{32f}}{4\Delta \eta}, \quad A_{be} = -\frac{D_{13e}}{4\Delta \gamma} - \frac{D_{31b}}{4\Delta \xi}$$

$$A_{bw} = \frac{D_{13w}}{4\Delta \gamma} + \frac{D_{31b}}{4\Delta \xi}, \quad A_{bn} = -\frac{D_{23n}}{4\Delta \gamma} - \frac{D_{32b}}{4\Delta \eta}, \quad A_{bs} = \frac{D_{23s}}{4\Delta \gamma} + \frac{D_{32b}}{4\Delta \eta} \qquad (12.192)$$

The $L\left[\hat{P}^\phi\right]_P$ term, appears when ϕ is u, v or w, and assumes the following form for the three components of the momentum equations,

12.5 Discretization of the Transformed Equations

$$L[\hat{p}^u]_P = \frac{(p_E - p_W)}{2\Delta\xi}(J\xi_x)_P + \frac{(p_N - p_S)}{2\Delta\eta}(J\eta_x)_P + \frac{(p_F - p_B)}{2\Delta\gamma}(J\gamma_x)_P \qquad (12.193)$$

$$L[\hat{p}^v]_P = \frac{(p_E - p_W)}{2\Delta\xi}(J\xi_y)_P + \frac{(p_N - p_S)}{2\Delta\eta}(J\eta_y)_P + \frac{(p_F - p_B)}{2\Delta\gamma}(J\gamma_y)_P \qquad (12.194)$$

$$L[\hat{p}^w]_P = \frac{(p_E - p_W)}{2\Delta\xi}(J\xi_z)_P + \frac{(p_N - p_S)}{2\Delta\eta}(J\eta_z)_P + \frac{(p_F - p_B)}{2\Delta\gamma}(J\gamma_z)_P \qquad (12.195)$$

The parameter $\bar{\alpha}$ has the same sign of the velocity components and varies from -0.5 to 0.5, while $\bar{\beta}$ varies from 0 to 1. These are parameters which enters in the interpolation functions and it possible to set $\bar{\beta}$ always equal to 1, letting $\bar{\alpha}$ changing according to the Peclet number. The WUDS scheme, as discussed in Chap. 5, ranges from upwind to central differencing schemes.

An interesting approach to avoid the 19 diagonals matrix is to treat explicitly the nonorthogonal terms in Eq. (12.174), leaving in the source term all coefficients D_{ij} with $i \neq j$. There are no simplifications in terms of the mathematical model, but only with the respect of the solution procedure. If the nonorthogonality is severe, the source term may have undesired effects on the convergence rate of the iterative procedure. As it is always attempted to generate grids orthogonal as possible, aiming minimizing the nonorthogonality, this can be a good choice. Doing this, the resulting equation resembles the equation written in Cartesian coordinates, as

$$\begin{aligned}
&\frac{M_P \phi_P - M_P^o \phi_P^o}{\Delta t} + \dot{M}_e \phi_e - \dot{M}_w \phi_w \\
&+ \dot{M}_n \phi_n - \dot{M}_s \phi_s + \dot{M}_f \phi_f - \dot{M}_b \phi_b \\
&= \left[D_{11}\frac{\partial \phi}{\partial \xi}\right]_e - \left[D_{11}\frac{\partial \phi}{\partial \xi}\right]_w + \left[D_{22}\frac{\partial \phi}{\partial \eta}\right]_n - \left[D_{22}\frac{\partial \phi}{\partial \eta}\right]_s \\
&+ \left[D_{33}\frac{\partial \phi}{\partial \gamma}\right]_f - \left[D_{33}\frac{\partial \phi}{\partial \gamma}\right]_b + L[\hat{p}^\phi]_P \Delta V + L[\hat{S}^\phi]_P \Delta V \qquad (12.196)
\end{aligned}$$

The stencil will be of 7 points, with the usual seven-diagonal matrix with well-ordered diagonals, since, as the curvilinear grid is structured, the numbering of the control volumes follows a regular pattern. In the limit, of course, it recovers the Cartesian coordinate system.

12.6 Comments on the Solution of the Equation System

All strategies and techniques using Cartesian grids are all extended to curvilinear grids, considering, logically, the particularities of the curvilinear system, being the most important the nonorthogonality. The solution of the coupled system of equations can be done simultaneously or segregated.

It is considered that the co-located grid is used, that is, all variables are stored and calculated at the same point, as this is the commonly used arrangement in modern software. How the coupling methods work is the principal issue when curvilinear nonorthogonal grids are employed, and this will be discussed embodied in the strategy for the solutions of the linear systems of equation. For this analysis let's consider a 2D incompressible, isothermal flow for simplicity.

12.6.1 Simultaneous Solution

The system of equations to be solved comprises the Navier–Stokes equations for the u and v and the mass conservation equation. The system is,

$$\frac{M_P u_P - M_P^o u_P^o}{\Delta t} + \dot{M}_e u_e - \dot{M}_w u_w + \dot{M}_n u_n - \dot{M}_s u_s$$
$$= \left[D_{11} \frac{\partial u}{\partial \xi} + D_{12} \frac{\partial u}{\partial \eta} \right]_e - \left[D_{11} \frac{\partial u}{\partial \xi} + D_{12} \frac{\partial u}{\partial \eta} \right]_w$$
$$+ \left[D_{21} \frac{\partial u}{\partial \xi} + D_{22} \frac{\partial u}{\partial \eta} \right]_n - \left[D_{21} \frac{\partial u}{\partial \xi} + D_{22} \frac{\partial u}{\partial \eta} \right]_s +$$
$$- \left(\frac{(p_E - p_W)}{2\Delta \xi} (J\xi_x)_P + \frac{(p_N - p_S)}{2\Delta \eta} (J\eta_x)_P \right) \Delta V \qquad (12.197)$$

$$\frac{M_P v_P - M_P^o v_P^o}{\Delta t} + \dot{M}_e v_e - \dot{M}_w v_w + \dot{M}_n v_n - \dot{M}_s v_s$$
$$= \left[D_{11} \frac{\partial v}{\partial \xi} + D_{12} \frac{\partial v}{\partial \eta} \right]_e - \left[D_{11} \frac{\partial v}{\partial \xi} + D_{12} \frac{\partial v}{\partial \eta} \right]_w$$
$$+ \left[D_{21} \frac{\partial v}{\partial \xi} + D_{22} \frac{\partial v}{\partial \eta} \right]_n - \left[D_{21} \frac{\partial v}{\partial \xi} + D_{22} \frac{\partial v}{\partial \eta} \right]_s +$$
$$- \left(\frac{(p_E - p_W)}{2\Delta \xi} (J\xi_y)_P + \frac{(p_N - p_S)}{2\Delta \eta} (J\eta_y)_P \right) \Delta V \qquad (12.198)$$

$$\frac{M_P - M_P^o}{\Delta t} + \dot{M}_e - \dot{M}_w + \dot{M}_n - \dot{M}_s = 0 \qquad (12.199)$$

or, identifying the contravariant velocity components, as

12.6 Comments on the Solution of the Equation System

$$\frac{M_P - M_P^o}{\Delta t} + (\rho U \Delta \eta)_e - (\rho U \Delta \eta)_w + (\rho V \Delta \xi)_n - (\rho V \Delta \xi)_s = 0 \quad (12.200)$$

In matrix form the system given by Eqs. (12.197–12.199) reads,

$$\begin{bmatrix} A^{uu} & 0 & A^{up} \\ 0 & A^{vv} & A^{vp} \\ A^{pu} & A^{pv} & 0 \end{bmatrix} \begin{bmatrix} u \\ v \\ p \end{bmatrix} = \begin{bmatrix} B^u \\ B^v \\ B^p \end{bmatrix} \quad (12.201)$$

Moreover, when the solution is simultaneous, the segregated pressure–velocity coupling (SPVC) doesn't exist, since the coupling among the equations is resolved automatically. If the grid arrangement is staggered [16], which would eliminate the co-located coupling (CPVC) the solution of the system could be done simultaneously as it is in Eqs. (12.197–12.199), represented by Eq. (12.201).

However, staggered grids are prohibitive in 3D problems, therefore, the following developments consider co-located grids.

The Rhie and Chow-like methods, already discussed in Chap. 7, are techniques which finds the velocity at the interfaces based on the neighbor momentum equations, creating a coupling among pressure and velocity. The method written for Cartesian grids is expanded to curvilinear coordinates. Besides calculating the advecting velocity at the interfaces, this helps to better conditioning the matrix of Eq. (12.190) by removing the zero entry in the diagonal of the matrix. This is true for this system of equation in any situation.

The momentum equations are written for the Cartesian components and the mass conservation equation demands the contravariant velocity components for mass flow computations. Thus, the discretized contravariant velocity components should be written in terms of the discretized Cartesian components. For 2D, Eqs. (12.154) and (12.155) give these relations, repeated here for convenience,

$$U = u y_\eta - v x_\eta \quad (12.202)$$

$$V = v x_\xi - u y_\xi \quad (12.203)$$

According to Fig. 12.38, the term $(\rho U \Delta \eta)_e$ is used as example, in which the U_e must be calculated at the interface e. This velocity will be found using the discretized momentum equation for the u and v component. The discretized momentum equation for u at control volumes P and E are,

$$(A_P)_E u_P = \sum (A_{nb} u_{NB})_P + L\left[\hat{S}^u\right]_P \Delta V + \frac{M_P^o u_P^o}{\Delta t} - L[\hat{p}^u]_P \Delta V \quad (12.204)$$

$$(A_P)_E u_E = \sum (A_{nb} u_{NB})_E + L\left[\hat{S}^u\right]_E \Delta V + \frac{M_E^o u_E^o}{\Delta t} - L[\hat{p}^u]_E \Delta V \quad (12.205)$$

Fig. 12.38 Nodes used to find U_e^*

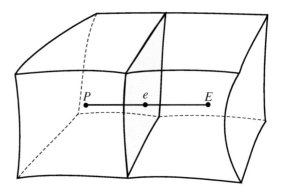

Using an average of Eqs. (12.204) and (12.205), keeping the consistent pressure gradient, as shown in Chap. 7, a pseudo-equation momentum equation for u_e is found. In a similar manner, using the discretized equations for v_e at control volumes P and E, as

$$(A_P)_P v_P = \sum (A_{nb} v_{NB})_P + L\left[\hat{S}^v\right]_P \Delta V + \frac{M_P^o v_P^o}{\Delta t} - L\left[\hat{p}^v\right]_P \Delta V, \quad (12.206)$$

$$(A_P)_E v_E = \sum (A_{nb} v_{NB})_E + L\left[\hat{S}^v\right]_E \Delta V + \frac{M_E^o v_E^o}{\Delta t} - L\left[\hat{p}^v\right]_E \Delta V, \quad (12.207)$$

it is obtained the pseudo-velocity for v_e. With u_e and v_e one can find the pseudo contravariant velocity which enters the mass conservation equation [17]. Repeating the procedure for the others control volume interfaces and, introducing all pseudo contravariant equations in Eq. (12.200), it is obtained a modified mass conservation equation which involves the neighbor velocities and pressures, promoting the required coupling among pressure and velocity. This modified mass conservation equation added to Eqs. (12.197) and (12.198) form the linear system to be solved,

$$\begin{bmatrix} A^{uu} & 0 & A^{up} \\ 0 & A^{vv} & A^{vp} \\ A^{pu} & A^{pv} & A^{pp} \end{bmatrix} \begin{bmatrix} u \\ v \\ p \end{bmatrix} = \begin{bmatrix} B^u \\ B^v \\ B^p \end{bmatrix} \quad (12.208)$$

in which it is possible to see that the diagonal is now complete. In Chap. 7 it is explained how this modified mass conservation equation rapidly reduces to the original equation as the grid is refined.

12.6.2 Segregated Solution

By solving the system of equations in a segregated fashion, each equation is solved independently, assuming all other variables as known. This translates into the need to handle the so-called segregated pressure-velocity coupling, SPVC, which requires an algorithm to be created to update the velocities and pressures for the next iterative cycle. These are the well-known SIMPLE-like methods for handling this coupling. When the mesh is co-located and the solution segregated, both couplings need to be handled. As just seen, the CPVC is treated with a Rhie and Chow-like method [17], and the SPVC with SIMPLEC, or similar. Just indications of how the velocity correction equations for curvilinear meshes should be constructed are now presented, since the combination of the Cartesian velocities to obtain the contravariant components follows what was seen for the Rhie and Chow-like method.

The mass conservation equation is again the main actor in the scheme for treating this coupling. At this time, it will be transformed in an equation for pressure, or pressure correction depending on the method, disappearing the velocities previously present in the mass conservation equation. Consider the SIMPLEC method, which uses the pressure correction P', to illustrate. Using Chap. 7, for the x component of the momentum conservation, the velocity correction has the following form,

$$u_e = u_e^* - \frac{L\left[\hat{P}'^{\,u}\right]_e \Delta\xi \Delta\eta \Delta\gamma}{A_p^u - \sum A_{nb}^u} \tag{12.209}$$

Again, what is required in the mass conservation equation is the contravariant velocity component, which can be obtained using the velocity correction for v_e, given by

$$v_e = v_e^* - \frac{L\left[\hat{P}'^{\,v}\right]_e \Delta\xi \Delta\eta \Delta\gamma}{A_p^v - \sum A_{nb}^v}, \tag{12.210}$$

which combined with Eq. (12.209) and help of Eq. (12.201) calculates the U_e velocity. Finding U_w, V_n and V_s, and substituting them in the mass conservation equation, Eq. (12.200), one finds an equation for p', which will involve, in this 2D case, 9 pressure points, instead of 4 when the grid is orthogonal, since $L\left[\hat{p}'^{\,u}\right]$ and $L\left[\hat{p}'^{\,v}\right]$ contain pressure derivatives in the ξ and η directions (see Eqs. (12.193–12.195)). The full iterative procedure is the same as seen in Chap. 7 and will not be repeated here.

Any SIMPLE-like method can be applied using general curvilinear coordinates, since, as already mentioned, all methods analyzed in the previous chapters using the Cartesian system are readily extended to any control volume with attention for the issues that were just presented. All speed flows, discussed in Chap. 9, were

12.7 Boundary Conditions

The application of boundary conditions of a given physical problem is one important part of numerical modeling. In real problems they are usually not clearly identified, and only detailed knowledge of the physics of the problem will allow choosing the best boundary condition. Knowing how to correctly choose the boundary conditions such that they do not destroy the quality of the solution, is part of the experience in solving engineering problems.

In Chap. 7, two ways of applying boundary conditions were presented: the use of fictitious points and the integration of the conservation equations for the boundary control volumes. In this section we will only recall the second procedure, trying to discuss only what concerns to the use of nonorthogonal coordinates.

The natural technique for applying boundary conditions in finite volume methods, and also for being consistent with the procedure adopted for internal volumes, is to perform balances of the physical quantity under consideration for the boundary volumes, incorporating the boundary condition into the approximate equation of the boundary volume.

The procedure is identical to that for the internal control volumes, i.e., integrate the differential equation in the conservative form over the boundary volume. Consider Fig. 12.39, where an eastern boundary is depicted in which it will be applied boundary conditions. Since it is being used co-localized variables all variables are at the center of the control volume. Writing the advective and diffusive terms of the general equation given by Eq. (12.167) in 2D for the ξ direction, one has

$$\frac{\partial}{\partial \xi}\left(\rho \tilde{U} \phi\right)$$

$$\frac{\partial}{\partial \xi}\left(\Gamma^\phi J \alpha_{11} \frac{\partial \phi}{\partial \xi} + \Gamma^\phi J \alpha_{12} \frac{\partial \phi}{\partial \eta}\right),$$

in which it should be recognized that in the diffusive flux the term in the parentheses is the normal derivative of ϕ in a ξ constant surface. To obtain the discretized equation for the volume centered at P, in this case a boundary volume, we must integrate the conservation equation in this volume. For example, integrating only the advective and diffusive terms, as shown, it is obtained, for the advective term,

$$\rho \tilde{U} \phi \bigg|_e - \rho \tilde{U} \phi \bigg|_w \qquad (12.211)$$

and for the diffusive,

12.7 Boundary Conditions

Fig. 12.39 Boundary control volume

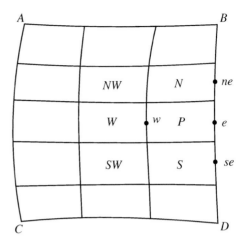

$$\left(\Gamma^\phi J \alpha_{11} \frac{\partial \phi}{\partial \xi} + \Gamma^\phi J \alpha_{12} \frac{\partial \phi}{\partial \eta} \right)_e$$
$$- \left(\Gamma^\phi J \alpha_{11} \frac{\partial \phi}{\partial \xi} + \Gamma^\phi J \alpha_{12} \frac{\partial \phi}{\partial \eta} \right)_w \tag{12.212}$$

For the terms evaluated in w face, the procedure is identical to that applied to the internal volumes, since this boundary is internal and an interpolation function applies. For the e face, one must substitute the advective and diffusive fluxes by the existing ones at the boundary. It is always helpful to remember that the application of boundary conditions in a finite volume methodology is always specifying the advective and diffusive fluxes, in any type of boundary condition. If the flux is not known, it should be found with the given values of the variable. The various types of boundary conditions are discussed below.

12.7.1 No-Flow Boundary ($\rho U = 0$). ϕ Prescribed

For this case, the diffusive flux, i.e. the value in parentheses in Eq. (12.212), on face e must be approximated by employing the values of prescribed ϕ at the boundary. For example, if a distribution of ϕ is prescribed on the eastern boundary ($\xi = $ constant), the derivatives that appear in Eq. (12.200) can be evaluated by,

$$\left. \frac{\partial \phi}{\partial \xi} \right|_e = \frac{(\phi_e - \phi_P)}{\frac{\Delta \xi}{2}} \tag{12.213}$$

$$\left. \frac{\partial \phi}{\partial \eta} \right|_e = \frac{\phi_{ne} - \phi_{se}}{2 \Delta \eta}, \tag{12.214}$$

in which ϕ_{ne}, ϕ_e and ϕ_{se} are at the boundary and, therefore, are known values. When Eqs. (12.213) and (12.214) are inserted in Eq. (12.212), which is part of the full equation, the discretized equation for the boundary control volume is obtained with the boundary conditions embedded.

12.7.2 No-Flow Boundary ($\rho U = 0$). Flux of ϕ Prescribed

Remembering that, in Eq. (12.212), the expression

$$\left(\Gamma^\phi J \alpha_{11} \frac{\partial \phi}{\partial \xi} + \Gamma^\phi J \alpha_{12} \frac{\partial \phi}{\partial \eta} \right)_e$$

is exactly the flux per unit of area crossing the east face of the control volume. So, it is enough to directly replace the term in parentheses by the prescribed value.

12.7.3 Bounday With Mass Flow ($\rho U \neq 0$). Mass Entering With ρU Known

Referring again to Fig. 12.39, consider a known mass influx on the east (e) face. In this case, there can be no diffusive flow on this face because, if there were, it would be affecting the upstream distribution, and therefore it would not be possible to prescribe the advective flow. In other words, the boundary of the calculation domain could not have been chosen in that position. As usual, if there is mass input at the boundary, for the value of ϕ to be prescribed, the value of the diffusive flux of ϕ must be prescribed zero. If the numerical analyst "senses" that the diffusive flux is important on that boundary, this means that the chosen location is not suitable for prescribing boundary conditions. The computational domain should be removed to a position where clearly boundary conditions are available, or one can apply the above condition knowing possible errors. It is an important task in mathematical modeling to properly choose the boundaries of the calculation domain and the corresponding boundary conditions.

12.7.4 Boundary With Mass Flow ($\rho U \neq 0$). Mass Leaving With ρU Unknown

In boundaries with mass output, the output velocity distribution is rarely known, and even the total mass flow rate may be unknown. In most cases, the prescribed condition is pressure. For these cases, the usual condition is to prescribe locally parabolic condition for the velocity. The values of ϕ at the outlet boundary are in general dictated by the solution of the problem and can be extrapolated from internal values. With respect to the diffusive flow, as the condition itself indicates, it must be prescribed equal to zero, and the comments made earlier again applies.

12.8 Conclusions

In this chapter, the coordinate transformation from the Cartesian system to a general curvilinear system was presented, linking its application in numerical methods. Emphasis was given to the geometric interpretation of the transformation relationships, always seeking to make clear points that, as we understand, cause considerable difficulty in learning numerical methods for general control volumes. The existing relationship between the physical components and the covariant and contravariant components of a vector was emphasized, since the intimacy with these representations is mandatory for learning finite volume methods in which balances are realized and, therefore, fluxes need to be calculated.

In curvilinear systems it is possible to write the conservation equations in the curvilinear system, allowing the solution of the equations to be obtained in the transformed domain, fixed in space and even in time. The computer code, therefore, is written for a fixed geometry, independent of the physical domain, which can change, whose real dimensions are informed to the transformed equations by the metrics of the transformation.

The transformation was done in its complete form, that is, involving the three spatial coordinates (ξ, η, γ) and t, allowing problems, in which the grid in the physical plane moves, to be solved also in the fixed computational plane. To write the conservation equations in the curvilinear system (or, in other words, in the fixed computational domain) has an important peculiarity from a didactic point of view, which is to show the automatic appearance of the grid velocity, and thus the relative velocity, in the advective terms.

It was also exercised to obtain the energy equation in an arbitrary domain using the coordinate transformation, as well as performing an energy balance in the arbitrary control volume. This is the strongest way of seeing the connection among the physics and the mathematical relations of the coordinate transformation.

The integration of the transformed equations was also realized in this chapter. It was observed that the resulting discretized equation differs from the Cartesian ones by the fact that the pressure and diffusive terms involve derivatives in all directions, leading to a 19 stencil for the discretized equations. A good alternative to avoid this large stencil is to carry explicitly the cross-derivative terms of the pressure and diffusive terms. In general, the nonorthogonality is not so severe, and the cross-derivative terms are small compared to the one in the principal direction.

Few words about boundary conditions were also given, pointing out that the same philosophy applies for all finite volume formulation, that is, applying balances of the quantity for the control volumes at the boundaries, incorporating the boundary condition in the approximate equation.

Curvilinear coordinates system, widely known by boundary-fitted grids, still have room in specific niches of engineering. However, nowadays, general CFD codes employs unstructured grids, due to its generality and flexibility, the topic of the coming chapter.

12.9 Exercises

12.1 Obtain Eqs. (12.12–12.14) and (12.18–12.20)
12.2 For the following orthogonal curvilinear systems, find the metrics given by Eq. (12.18–12.20):

a. Cylindrical parabolic

$$x = 0,5(\xi^2 - \eta^2)$$
$$y = \xi\eta$$
$$z = \gamma$$

b. Cylindrical elliptic

$$x = a\cosh\xi \ \cos\eta$$
$$y = a\,\text{senh}\,\xi \ \text{sen}\,\eta$$
$$z = \gamma$$

12.3 Find the expression for the volume of the element in the cylindrical elliptic coordinate system.
12.4 Show that for the cylindrical parabolic and elliptic coordinate systems the components of the metric tensor g_{ij} with $i \neq j$ are equal to zero.
12.5 Obtain, for a general curvilinear system the expressions of $\nabla\phi$, $\nabla \cdot \mathbf{F}$ and $\nabla^2\phi$.
12.6 Show that it is not possible to calculate the mass flow rate crossing an element of length over a coordinate line using only one component of the covariant velocity vector.
12.7 Using the following nonorthogonal transformation

$$\xi = 2,5x - 5y$$
$$\eta = 2x + 4y,$$

obtain:

a. the inverse of the transformation;

12.9 Exercises

b. g_{ij} and g^{ij};
c. the Jacobian of the transformation and of the inverse;
d. the lengths on the coordinate lines
e. the distance between the origin of the system and the point (0.4;0.2);
f. the covariant and covariant base vectors;
g. the angle between the lines ξ and η using the g_{ij} components.

12.8 Consider the vector $\mathbf{V} = 0.2\mathbf{i} + 0.4\mathbf{j}$ and the coordinate system given in the previous exercise, and find:

a. the Cartesian, Contravariant and Covariant components of the given vector;
b. the expressions of $\mathbf{V}$ in the covariant and contravariant base;
c. the value of the normal of $\mathbf{V}$ to the lines ξ and η in a generic point (x, y).

12.9 Multiply Eq. (12.96) by (-1), recalculate all metrics as done in Sect. 12.2.7 and see that the Jacobian and the value of β result negatives. Explain why this happens. Draw this new coordinate system and compare with the one of Sect. 12.2.7.

12.10 For a 2D nonorthogonal system (ξ, η), find $(\partial\phi/\partial\mathbf{n})_{\xi=cte}$ and $(\partial\phi/\partial\mathbf{n})_{\eta=cte}$.

12.11 Draw the computational domain for the grid shown in Fig. 12.40. If a heat conduction problem is solved in the domain, with boundary conditions of constant temperature prescribed at $\overline{3456}$ and different temperatures on the lower and upper sides of the rhombus, how the boundary conditions will be in the computational domain? Think about the numbering of the control volumes in the computer code and about the storage of the boundary condition information on the rhombus.

12.12 Consider an incompressible isothermal fluid flow coming from the left in the geometry with the grid shown in Fig. 12.41. Draw the grid in the computational domain and specify the boundary conditions in this domain.

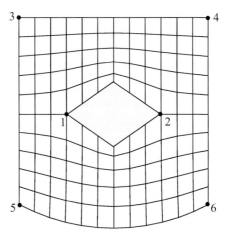

Fig. 12.40 Figure for Problem 12.11

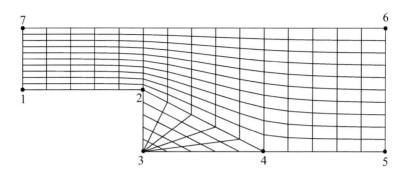

Fig. 12.41 Figure for Problem 12.12

12.13 Draw three different grids fitting in a square with their respective transformed plane.

12.14 For the geometry shown in Fig. 12.42, draw 5 lines ξ and 4 lines η and give the discrete correspondence between the coordinate points (x, y) and (ξ, η), in the form given by Eq. (12.115).

12.15 Use the elliptic coordinate system (2D) presented in Prob. 12.2 (b), with ξ ranging from 1 to M and η from 1 to N, where M and N are integers. With the analytic transformation, calculate analytically all the metrics and information of interest, such as g_{11}, g_{22} g_{12}, J, lengths along the coordinate lines ξ and η, areas etc. Now, using the points (x, y) obtained from the analytic transformation, calculate numerically the metrics and parameters of interest. Compare the results. Vary M and N.

12.16 When solving flow problems inside straight ducts of variable cross-section, as shown in Fig. 12.43, it is possible to march parabolically along the duct axis. Thus, only two calculation sections need to be stored simultaneously.

Fig. 12.42 Figure for Problem 12.14

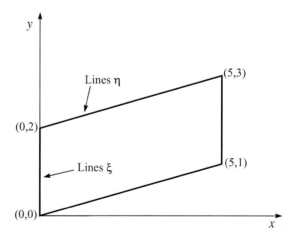

12.9 Exercises

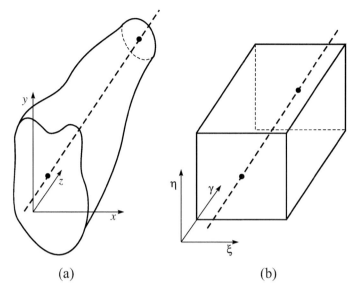

Fig. 12.43 Figure for Problem 12.16

If the marching planes are parallel, even if the cross-section changes with z, the following coordinate transformation applies,

$$\xi = \xi(x, y, z)$$
$$\eta = \eta(x, y, z)$$
$$\gamma = z$$

Considering the three-dimensional parabolic, laminar, incompressible flow inside a duct with the above characteristics, obtain the transformed equation for a generic variable ϕ, remembering that the mesh does not change with time. In the general equation for ϕ consider the advective and diffusive fluxes without source term. Show that the expressions for the contravariant components, without metric normalization, U, V and W will have the form,

$$U = y_\eta u - x_\eta v + (y_\gamma x_\eta - x_\gamma y_\eta)w$$
$$V = x_\xi v - y_\xi u + (x_\gamma y_\xi - y_\gamma x_\xi)w$$
$$W = wJ$$

12.17 The heat flux vector **q** is known from its Cartesian components q_x'' and q_y''. Determine the amount of heat per unit of time that passes through the area $\overline{AB}$. Then, using central differences, relate this flux as function of the temperatures at the points shown in Fig. 12.44.

Fig. 12.44 Figure for problem 12.17

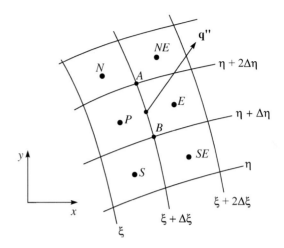

Fig. 12.45 Figure for problem 12.18

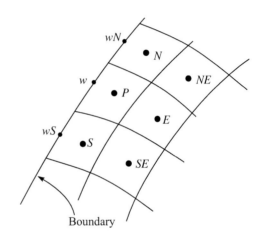

12.18 Using the two-dimensional form of Eq. (12.167) without pressure and source terms, obtain the approximate equation for the boundary volume P of Fig. 12.45. At the boundary, which has no mass flow, ϕ is a prescribed function, i.e., with known values in w, wN and wS.

References

1. Aris R (1989) Vectors, tensors, and the basic equations of fluid mechanics. Unabridged, October 1989
2. Korn GA, Korn TM (1972) Mathematical handbook for scientist and engineers. McGraw-Hill
3. Thompson JF, Warsi ZUA, Mastin CW (1985) Numerical grid generation foundations and applications. Elsevier Science Publishing Co., EUA

4. Milne-Thompson LM (1950) Theoretical hydrodynamics, 2a. d. MacMillan, Nova York
5. Winslow AM (1967) Numerical solution of quasi-linear Poison equation in a non-uniform triangle mesh. J Comput Phys 2:149–172
6. Chu WH (1971) Development of a general finite difference approximation for a general domain-Part I: machine transformation. J Comput Phys 8:392–408
7. Barfield WD (1970) An optimal mesh generator for Lagrangean hydrodynamics calculation in two-space dimensions. J Comput Phys 6:417–429
8. Amsden AA, Hirt CW (1973) A simple scheme for generating general curvilinear grids. J Comput Phys 11:348–359
9. Godunov SK, Prokopov GP (1968) The solution of differential equations by the use of curvilinear difference networks. USSR Comput Math Phys 8:1
10. Thompson JF, Thames FC, Mastin WC (1974) Automatic numerical generation of body-fitted curvilinear coordinate system for field containing any number of arbitrary two-dimensional bodies. J Comput Phys 15:299–319
11. Thompson JF, Thames FC, Mastin WC (1976) Boundary-fitted curvilinear coordinate system for solution of partial differential equations on fields containing any number of arbitrary two-dimensional bodies. NASA Langley Research Centre CR-2729, EUA
12. Thompson JF, Thames FC, Mastin WC (1977) TOMCAT–A code for numerical generation of boundary-fitted curvilinear coordinate systems on fields containing any number of arbitrary two-dimensional bodies. J Comput Phys 24:274–302
13. Anderson DA, Tannehill JC, Pletcher RH (1984) Computational fluid mechanics and heat transfer. Hemisphere Pub. Corporation
14. Thomas PD, Lombard CK (1978) The geometric conservation law—a link between finite-difference and finite-volume methods of flow computation in moving grids. AIAA Paper 78-1208, Seattle, Washington
15. Maliska CR (1981) A solution method for three-dimensional parabolic fluid flow problems in nonorthogonal coordinates. PhD Thesis, University of Waterloo, Waterloo, Canada
16. Maliska CR, Raithby GD (1984) A method for computing three dimensional flows using non-orthogonal boundary-fitted co-ordinates. Int J Numer Methods Fluids 4(4):519–537
17. Marchi CH, Maliska CR (1994) A nonorthogonal finite volume method for the solution of all speed flows using co-located variables. Numer Heat Transfer 26:293–311

Chapter 13
Unstructured Grids

13.1 Introduction

When the computational domain is complex, with holes and protuberances, structured grids can still be used subdividing the domain into blocks, each one being structured. However, when the geometry is too complex it generates excessive number of blocks, requiring special algorithms to deal with the transfer of information from block-to-block. The number of blocks can be so large that it resembles an unstructured grid. Thus, the alternative is to use full unstructured grids with control volumes of different shapes, taking advantage of the inherent flexibility. Adaption and grid refinement in specific regions of the domain are more easily achieved, and highly complex geometries can be handled.

On the other hand, the algorithms for solving the discretized equations grow a bit in complexity. The methods for solving linear systems are more elaborate, since ordering, which is a trivial procedure in structured meshes, assumes some importance in unstructured meshes, because the bandwidth of the coefficient matrix is dependent on the nature of this ordering. In multigrid methods, the algorithm for clustering the volumes is also more involved if an algebraic method is employed. However, great flexibility and generality is attained, and unstructured grids are, nowadays, the discretization choice in the major CFD applications in the market.

An unstructured discretization can consist, usually, of triangular and quadrangular elements in 2D, tetrahedra, hexahedra, prims and pyramids in 3D, as shown in Fig. 13.1. The elements can be used as control volume and agglomerated to obtain polyhedral grids in the context of cell-center finite volume methods, or using part of them in the construction of the control volume, as in cell-vertex finite volume methods, which will be seen in detail for the Element-based Finite Volume Method—EbFVM.

Historically, the use of unstructured meshes has always been associated with the finite element method with application in structural analysis, since this is the predominant area of its application. This experience has led the developers of finite element be pioneers in the solution of fluid flow problems [1] in unstructured grids.

© The Author(s), under exclusive license to Springer Nature Switzerland AG 2023, corrected publication 2023
C. R. Maliska, *Fundamentals of Computational Fluid Dynamics*, Fluid Mechanics and Its Applications 135, https://doi.org/10.1007/978-3-031-18235-8_13

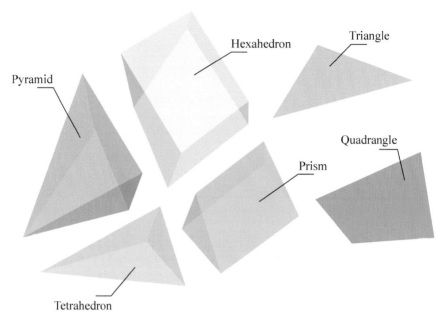

Fig. 13.1 Elements in 2D and 3D

Unstructured grid strategies, however remained dormant in the aerodynamics area until the schemes were cast in the framework of finite volume [1].

The use of unstructured grids, predominantly with triangular and tetrahedral meshes, has a clear beginning in aerodynamics with the pioneering works described in [2–5], among others. In solving low speed flows with heat transfer the developments started a little later, with the works reported in [6–10]. The pioneering work of Baliga and Patankar [6], already using the concept of elements, gained prominence with the works of Raw and Schneider extending to quadrangular elements and using the shape functions as in finite element. A summary of these procedures can be found in [11] (Sect. 11.4) and a review of unstructured grid adaption can be found in [12]. In the last three decades the developing pace continued, making available to the numerical analyst a myriad of techniques for fluid flow calculations using finite volumes, with some of them fully validated by commercial applications.

This chapter aims bringing to the reader the finite volumes method for unstructured meshes, covering the cell-center and cell-vertex methods. In the class of cell-vertex methods, the EbFVM will be the focus. Voronoi diagrams, the dual of the Delaunay triangulation, because they have local orthogonality, offer the possibility of developing a formulation where only two nodal points need to be employed for fluxes calculation. Formulation using Voronoi diagrams will be seen in the context of cell-center methods.

Again, we will try to describe the methodologies in a way that they can serve as a guideline for those who begin the studies in unstructured meshes having the

13.1 Introduction

knowledge of the previous fundamental chapters, without the concern to review and cite the variants of existing methods in the literature, which are so many and very easy to find online.

Before entering in details of the cell-center and cell vertex methods, let's point out two main characteristics of them, already seen in Chap. 3, but that is worth to recap. Inspecting Fig. 13.2a, in which the element is taken as control volume, for a quadrangular element, there will be four fluxes for the balance. If the elements are used to build the control volume, as in Figs. 11.9 and 13.2b, more fluxes will be involved, what improves accuracy. The second point is about boundary conditions. In the cell-center there is no grid nodes on the boundary and all control volumes are integral, and balances will be performed for all of them. The cell-vertex variant has grid nodes on the boundary, with the appearance of half volumes at the edges and a quarter of a volume at the corners in two dimensions. Dirichlet boundary conditions are easy to apply, but conservation will be not obeyed for the control volumes at the boundaries, unless some special scheme is constructed.

The discretization of the equation will be done for a general scalar ϕ, given by

$$\frac{\partial}{\partial t}(\rho\phi) + \nabla.(\rho V \phi) = \nabla.(\Gamma^\phi \nabla \phi) + S^\phi, \qquad (13.1)$$

Integrating in the control volume, despite of its shape,

$$\int_V \left\{ \frac{\partial}{\partial t}(\rho\phi) + \nabla.(\rho V \phi) = \nabla.(\Gamma^\phi \nabla \phi) + S^\phi \right\} dV, \qquad (13.2)$$

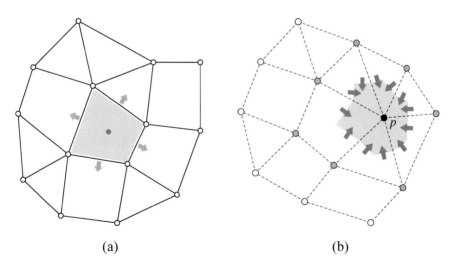

Fig. 13.2 Fluxes in the cell-center (**a**) and cell-vertex (**b**)

it is obtained, after using a first order approximation (interpolation in time) for the evaluation of the transient term,

$$\frac{M_P \phi_P - M_P^o \phi_P^o}{\Delta t} = -\sum_{ip} \left(\rho (V.n) \underline{\phi} \Delta s \right)_{ip}$$

$$+ \sum_{ip} \left(\Gamma^\phi \underline{\nabla \phi}.n \Delta s \right)_{ip} + (S_P \phi_P + S_C) \Delta V \qquad (13.3)$$

The reader is referred to Chap. 3 to review the basic procedures of finite volume methods. The interpolation required in space in Eq. 13.3 are for ϕ_{ip} and $(\nabla \phi)_{ip}$, the terms underlined in Eq. (13.3). The interpolation for ϕ_{ip} was extensively discussed in Chap. 5, and the outcome of those considerations applies entirely here. The interpolation for $(\nabla \phi)_{ip}$ was exercised in Chap. 3, when the introduction to finite volume method was presented using the 1D heat diffusion equation. As the Cartesian coordinate system was used there, the calculation of the gradient went almost unnoticed, since the components of the gradient of ϕ, $(\partial \phi / \partial x)_{ip}$ and $(\partial \phi / \partial y)_{ip}$ was easily found. In unstructured grids this calculation is a little bit more elaborated, particularly in the cell-center methods. How to do this will be seen in the following sections.

For the sake of simplicity, in this chapter, all pictures used in the mathematical developments will be done for 2D situations, despite the fact that most of the developments done by the author and colleagues were done for 3D. Few results will be shown in Chap. 15. Equation (13.3) will be followed, term by term, seeking its full discretization, starting with the cell-center methods.

13.2 Cell-Center Methods

13.2.1 Conventional Finite Volume Method

13.2.1.1 Determination of $(\nabla \phi)_{ip}$

Figure 13.3 shows an unstructured grid with evidence of a control volume coincident with the element, that is, a cell-center approach. The grid node P and its neighbors should be used for finding the gradient of ϕ at point P, which is, then, used for finding the gradient at point ip. This figure also shows that in unstructured grids the nodes are not aligned with the axis, precluding the calculation as done in Cartesian meshes. It could be possible, of course, interpolating the neighbors grid nodes on the x and y axis passing on point P and calculating $(\partial \phi / \partial x)_{ip}$ and $(\partial \phi / \partial y)_{ip}$, but it would be a cumbersome calculation with lots of interpolations. Another alternative would be to calculate the direct part of the gradient in the direction joining P and N, and using the neighbor grid nodes to find the remaining part of the gradient in the cross direction [13, 14]. The basis vector seen in Chap. 12 are helpful in this calculation,

13.2 Cell-Center Methods

Fig. 13.3 Gradient calculation. Cell-center method

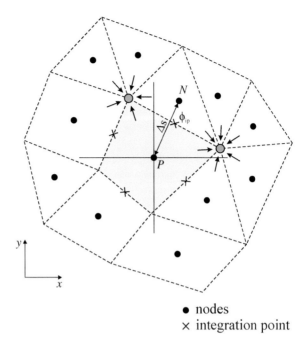

- nodes
× integration point

since a local nonorthogonal coordinate system (ξ, η) (not shown in Fig. 13.3) can be set at the integration point. The derivatives $(\partial\phi/\partial\xi)_{ip}$ and $(\partial\phi/\partial\eta)_{ip}$ could be determined, and then, $(\partial\phi/\partial\mathbf{n})_{ip}$. This form of calculation recovers the two-point flux approximation when the grid is orthogonal, since the cross derivatives are multiplied by a component of the metric tensor which is zero for orthogonal systems.

Probably, the best way and easier to automate its calculation inside the full algorithm, is to apply one of the gradient recovering methods available. Three methods will be described, the cell-based, node-based and least-square cell-based.

One form of the Green-Gauss's theorem is given by,

$$\int_V \nabla\phi \, dV = \int_S \phi \, dS, \qquad (13.4)$$

in which the surface S defines a closed region in space with volume V. Applying the mean value theorem and replacing the integral by its discrete counterpart, one gets

$$(\nabla\phi)_P \approx \frac{1}{\Delta V_P} \sum_{ip=1}^{N} \phi_{ip} \Delta S_{ip}, \qquad (13.5)$$

in which ϕ_{ip} is the value at the centroid of the face (at the integration point) and N is the number of faces of the respective control volume. The values of ϕ_{ip} remain to be found based on the values of ϕ at the grid nodes. Two approaches are available for this task, the cell-based and the vertex-based [15, 16].

Cell-Based Method

For a general integration point ip, the value of ϕ_{ip} is found as an averaging of the values of ϕ at grid nodes P and N, considering the distance of these points to the point ip. . If the grid is fairly homogenous a simple arithimetic average suffices (see Fig. 13.3).

Knowing all ϕ_{ip} belonging to the volume, Eq. (13.5) can be applied and the gradient at point P can be determined. The gradient of ϕ, $(\nabla \phi)_{ip}$, at the integration points can be found by an averaging of the gradients of the grid nodes. This averaging must take in account the non-uniformity of the grid.

Vertex-Based Method

The vertex-based method uses the value of ϕ at the vertices of the face to obtain the value of ϕ at the centroid of the face, ϕ_{ip}. As there are no values available at the vertices of the face, they are found from an interpolation of neighbor volumes that shares that vertex, as depicted in Fig. 13.3. The small arrows points which grid nodes feed the vertex. The value of ϕ_{ip} is found through an average of the values at the vertices. To find the gradient at the center of the control volume Eq. (13.5) applies again.

In three dimensions, it is common the grid generators to deliver tetrahedral elements, which, if a cell-center method is used, means to have one unknown for each tetrahedral element, what implies in a large linear system to be solved. To reduce the number of unknowns, the tetrahedral elements can be cast in a polyhedral grid. This procedure increases the number of surfaces in which fluxes will be calculated, improving accuracy. Figures 13.4 and 13.5, respectively, show a polygonal grid constructed with triangles for 2D, and a polyhedral grid constructed with pyramids in 3D. The commercial package [16] has the option of converting grids of tetrahedra and other elements in polyhedral grids. On the grids shown in Figs. 13.4 and 13.5, for 2D and 3D the gradient recovery methods just described were applied [17].

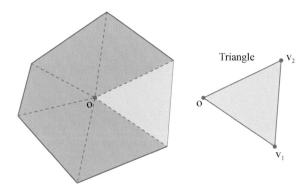

Fig. 13.4 Polygonal grid (made of triangles)

13.2 Cell-Center Methods

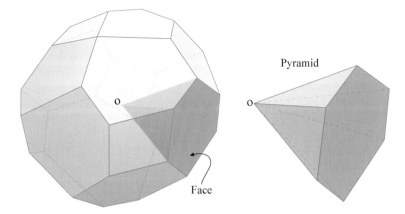

Fig. 13.5 Polyhedral grid (made of pyramids)

Least Square Minimization

An alternative way to reconstruct the gradient from a set of discrete values of a scalar variable may be done when a linear variation is assumed in the region around a central point. Then, a set of linear equations is obtained in which the unknowns are the components of the gradient. In general, there will be more equations than unknowns, because the unknowns are the three components of the gradient and the number of neighbors is almost always more than that. Therefore, the components of the gradient can be determined by considering an auxiliary least square problem.

The reconstruction of the gradient with this alternative methodology is generally more robust than those applying Gauss's theorem. With the least square technique, the gradient reconstruction of linear functions is exact and with nonlinear functions the approximation error tends to zero as the meshes are refined, even when they present severe distortions [5]. Another important aspect of this method is that for its application, geometric information of the mesh cells is not necessary. It is only necessary to know the coordinates of the points where the discrete values of the variable were defined. For cells adjacent to the boundaries, it is not necessary to consider values of the variable on the boundaries either, as happens with methods based on the Gauss theorem.

To demonstrate the method, a polygonal grid, shown in Fig. 13.6, constructed with triangles will be used. Applying a Taylor series expansion not considering terms with second or higher order around point P for all its neighboring points,

$$\phi_{NB} = \phi_P + (\nabla\phi)_P \Delta r_{P,NB}, \tag{13.6}$$

or,

$$\Delta r_{P,NB}(\nabla\phi)_P = \phi_{NB} - \phi_P, \tag{13.7}$$

Fig. 13.6 Least square method in a polygonal grid

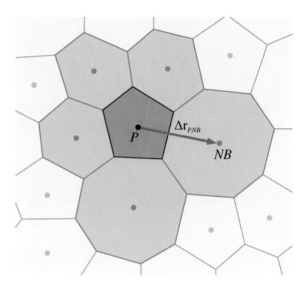

in which $\Delta \boldsymbol{r}_{P-NB}$ is the difference vector among the position vectors of the centroid of the control volume P and the centroid of the neighbor control volume NB, and $(\nabla \phi)_P$ is the gradient of ϕ, in fact, the unknown of the equation. Applying Eq. (13.7) for all neighbor cells, and considering a 3D problem, the linear system becomes,

$$\begin{pmatrix} \Delta x_1 & \Delta y_1 & \Delta z_1 \\ \Delta x_2 & \Delta y_2 & \Delta z_2 \\ . & . & . \\ . & . & . \\ \Delta x_m & \Delta y_m & \Delta z_m \end{pmatrix} \begin{pmatrix} \frac{\partial \phi}{\partial x} \\ \frac{\partial \phi}{\partial y} \\ \frac{\partial \phi}{\partial z} \end{pmatrix} = \begin{pmatrix} \phi_{NB_1} - \phi_P \\ \phi_{NB_2} - \phi_P \\ . \\ . \\ \phi_{NB_m} - \phi_P \end{pmatrix}, \qquad (13.8)$$

in which Δx, Δy and Δz are the cartesian components of the vector $\Delta \boldsymbol{r}$. The linear system (13.8) is overdetermined, that is, there are infinite solutions, because one has three unknowns and five equations. There is not a single solution for this system. The linear system can be written as,

$$\boldsymbol{A}\boldsymbol{x} = \boldsymbol{b}, \qquad (13.9)$$

and a minimization should be found, as

$$\min \|\boldsymbol{A}\boldsymbol{x} - \boldsymbol{b}\|_2, \qquad (13.10)$$

in which the subscript in Eq. (13.10) means the Euclidian norm. This least square minimization problem is usually weighted by the inverse of the distance among the neighboring points and point P, as

13.2 Cell-Center Methods

$$\begin{pmatrix} w_1 \Delta x_1 & w_1 \Delta y_1 & w_1 \Delta z_1 \\ w_2 \Delta x_2 & w_2 \Delta y_2 & w_2 \Delta z_2 \\ \cdot & \cdot & \cdot \\ \cdot & \cdot & \cdot \\ w_m \Delta x_m & w_m \Delta y_m & w_m \Delta z_m \end{pmatrix} \begin{pmatrix} \frac{\partial \phi}{\partial x} \\ \frac{\partial \phi}{\partial y} \\ \frac{\partial \phi}{\partial z} \end{pmatrix} = \begin{pmatrix} w_1 (\phi_{NB_1} - \phi_P) \\ w_2 (\phi_{NB_2} - \phi_P) \\ \cdot \\ \cdot \\ w_m (\phi_{NB_m} - \phi_P) \end{pmatrix} \quad (13.11)$$

Solving this minimization problem, the components of the gradient are determined.

13.2.1.2 Determination of $\phi)_{ip}$

Perhaps it would be elucidative to recall that the need of calculating ϕ_{ip} is due to the advective flux calculation at the interfaces of a control volume, a characteristic of finite volume methods. This procedure requires the interpolation among the neighbor nodal values of ϕ to obtain ϕ_{ip}. This topic was seen in detail in Chap. 5 and most of the schemes already discussed for Cartesian grids can be applied for unstructured grids. Two of the methods will be repeated here as they are more suitable for unstructured meshes. Recall that the key feature of an interpolation function is to represent, as better as possible, the physics happening from an upstream point and the integration point ip, where the variable should be calculated. Any interpolation function, independently of its fidelity with the physics, could be used, being aware of possible numerical diffusion or oscillations. The CDS, central differencing scheme [16] can be constructed, according Fig. 13.7, as

$$\phi_{ip} = \phi_P + (\nabla \phi)_P \Delta s_P \quad (13.12)$$

$$\phi_{ip} = \phi_N + (\nabla \phi)_N \Delta s_N \quad (13.13)$$

Equations (13.12) and (13.13) represent the SOU (Second Order Upwind), when Eq. (5.67) is applied from P to ip and from NB to ip, respectively. These two values can be used to obtain a CDS scheme, as

$$\phi_{ip} = \phi_{ip}^{CDS} = \frac{1}{2}(\phi_P + \phi_N) + \frac{1}{2}(\nabla \phi_P \Delta s_P + \nabla \phi_N \Delta s_N), \quad (13.14)$$

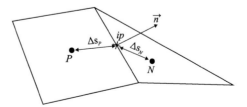

Fig. 13.7 Construction of interpolation schemes using two grid nodes

in which the $\nabla\phi$ at points P and N are calculated as in Sect. 13.2.1. As stated in Chap. 5, this interpolation function doesn't guarantee positivity of the coefficients

The upwind interpolation, which is stable but may introduces severe numerical diffusion can be calculated, considering the direction of the velocity vector from P to ip, as

$$\phi_{ip} = \phi_{ip}^{UDS} = \phi_P + (\nabla\phi)_P \Delta s_P \qquad (13.15)$$

These schemes can be blended using Eq. (5.44), Chap. 5, resulting in

$$\phi_{ip} = \phi_{ip}^{UDS} + \left(\phi_{ip}^{CDS} - \phi_{ip}^{UDS}\right)^o \qquad (13.16)$$

As indicated in Eq. (13.16), the second term in RHS is treated explicitly, that is, it is embedded in the independent term in the linear system, while the first term is treated implicitly, that is, takes part in the matrix coefficients. During the iterative procedure the UDS scheme stabilizes the solution and, when it converges, the explicit term cancels out, and the interpolation function applied was, in fact, CDS, therefore, absent of numerical diffusion. Oscillations, in the other hand, may appear.

The MUSCLE scheme [18] also reported in Chap. 5, can also be applied to unstructured grids, as

$$\phi_{ip}^{MUS} = \theta\phi_{ip}^{CDS} + (1-\theta)\left(\phi_{ip}^{SOU}\right), \qquad (13.17)$$

in which ϕ_{ip}^{SOU} is given by Eq. (13.12). The schemes reported, as they have no direct connection with the Cartesian system in its structure, they apply to three-dimensional problems with unstructured grids too. The general interpolation function along a streamline seen in Chap. 5 generalizes several interpolation schemes already described. A Taylor series expansion along s for the determination of ϕ_{ip} is, referring to Fig. 13.8,

$$\phi_{ip} = \phi_u + \left.\frac{\partial\phi}{\partial s}\right|_{ip} \Delta s + \left.\frac{\partial^2\phi}{\partial s^2}\right|_{ip} \frac{1}{2}\Delta s^2 + O(\Delta s^3) + \cdots +, \qquad (13.18)$$

in which, $(\partial\phi/\partial s)_{ip}$ replaces the $(\nabla\phi)_u$ in the similar equation, Eq. (13.12). It is obtained discretizing the partial differential equation for the ip point, as

$$\left.\frac{\partial\phi}{\partial s}\right|_{ip} = \frac{1}{\rho V}\left[\Gamma_{ip}^\phi\left(\frac{\partial^2\phi}{\partial x^2} + \frac{\partial^2\phi}{\partial y^2}\right)_{ip}\right] + B^\phi \qquad (13.19)$$

in which V is the magnitude of the velocity vector in the s direction. The right-hand side of Eq. (13.19) contain all the local physics involved in the determination of ϕ_{ip} and can be discretized using finite differences, since in this equation there is no need of obeying conservation principles. In Eq. (13.18), the key issue is the determination of ϕ_u (point u in Fig. 13.8) which could be found by a weighted average of ϕ_A and

13.2 Cell-Center Methods

ϕ_B. This approach, however, doesn't create a scheme with positive coefficients. To obtain positive coefficients, ϕ_u should be the donor cell, that is the cell with donates the physical information to the integration point. Mass weighted upstream schemes, as discussed in Chap. 7, does this job.

Inserting the values of ϕ_{ip} and $(\nabla\phi)_{ip}$ as functions of the grid nodes in Eq. 13.3, the linear system for ϕ can be written as

$$[A][\phi] = [B] \qquad (13.20)$$

The methods for solving the linear system given by Eq. 13.20 can be seen in Chap. 4.

In a fluid flow problem, the delicate coupling is among pressure and velocity, and the strategies for handling it can be in the context of a segregated or a simultaneous solution. These approaches were deeply discussed in Chap. 7 and flashes of the key issues will be repeated here.

Considering co-located variables with segregated solution one is faced with the two types of coupling, the segregated (SPVC) and the co-located (CPVC) couplings. The former is tackled with the SIMPLE-like methods, while the latter with the RC-like or PIS-like schemes.

If a simultaneous solution is chosen, still with co-located variables, the segregated coupling (SPVC) no longer exists, remaining the co-located coupling (CPVC), which always exists when it is necessary to calculate the velocity at the interface of a control volume for mass conservation. The simultaneous solution, using a proper method for obtaining the advecting velocity at the interfaces (PIS or RC-like schemes), results in the following linear system,

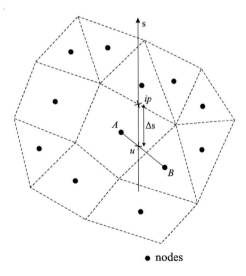

Fig. 13.8 General interpolation function

$$\begin{bmatrix} A^{uu} & A^{uv} & A^{uw} & A^{up} \\ A^{vu} & A^{vv} & A^{vw} & A^{vp} \\ A^{wu} & A^{wv} & A^{ww} & A^{wp} \\ A^{pu} & A^{pv} & A^{pw} & A^{pp} \end{bmatrix} \begin{bmatrix} u \\ v \\ w \\ p \end{bmatrix} = \begin{bmatrix} B^u \\ B^v \\ B^w \\ B^p \end{bmatrix} \qquad (13.21)$$

which can be solved using an iterative solver, with a recommendation that a multigrid method, or other strategy for accelerating the process, should be used. A direct solver with a strategy for handling the non-zeros can also be applied. The reason why this matrix may contain all entries non-zeros depends on which velocity terms are left implicitly in the momentum equations. This will be discussed when the matrix construction for EbFVM will be shown. The term A^{pp} appears due to the application of a RC-like method.

13.2.1.3 Boundary Conditions

In Chap. 12 it was dealt with cell-center methods using boundary-fitted grids, or general curvilinear grids, in which the control volume was of distorted arbitrary rectangular (or hexahedral in 3D) shape. Boundary conditions for cell-center methods using unstructured grids follow the same rationale, that is, the application of balances for the entities for the volumes at the boundaries. Prescribed temperature or fluxes with no boundary flow, as well as inflow and outflow boundary conditions, receive the same treatment as shown in Chap. 12. It is our understanding that the cell-center method, or conventional finite volume method was sufficiently covered when Cartesian and curvilinear grids were treated. The knowledge gained in the previous chapters and the points raised in this section are enough ingredients to start to conceive and write a code for fluid flow calculations using a cell-center method using unstructured grids.

13.2.2 Voronoi Diagrams

The Voronoi diagrams, used as control volumes, are created based on the Delaunay triangulation. Figure 13.9 shows a Delaunay triangulation with its dual, the Voronoi diagrams, which exhibit a very special and desired characteristics for a grid, specifically, local orthogonality. Thus, the conventional finite volume formulation for orthogonal grids can be easily applied to the Voronoi diagrams. Figure 13.10 shows a Voronoi diagram in which the following properties can be observed:

1. The line segment $\overline{AB}$ is perpendicular to the line segment $\overline{P2}$ and cuts $\overline{P2}$ exactly in its middle. These two properties are of fundamental importance in the discretization. The former means that a two-point flux approximation calculates the flux exactly, and the latter, the gradient is calculated in the middle of $\overline{P2}$, not

13.2 Cell-Center Methods

Fig. 13.9 Delaunay triangulation and its dual

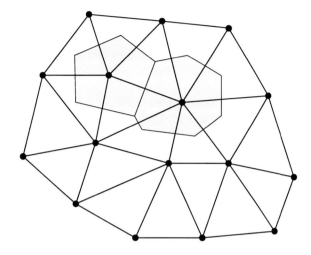

Fig. 13.10 Voronoi control volume

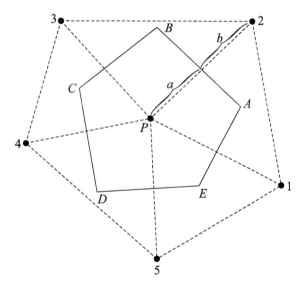

requiring a weighted approximation. In addition, the interpolation of physical properties is done as in a homogenous grid, since $a = b$;

2. Any point inside the Voronoi diagram centered on P is closer to P than any neighboring nodal point. This characteristic is like grid uniformity;
3. By the vertices of the triangle passes a circle. This is a construction characteristic.

Voronoi diagrams combine the advantages of the simplicity of numerical approximations existing in a structured orthogonal system and the flexibility of unstructured meshes. These are all very interesting characteristic for implementing the code. However, as always, there are drawbacks, and the crucial one is the difficulty in generating the Delaunay triangulation. In 2D the issue is not too difficult to face, but in 3D, the efforts required are considerable. To exercise the use of the Voronoi diagrams a pure diffusion and a diffusion/advection problem will be discretized.

13.2.2.1 Diffusion of a Scalar ϕ

Taking as example the 2D transient diffusive equation given by,

$$\frac{\partial}{\partial t}(\rho\phi) = \frac{\partial}{\partial x}\left(\Gamma^\phi \frac{\partial \phi}{\partial x}\right) + \frac{\partial}{\partial y}\left(\Gamma^\phi \frac{\partial \phi}{\partial y}\right) + S, \tag{13.22}$$

or

$$\frac{\partial}{\partial t}(\rho\phi) = \nabla \cdot \left(\Gamma^\phi \nabla \phi\right) + S, \tag{13.23}$$

in which, for recovering the heat conduction equation, the transport coefficient Γ^ϕ will be k/c_P and $\phi = T$. Figure 13.11 depicts the control volume P and its neighbor points denote by NB. Linearizing the source term by $S = S_P \phi_P + S_C$, adopting a fully implicit formulation and integrating Eq. (13.23) over the control volume shown in Fig. 13.11 and over time, as

$$\int_{V,t} \frac{\partial}{\partial t}(\rho\phi)\, dV\, dt = \int_{V,t} \nabla \cdot \left(\Gamma^\phi \nabla \phi\right) dV\, dt + \int_{V,t} (S_P \phi + S_C)\, dV\, dt \tag{13.24}$$

it reads, after applying the divergence theorem,

$$\frac{M_P \phi_P - M_P \phi_P^o}{\Delta t} = \sum_{ip} \left(\Gamma^\phi \nabla \phi \cdot \mathbf{n} \Delta S\right)_{ip} + S_P^\phi \phi_P \Delta V + S_C^\phi \Delta V, \tag{13.25}$$

in which M_P is the mass inside the control volume.

Note that the line joining the center of the control volume P to the centers of the neighboring volumes NB is orthogonal to the surfaces of the control volumes, what allows the use of two grid nodes to calculate the flows numerically exact. The integration point, ip is at the intersection of the lines connecting points P and NB, and the face of the control volume. Note that this point is not always in the middle of the face, and may even be outside the face, a situation in which the Voronoi diagram is called degenerated, which happens when the triangulation is not well behaved (very flat triangles, for example). Recognizing that $\Gamma^\phi \nabla \phi \cdot \mathbf{n}$ in Eq. (13.25) is exactly the derivative in the normal direction, and using an approximation in central differences, since the problem is of pure diffusion, Eq. (13.25) becomes,

13.2 Cell-Center Methods

Fig. 13.11 Local orthogonality

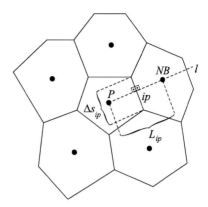

$$\frac{M_P \phi_P}{\Delta t} + \sum_{ip}\left(\Gamma^\phi(\phi_P - \phi_{NB})\frac{\Delta S}{\Delta L}\right)_{ip} - S_P^\phi \phi_P \Delta V = \frac{M_P^o \phi_P^o}{\Delta t} + S_C^\phi \Delta V, \quad (13.26)$$

or

$$A_P \phi_P = \sum_{pi} A_{nb}\phi_{NB} + B, \quad (13.27)$$

in which

$$A_{nb} = \left(\Gamma^\phi \frac{\Delta S}{\Delta L}\right)_{ip} \quad (13.28)$$

$$A_P = \sum A_{nb} - S_P^\phi \Delta V + \frac{M_P}{\Delta t} \quad (13.29)$$

$$B = \frac{M_P^o \phi_P^o}{\Delta t} + S_C \Delta V \quad (13.30)$$

Equation (13.29) and (6.39) are the same, the later developed in the contest of conventional finite volume method (cell-center) using structured grids. Now, with unstructured grids but locally orthogonal the formulation is the same.

13.2.2.2 Advection/Diffusion of a Scalar ϕ

In this section, problems in which a scalar is transported by advection and diffusion will be considered, assuming that the velocity field is known. This is the situation in which, after the velocity field has been calculated, the scalar fields are advanced. The conservation equation for ϕ is given by,

$$\frac{\partial}{\partial t}(\rho\phi) + \nabla \cdot (\rho \mathbf{V}\phi) = \nabla \cdot \left(\Gamma^\phi \nabla \phi\right) + S^\phi \tag{13.31}$$

The process now is the same as in the previous section, that is, we must integrate Eq. (13.31) over the control volume and in time. Using the divergence theorem, the integration results,

$$\frac{M_p \phi_P - M_p^o \phi_P^o}{\Delta t} + \sum_{ip} (\rho \mathbf{V} \cdot \mathbf{n}\Delta S)_{ip}^o \phi_{ip}$$

$$= \sum_{ip} \left(\Gamma^\phi \frac{\partial \phi}{\partial \mathbf{n}} \Delta S\right)_{ip} + \left(S_P^\phi \phi_P + S_C^\phi\right) \Delta V, \tag{13.32}$$

in which $\nabla\phi \cdot \mathbf{n}$ can readily be substituted by $\frac{\partial \phi}{\partial \mathbf{n}}$, easily computed due to the local orthogonality of the grid. This term, being a diffusion term, can be approximated by central differencing with no difficulties in convergence. The dependent variables u and v are stored at the center of the control volume (grid nodes) and u_{ip} and v_{ip} are obtained averaging the velocities at the grid nodes. The superscript in the second term means a value known from the previous time level. In this case the velocity is known, therefore, it has a fixed value along the iterations. Defining the velocity normal to the face of the control volume by V_{ip}, positive pointing out of the surface of the control volume, it can be found by,

$$V_{ip} = u\Delta y - v\Delta x, \tag{13.33}$$

in which the signal of Δx and Δy for each segment of the surface of the control volume is obtained by circling the area counterclockwise. The mass flow rate is given by,

$$(\rho \mathbf{V} \cdot \mathbf{n}\Delta S)_{ip} = (\rho V \Delta S)_{ip} \tag{13.34}$$

For the interpolation function for ϕ_{ip} the WUDS [19] is used,

$$\phi_{ip} = \left(\frac{1}{2} + \alpha_{ip}\right)\phi_P + \left(\frac{1}{2} - \alpha_{ip}\right)\phi_{NB} \tag{13.35}$$

in which α_{ip} is given by,

$$\alpha_{ip} = \frac{\text{Pe}^2}{10 + 2\text{Pe}^2} \tag{13.36}$$

with the Peclet number given by

$$\text{Pe} = \frac{V_{ip} L_{ip}}{\Gamma^\phi} \tag{13.37}$$

13.2 Cell-Center Methods

Observe that in Eq. (13.35) α_{ip} is positive when the flow leaves the control volume and negative in the opposite case. For example, when α_{ip} is equal to 0.5, ϕ_{ip} will be equal to ϕ_P, what characterizes an upwind scheme. The value α_{ip} calculated using by Eq. (13.36) is always positive and, thus, when programming it must be remembered that the signal of α is always the same of the velocity. In this case the upwind scheme is not applied on the direction of the velocity vector, what may cause numerical diffusion.

The developments that has been done so far, have gone through the Cartesian, curvilinear and now Voronoi grids. The interpolation functions have always the same philosophy, and it just need to be adjusted to the schemes for the different grids.

For example, it is possible to use an interpolation function along l(l is the line in the $\overline{P-NB}$ direction) for the Voronoi meshes, as

$$\phi_{ip} = \phi_P + \left.\frac{\partial \phi}{\partial l}\right|_P \left(\frac{L_{ip}}{2}\right) \tag{13.38}$$

The same discussion arises again, that is, ϕ_P is used in Eq. (13.38) in place of ϕ_u, which should be the donor cell in order to have positive coefficients. It is also possible to find the streamline passing on ip, but it will not pass on P (it may, but by an enormous coincidence) and the upwind point, ϕ_u, would need to be found. The term $(\partial \phi / \partial l)_P$ would be found applying the differential equation in analysis at the point P.

Introducing the normal derivatives of ϕ using a central differencing scheme and the values of ϕ_{ip} according to Eq. (13.35), one finds

$$\left[\frac{M_P}{\Delta t} + \sum_{ip}\left(\left(\frac{\Gamma^\phi}{L} + \rho V\left(\frac{1}{2} + \alpha\right)\right)\Delta S\right)_{ip} - S_P^\phi \Delta V\right]\phi_P$$
$$= \sum_{ip}\left(\left(\frac{\Gamma^\phi}{L} - \rho V\left(\frac{1}{2} - \alpha\right)\right)\Delta S\right)_{ip} \phi_{NB} + \frac{M_P^o \phi_P^o}{\Delta t} + S_C^\phi \Delta V \tag{13.39}$$

Introducing the mass conservation equation, as

$$-\sum_{ip}(\rho V \Delta S)_{ip} - \frac{M_P}{\Delta t} + \frac{M_P^o}{\Delta t} = 0, \tag{13.40}$$

into the brackets, one finds

$$\left[\sum_{ip}\left(\left(\frac{\Gamma^\phi}{L} - \rho V\left(\frac{1}{2} - \alpha\right)\right)\Delta S\right)_{ip} - S_P^\phi \Delta V + \frac{M_P^o}{\Delta t}\right]\phi_P$$
$$= \sum_{ip}\left(\left(\frac{\Gamma^\phi}{L} - \rho V\left(\frac{1}{2} - \alpha\right)\right)\Delta S\right)_{ip} \phi_{NB} + \frac{M_P^o \phi_P^o}{\Delta t} + S_C^\phi \Delta V, \tag{13.41}$$

or, in the compact form,

$$A_P \phi_P = \sum_{ip} A_{nb} \phi_{NB} + B \qquad (13.42)$$

in which,

$$A_{nb} = \left(\left(\frac{\Gamma^\phi}{L} - \rho V \left(\frac{1}{2} - \alpha \right) \right) \Delta S \right)_{ip} \qquad (13.43)$$

$$A_P = \sum A_{nb} - S_p^\phi \Delta V + \frac{M_P}{\Delta t} \qquad (13.44)$$

$$B = \frac{M_P^o \phi_P^o}{\Delta t} + S_C^\phi \Delta V \qquad (13.45)$$

Note that Eqs. (13.42) and (13.44) are identical to Eqs. (6.38) and (6.39). The expression of the coefficient A_{nb} is the same for all neighboring volumes, while for structured meshes defined by a system of global axes, the coefficients for the south, west and back faces change sign relative to the north, east and front coefficients, as can be seen in Eqs. (6.26)–(6.28). It is easy to see that for unstructured meshes, in 2D, all the coefficients are as if they were coefficients A_e and A_n, since our local axis is the normal, always pointing outward, thus coincident on these two faces, with the x and y axes, both pointing outward from the faces. In Eq. (13.43) for each ip there will be a A_{nb} calculated with information at that integration point connecting the control volume P with its neighboring nodal point NB.

13.2.2.3 Calculating the Velocity Field

To solve the discretized equations for ϕ, as described in the two previous sections the velocity field is required. We continue using 2D formulation for the sake of simplicity, and the system of equations to be solved, is, therefore, considering incompressible flow and constant viscosity,

$$\frac{\partial \rho}{\partial t} + \frac{\partial}{\partial x}(\rho u) + \frac{\partial}{\partial y}(\rho v) = 0 \qquad (13.46)$$

$$\frac{\partial}{\partial t}(\rho u) + \frac{\partial}{\partial x}(\rho u u) + \frac{\partial}{\partial y}(\rho v u) = \\ -\frac{\partial P}{\partial x} + \frac{\partial}{\partial x}\left(\mu \frac{\partial u}{\partial x}\right) + \frac{\partial}{\partial y}\left(\mu \frac{\partial u}{\partial y}\right) + S^u \qquad (13.47)$$

13.2 Cell-Center Methods

$$\frac{\partial}{\partial t}(\rho v) + \frac{\partial}{\partial x}(\rho u v) + \frac{\partial}{\partial y}(\rho v v) = \\ -\frac{\partial P}{\partial y} + \frac{\partial}{\partial x}\left(\mu \frac{\partial v}{\partial x}\right) + \frac{\partial}{\partial y}\left(\mu \frac{\partial v}{\partial y}\right) + S^v \quad (13.48)$$

The discretization of the mass conservation equation, Eq. (13.46), reads

$$-\sum_{pi}(\rho V \Delta S)_{pi} = \frac{M_P - M_P^o}{\Delta t}, \quad (13.49)$$

in which, repeating, the velocity V is the normal velocity at the boundaries of the control volume pointing outwards. The integration of Eqs. (13.47) and (13.48) follows the same steps as done with Eq. (13.31), since these equations differ only in the source terms. Therefore, these equations when discretized, read

$$A_P \phi_P = \sum_{ip} A_{nb} \phi_{NB} + B, \quad (13.50)$$

with coefficients given by Eqs. (13.43) and (13.44). The source terms now contain the pressure gradient,

$$B = -\frac{\Delta p}{\Delta x_i} \cdot \Delta V + S_C \Delta V + \frac{M_P^o \phi_P^o}{\Delta t}, \quad (13.51)$$

in which i represents x or y, when ϕ is equal to u or v, respectively.

Now comes again the decision on how to solve the equations for (u, v, p), segregated or simultaneously. Following the SIMPLEC [20] approach in a simplified format, since details of this procedure is given in Chap. 7, and remembering that the co-located grid is used, pseudo-equations for the u, v velocity at the integration point (ip) are given by,

$$u_{ip} = u_{ip}^* - \frac{\Delta p'}{\Delta x} \frac{\overline{\Delta V}}{\left(A_P - \sum A_{nb}\right)_{ip}} \quad (13.52)$$

$$v_{ip} = v_{ip}^* - \frac{\Delta p'}{\Delta y} \frac{\overline{\Delta V}}{\left(A_P - \sum A_{nb}\right)_{ip}}, \quad (13.53)$$

in which $\overline{\Delta V}$ is the average of the control volumes that shares the same integration point, and $A_P - \sum A_{nb}$ is given by Eq. (13.60). It is proposed a correction equation for the normal velocity at the interface. Remember, once again, that the equation for correcting the velocities does not interfere with the solution of the problem. Taking advantage of the fact that the pressures are stored along the line normal to the interface, the following correction equation is adequate,

$$V_{ip} = V_{ip}^* - \frac{(p'_P - p'_{NB})}{\left(A_P - \sum A_{nb}\right)_{ip}} \overline{\frac{\Delta V}{L_{ip}}} \tag{13.54}$$

Inserting Eq. (13.54) in the mass conservation equation, one gets

$$-\sum_{ip}\left(V_{ip}^* - \frac{(p'_P - p'_{NB})}{\left(A_P - \sum A_{nb}\right)_{ip}} \overline{\frac{\Delta V}{L_{ip}}}\right)\Delta S_{ip} = 0, \tag{13.55}$$

which, when expanded to all ip and rearranging the terms, it is obtained the pressure correction as,

$$A_P p'_P = \sum A_{nb} p'_{NB} + B, \tag{13.56}$$

in which

$$A_{nb} = \frac{\Delta S_{ip} \overline{\Delta V}}{\left(A_P - \sum A_{nb}\right)_{ip} L_{ip}} \tag{13.57}$$

$$A_P = \sum_{ip} A_{nb} \tag{13.58}$$

$$B = \sum \left(V^* \Delta S\right)_{ip} \tag{13.59}$$

The term $\overline{A_P - \sum A_{nb}}$ need to be computed at the interfaces (ip). However, the coefficients A_P and $\sum A_{nb}$ are available at the nodal points. The strategy is to perform an average among the values at nodal point P and NB which shares the same ip, as

$$\overline{\left(A_P - \sum A_{nb}\right)}_{ip} = \frac{(A_P)_P + (A_P)_{NB} - \sum(A_{nb})_P - \sum(A_{nb})_{NB}}{2} \tag{13.60}$$

It was shown in [21] that the convergence rate increases if the following weighted average is employed,

$$\overline{\left(A_P - \sum A_{nb}\right)}_{ip} = \frac{2\left((A_P)_P - \sum(A_{nb})_P\right)\left((A_P)_{NB} - \sum(A_{nb})_{NB}\right)}{(A_P)_P - \sum(A_{nb})_P + (A_P)_{NB} - \sum(A_{nb})_{NB}} \tag{13.61}$$

Inspecting Eq. (13.55) one sees that the V_{ip}^* velocity are required at the interfaces. This means that a RC-like method is required. In [22] the following discretized equations are used,

$$\left(A_P\right)_P u_P^* = \sum \left(A_{nb} u_{NB}^*\right)_P + \frac{M_P^o u_P^o}{\Delta t} + (S_C \Delta V)_P - \frac{\Delta p^*}{\Delta x}\Delta V \tag{13.62}$$

13.2 Cell-Center Methods

$$(A_p)_{NB} u^*_{NB} = \sum (A_{nb} u^*_{NB})_{NB} + \frac{M^o_{NB} u^o_{NB}}{\Delta t} + (S_C \Delta V)_{NB} - \frac{\Delta p^*}{\Delta x} \Delta V, \quad (13.63)$$

to obtain the equation for the Cartesian velocity u^*_{ip} at the interface, as

$$u^*_{ip} = \frac{\sum (A_{nb} u^*_{NB})_P + \sum (A_{nb} u^*_{NB})_{NB} + (S_C \Delta V)_P + (S_C \Delta V)_{NB}}{(A_p)_P + (A_p)_{NB}}$$

$$+ \frac{1}{((A_p)_P + (A_p)_{NB})} \left(\left(\frac{M^o_P + M^o_{NB}}{\Delta t}\right) u^o_{ip} - \left(2\frac{\Delta p^*}{\Delta x}\right)_{ip} \overline{\Delta V} \right) \quad (13.64)$$

In a similar manner,

$$v^*_{ip} = \frac{\sum (A_{nb} v^*_{NB})_P + \sum (A_{nb} v^*_{NB})_{NB} + (S_C \Delta V)_P + (S_C \Delta V)_{NB}}{(A_p)_P + (A_p)_{NB}}$$

$$+ \frac{1}{((A_p)_P + (A_p)_{NB})} \left(\left(\frac{M^o_P + M^o_{NB}}{\Delta t}\right) v^o_{ip} - \left(2\frac{\Delta p^*}{\Delta y}\right)_{ip} \overline{\Delta V} \right), \quad (13.65)$$

in which, as already seen,

$$\overline{\Delta V} = \frac{[(\Delta V)_P + (\Delta V)_{NB}]}{2} \quad (13.66)$$

Using Eqs. (13.64) and (13.65), one finds

$$V^*_{ip} \Delta S_{ip} = u^*_{ip} \Delta y - v^*_{ip} \Delta x \quad (13.67)$$

Note that the V^*_{ip} velocities enter in the term B in Eq. (13.56). The signal of Δx and Δy need to be according to the convention adopted (sweeping the surface counterclockwise). For example, considering Fig. 13.12, for the segment $\overline{AB}$, Δx is negative, and Δy, positive. For the segment $\overline{BC}$, both negatives, and for $\overline{EA}$, both positives. V^*_{ip} will result positive when leaving the control volume, as the convention adopted in this section.

Solving the linear system given by Eq. (13.56), the pressure can be correct through $p = p^* + p'$, and the velocities using Eq. (13.54). At this point a new pressure field and a new V_{ip} field which satisfy mass conservation is known. The Cartesian components, the dependent variables of the problem, should be now determined. It is possible to obtain them using Eqs. (13.52) and (13.53), but it is recommended to find them as average of the normal velocities which satisfy mass, remembering that there are infinite groups of u and v which satisfy the normal velocity.

The last quantity to be determined is the components of the pressure gradient, $(\partial p/\partial x)_P$ and $(\partial p/\partial y)_P$, which can be determined using the methods seen in Sect. 13.2.1.1.

Fig. 13.12
Counterclockwise sweep.
Signal convention

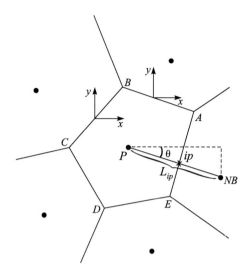

A possible iterative scheme for solving Eqs. (13.46–13.48), is

1. Give the initial conditions of the problem;
2. Advance a time step Δt. Estimate the variables at that time level. The initial estimated values are generally used;
3. Using Eq. (13.50) for u and v, with the source term B given by Eq. (13.51), and using $p*$, determine, by solving the two linear systems, the velocities $u*$ and $v*$ in the center of the control volume;
4. Having $u*$ and $v*$ at the center of the control volume (Voronoi volume), we can determine u_{ip}^* and v_{ip}^* using Eqs. (13.64) and (13.65);
5. Solve Eq. (13.56) and obtain p';
6. Correct the normal velocity at the interfaces using Eq. (13.54);
7. Obtain the pressure using $p = p* + p'$;
8. Obtain u and v at the center of the volumes.

Voronoi diagrams have also been shown to be suitable for solving free boundary problems [21] and multiphase flow in petroleum reservoirs, among other applications. Some results for these classes of problems will be shown in Chap. 15.

13.3 EbFVM—Element-based Finite Volume Method

All cell-vertex methods build the control volume using parts of the elements which shares the same grid node, as depicted in Fig. 13.13. The name cell-vertex comes from the fact that the unknowns are in the vertices of the elements.

This method, with pioneering works developed in [6, 9], was named CVFEM-Control Volume Finite Element Method, probably based on the facts that it uses

13.3 EbFVM—Element-based Finite Volume Method

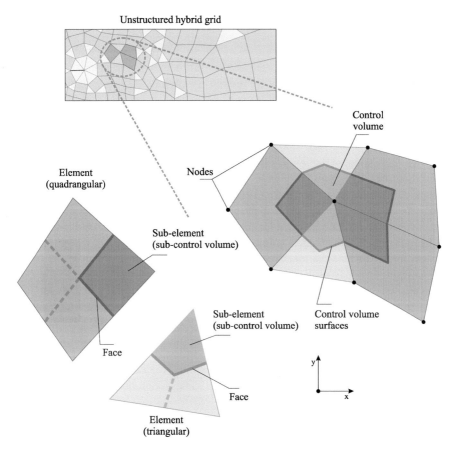

Fig. 13.13 Entities of the Element-based Finite Volume Method

concepts of the finite element technique. We feel that few words about this denomination is opportune. This denomination conveys the reader the idea that it is a finite element method that uses control volumes, that is, a conservative finite element method. In fact, it is not a finite element method, but a finite volume method which borrows from the finite element methodology the concept of element, its geometric representation via shape functions and the element-by-element sweep of the domain for obtaining the global matrix. Therefore, it is our feeling that the method is better defined as Element-based Finite Volume Methods—EbFVM, name that will be used in this text and which is already found in the literature and used in [23]. The denomination is strongly founded on the use of the elements to perform all calculations required for obtaining the approximated equations.

13.3.1 Geometrical Entities

Figure 13.13, shows a sample of an unstructured grid in which five elements are identified in the zoom. The following entities are depicted, the element, the sub-element (sub-control volume), faces of the control volume, control volume and grid nodes. The element is defined by the coordinates of their vertices and they can be of triangular or quadrangular shape in 2D, or 3D elements as shown in Fig. 13.1. The control volume is constructed by joining the centroid of the elements with the centroid of the faces which shares the same node, the latter being the position in which the dependent variable is located. Each element is subdivided in sub-elements or sub-control volumes.

This construction, in 2D, creates in each sub-control volume two faces where the fluxes are calculated for the balance of the conservative property. Regardless if the element is triangular or quadrangular, there will be two surfaces of each element that forms the surface of the control volume, as can be seen in Fig. 13.2b, whose control volume contains triangular and quadrangular elements in its construction.

For tridimensional grids, control volumes can be constructed with different elements, what originates the so-called hybrid grids, or with the same element, as shown in Fig. 13.14, in which just tetrahedra are used. In this picture 8 tetrahedral elements were combined forming the control volume. Each sub-control volume has 3 integration points, which totals 24 integration points for the whole control volume balance. If we were using 8 hexahedra, the resulting control volume would be a hexahedron, of course, with, again, 24 integration points. Regardless the element used, the sub-control volume in three dimension has 3 integration points contributing for the balance.

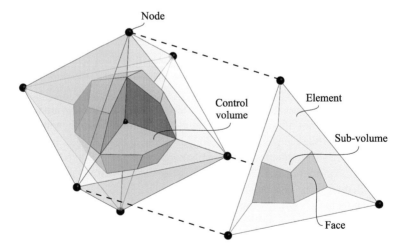

Fig. 13.14 Tetrahedral elements and the control volume

13.3.2 Local Coordinates. Shape Functions

In unstructured grids, in which control volumes of arbitrary shapes and with different number of neighbors are used, it is convenient to have a local coordinate system, such that all geometric representation of the elements can be done in a concise and general mathematical form, despite the physical size of the element, in a similar way as was done in Chap. 12 for curvilinear coordinates. All developments done in that chapter for a global coordinate transformation applies directly to the local coordinate transformation.

The idea, following the techniques used in finite element methods, is to have all the calculations performed at element level, and the assemble of the equations using the contribution of each element in the control volume through a sweep in the whole domain in an element-base. In Fig. 13.15, it is in evidence two elements, each element contributing with two fluxes for the global balance at the control volume.

Figure 13.16 shows, in two dimensions, a quadrangular and a triangular element in the physical and transformed domain with their respective local axis (ξ, η). As the variables of the problem are located at the vertices of the element (cell-vertex model), a suitable way to mathematically represent all quantities inside the element, as function of the values at the vertices, are through the shape functions. In Chap. 3 briefly comments were given with respect to the order of the elements, since one can have a triangular or quadrangular element of higher order, which means to have more nodes defined on the element. First order elements are easy to work with, and they suffice, most of the time, for numerical simulation, grounded on the fact that grid refinement is easier to apply than having complex elements for integration. Additionally, fine grids are usually required to capture the physics of the problem.

Moreover, the mathematics could be done using all equations in vector form. For the sake of compactness, however, it is preferable, for being didactic to present in 2D with all terms expanded to facilitate the understanding.

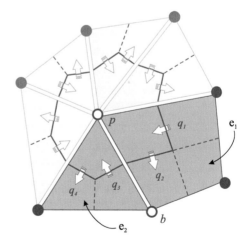

Fig. 13.15 Fluxes contribution from two elements

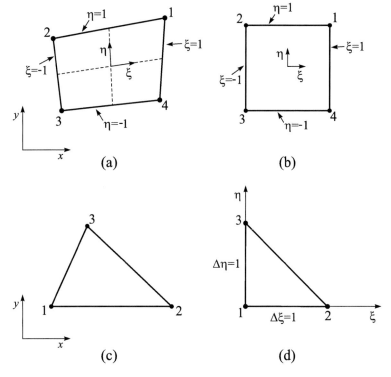

Fig. 13.16 Physical and transformed domains

The shape functions, or interpolation functions, for the quadrangular element of first order shown in Fig. 13.16 are bilinear functions of the type,

$$N_1(\xi, \eta) = \frac{1}{4}(1+\xi)(1+\eta) \tag{13.68}$$

$$N_2(\xi, \eta) = \frac{1}{4}(1-\xi)(1+\eta) \tag{13.69}$$

$$N_3(\xi, \eta) = \frac{1}{4}(1-\xi)(1-\eta) \tag{13.70}$$

$$N_4(\xi, \eta) = \frac{1}{4}(1+\xi)(1-\eta) \tag{13.71}$$

It is easy to infer that the shape functions are nothing more than a practical and clean form of the interpolation procedure for finding the values of quantities inside the element based on the values at its nodes. Therefore, there will be as many shape functions as the number of nodes of the element, and the summation of all shape function at any point of the element must be equal to unity [24]. This is the

13.3 EbFVM—Element-based Finite Volume Method

condition to be obeyed when finding the shape functions for the element. The shape functions are specified for a fixed computational element in the (ξ, η) domain, with constant dimensions irrespective of the physical dimensions of the element, similar to the computational domain in curvilinear coordinates. A coordinate transformation, relating the local with the global coordinates, is required. Using 2D for simplicity,

$$x = x(\xi, \eta)$$
$$y = y(\xi, \eta) \tag{13.72}$$

$$\xi = \xi(x, y)$$
$$\eta = \eta(x, y) \tag{13.73}$$

To better illustrate the behavior of the shape function, Fig. 13.17 plots the N_1 shape function inside the quadrangular element. For $\xi = 1$, Eq. 13.70 is a linear function with η, while for $\eta = 1$ it is a linear function with ξ, what originates the straight lines in all boundaries of the element. Inside the domain it is a bilinear function, represented by the curved line joining vertex 1 to vertex 3. It can be checked that for the center of the element ($\xi = \eta = 0$), $N_1 = 0.25$, and not 0.5 that it would be if it was linear. The other three shape functions have the same behavior in each vertex.

The shape functions, therefore, weights the contribution of the nodes 1, 2, 3 and 4 in the calculation of any quantity inside the element. This is the practical feature of the element, being possible to perform all calculations without the interference of the neighbor elements. In the conventional cell-center finite volume, a flux at an interface will involve the neighbor control volume, precluding an element-by-element sweep in the domain to obtain the global linear system. In the EbFVM, as is in finite element methods, local matrices for each element are constructed, and the global linear system is obtained by assembling the local matrices in an element-by-element approach.

To illustrate that the local coordinate transformation has all procedures as in a global transformation, the right upper quadrant of Fig. 13.18 should be compared with the same geometric element analyzed in Fig. 12.3. For example, ΔS_{ip2} is a length along the η axis, the same as given by Eq. (12.27), while ΔS_{ip1} is a length

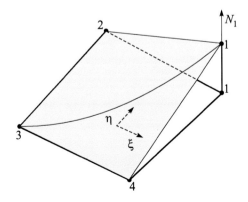

Fig. 13.17 N_1 shape function in the transformed domain

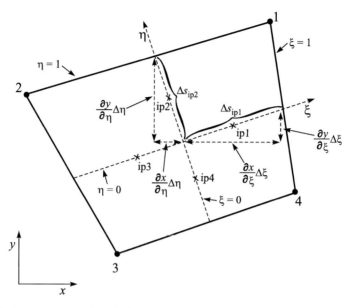

Fig. 13.18 The element and the sub-elements (sub-control volumes)

along the ξ axis, given by Eq. (12.26). The Jacobian, calculated, for example, at $\xi = \eta = 0.5$ using Eq. (12.16) gives the area (or volume in 3D) of the upper right quadrant of the element. These quantities will be used for the balance of the property in the control volume centered at node 1, the control volume which the upper right quadrant of element 1234 belongs to.

Figure 13.16c shows a general triangular element in the physical plane with its mapping in Fig. 13.16d. In this case, the shape functions are given by

$$N_1 = 1 - \xi - \eta$$
$$N_2 = \xi$$
$$N_3 = \eta \qquad (13.74)$$

and all features already pointed out, and valid for quadrangular elements, apply here. In this case the plot of the shape function in the element will be a plane passing by the three vertices.

Being all quantities inside the element represented by an interpolation of the values at the nodes, one can write, keeping the developments in two dimensions.

13.3 EbFVM—Element-based Finite Volume Method

$$x(\xi, \eta) = \sum_{i=1}^{N} N_i(\xi, \eta) x_i$$

$$y(\xi, \eta) = \sum_{i=1}^{N} N_i(\xi, \eta) y_i$$

(13.75)

in which N is the number of nodes of the element. The metrics of the transformation are, using the relations given by Eq. (13.75)

$$\frac{\partial x}{\partial \xi} = \sum_{i=1}^{4} \frac{\partial N_i}{\partial \xi} x_i \qquad (13.76)$$

$$\frac{\partial x}{\partial \eta} = \sum_{i=1}^{4} \frac{\partial N_i}{\partial \eta} x_i \qquad (13.77)$$

$$\frac{\partial y}{\partial \eta} = \sum_{i=1}^{4} \frac{\partial N_i}{\partial \eta} y_i \qquad (13.78)$$

$$\frac{\partial y}{\partial \xi} = \sum_{i=1}^{4} \frac{\partial N_i}{\partial \xi} y_i \qquad (13.79)$$

As stated, the value of any function inside the element can be interpolated using the shape functions, as

$$\phi(\xi, \eta) = \sum_{i=1}^{N} N_i(\xi, \eta) \phi_i, \qquad (13.80)$$

in which ϕ_i is the value of the variable at the nodes of the element. As in cell-center methods, the gradient of the variable must be calculated for the diffusive fluxes.

13.3.3 Determination of $(\nabla \phi)_{ip}$

The gradient of ϕ at the integration points is easily found by

$$\frac{\partial \phi}{\partial x} = \sum_{i=1}^{N} \left.\frac{\partial N_i}{\partial x}\right|_{\xi, \eta} \phi_i \qquad (13.81)$$

$$\frac{\partial \phi}{\partial y} = \sum_{i=1}^{N} \frac{\partial N_i}{\partial y}\bigg|_{\xi,\eta} \phi_i \qquad (13.82)$$

The coordinate transformation is used for finding the derivatives in Eqs. (13.81) and (13.82), since the shape functions are function of (ξ, η).

Using the chain rule,

$$\frac{\partial N_i}{\partial \xi} = \frac{\partial N_i}{\partial x}\frac{\partial x}{\partial \xi} + \frac{\partial N_i}{\partial y}\frac{\partial y}{\partial \xi} \qquad (13.83)$$

$$\frac{\partial N_i}{\partial \eta} = \frac{\partial N_i}{\partial x}\frac{\partial x}{\partial \eta} + \frac{\partial N_i}{\partial y}\frac{\partial y}{\partial \eta}, \qquad (13.84)$$

or in matrix form as,

$$\begin{bmatrix} \dfrac{\partial N_i}{\partial \xi} \\ \dfrac{\partial N_i}{\partial \eta} \end{bmatrix} = \begin{bmatrix} \dfrac{\partial x}{\partial \xi} & \dfrac{\partial y}{\partial \xi} \\ \dfrac{\partial x}{\partial \eta} & \dfrac{\partial y}{\partial \eta} \end{bmatrix} \begin{bmatrix} \dfrac{\partial N_i}{\partial x} \\ \dfrac{\partial N_i}{\partial y} \end{bmatrix} \qquad (13.85)$$

whose solution is

$$\begin{bmatrix} \dfrac{\partial N_i}{\partial x} \\ \dfrac{\partial N_i}{\partial y} \end{bmatrix} = \frac{1}{J}\begin{bmatrix} \dfrac{\partial y}{\partial \eta} & -\dfrac{\partial y}{\partial \xi} \\ -\dfrac{\partial x}{\partial \eta} & \dfrac{\partial x}{\partial \xi} \end{bmatrix} \begin{bmatrix} \dfrac{\partial N_i}{\partial \xi} \\ \dfrac{\partial N_i}{\partial \eta} \end{bmatrix}, \qquad (13.86)$$

in which the Jacobian reads,

$$J = x_\xi y_\eta - x_\eta y_\xi, \qquad (13.87)$$

The local derivatives of the shape functions are directly obtained from Eqs. (13.68–13.71), as

$$\begin{aligned}
\frac{\partial N_1}{\partial \xi} &= \frac{1}{4}(1+\eta) & \frac{\partial N_1}{\partial \eta} &= \frac{1}{4}(1+\xi) \\
\frac{\partial N_2}{\partial \xi} &= -\frac{1}{4}(1+\eta) & \frac{\partial N_2}{\partial \eta} &= \frac{1}{4}(1-\xi) \\
\frac{\partial N_3}{\partial \xi} &= -\frac{1}{4}(1-\eta) & \frac{\partial N_3}{\partial \eta} &= -\frac{1}{4}(1-\xi) \\
\frac{\partial N_4}{\partial \xi} &= \frac{1}{4}(1-\eta) & \frac{\partial N_4}{\partial \eta} &= -\frac{1}{4}(1+\xi),
\end{aligned} \qquad (13.88)$$

13.3 EbFVM—Element-based Finite Volume Method

Using the metrics of the coordinate transformation given by Eqs. (13.76–13.79), the derivatives of the shape functions with respect to x and y can be determined through Eq. (13.86) and, finally, the components of the gradient of ϕ, $\partial\phi/\partial x$ and $\partial\phi/\partial y$ can be found using Eqs. (13.81) and (13.82).

This gradient calculation is considerably easier when compared with the gradient reconstruction strategies required in the conventional cell-center finite volume method. The diffusive part of the flux which demands the gradient of ϕ is resolved using the gradient calculation just outlined. Later, we will discuss about the assemble of the global linear system. By now, it is enough to keep in mind, referring to Fig. 13.15, that each element will contribute with the balance in the control volume with two fluxes.

Connectivity among grid nodes and elements in the global coordinate system is required for later assembling the equations for the control volumes in terms of the elements. Figure 13.19 brings this connectivity for a part of a grid, for illustration.

Next step of the formulation is the determination of ϕ at the integration points, that is, ϕ_{ip}. At a first glance, we could be tempted to find ϕ_{ip} using the shape functions, but this would be equivalent of using central differencing schemes, which leads to numerical oscillations if the grid is not refined enough. Therefore, for avoiding this drawback, the same strategies used in the formulations previously seen are applicable, with the major advantage that the critical issue of finding the donor cell to obtain positive coefficients is now possible without difficulties, due to the construction of the control volume based on the elements in the EbFVM.

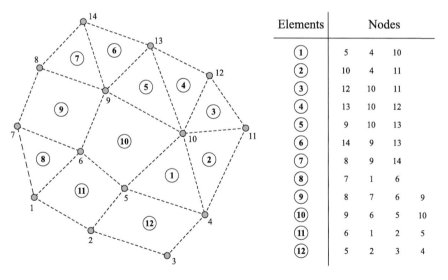

Fig. 13.19 Elements and grid nodes connectivity

13.3.4 Determination of ϕ_{ip}

Alternatives for interpolating the nodal values of ϕ was deeply discussed Chap. 5, mainly connected with Cartesian grids. In the first part of this chapter, the findings were extended to unstructured grids in the context of cell-center methods. Now, the task is to find the interpolation functions for methodologies which uses unstructured grids with cell-vertex methods. It is not surprising that all schemes already seen apply also here with subtle changes.

It is also a key point recognize that almost all schemes can be derived using a Taylor series expansion around the integration point along a coordinate axis. It is well-known that the pure advective scheme in Cartesian coordinates, which says that $\phi_w = \phi_W$, in which w stands for ip and ϕ_W for ϕ_u, is a Taylor series in which the second term in the RHS was dropped, as can be recognized inspecting Eq. (13.89). Therefore, to minimize numerical diffusion, the Taylor series expansion works better when done along the direction s of the velocity vector. An increased order of approximation can be obtained accounting for the term $(\partial \phi / \partial s)_{ip}$, which carries information of the local physics. One is talking, in fact, about Eq. (13.18), repeated her for completeness

$$\phi_{ip} = \phi_u + \left.\frac{\partial \phi}{\partial s}\right|_{ip} \Delta s + \left.\frac{\partial^2 \phi}{\partial s^2}\right|_{ip} \frac{1}{2}\Delta s^2 + O(\Delta s^3) + \cdots +, \qquad (13.89)$$

In this equation the derivatives of ϕ with respect to s is taken in the point ip, instead of u, since it is a locally defined point in the element. Considering a second order interpolation, one has

$$\phi_{ip} = \phi_u + \left(\frac{\Delta \phi}{\Delta s}\right)_{ip} \cdot \Delta s \qquad (13.90)$$

It was already pointed out that in Eq. (13.90), $(\partial \phi / \partial s)_{ip}$ introduces in the interpolation function all the physics involved, since this term is obtained through a discretization of the partial differential equation in analysis. The family of schemes which uses Eq. (13.90) has its roots in the SUDS—Skew Upstream Differencing Scheme [25], a pioneering work recognizing the need of considering the interpolation function along the direction of the velocity vector. Initially, let's bring the application of Eq. (13.90) to the environment of EbFVM using the elements.

Consider Fig. 13.20, in which an element is depicted with its local numbering 1-2-3-4, the grid nodes where the unknown variables are located. Using Eq. (13.90) one can start obtaining the two most known approximations, upwind and central differencing. Neglecting the second term in the RHS it is obtained a pure upwind interpolation function along s, direction of the flow, that is, $\phi_{ip} = \phi_u$, in which the upstream value is found through and interpolation among ϕ_2 and ϕ_3, as shown in Fig. 13.20. It is known that the upwind interpolation creates a robust scheme, but, on the contrary, generates excessive numerical diffusion.

13.3 EbFVM—Element-based Finite Volume Method

Fig. 13.20 Interpolation function along s [9]

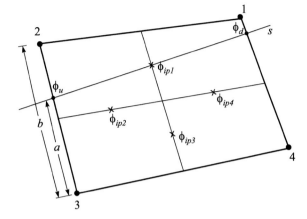

A central differencing scheme can also be constructed using Eq. (13.90), finding ϕ_d through an interpolation among grid nodes ϕ_1 and ϕ_4. Knowing that,

$$\left.\frac{\Delta\phi}{\Delta s}\right|_{ip1} \approx \frac{\phi_d - \phi_u}{2\Delta s}, \tag{13.91}$$

one finds,

$$\phi_{ip1} = \frac{\phi_d + \phi_u}{2}, \tag{13.92}$$

in which the upstream and downstream values are a function of the four grid nodes of the element. Of course, this is equivalent of calculating ϕ_{ip1} using the shape functions, and these procedures are not recommended since they are only applicable for diffusive dominant problems, otherwise, oscillations will appear. Other schemes, as WUDS, QUICK and SOU, already discussed in Chap. 5, can be obtained with a proper evaluation of $(\Delta\phi/\Delta s)_{ip1}$ in Eq. (13.90).

Whenever ϕ_u is calculated by employing an interpolation of only nodal points, as shown in Fig. 13.20 involving ϕ_2 and ϕ_3, the value of ϕ_u will be explicitly determined as a function of these nodal points, which are part of the linear system. When the inclination of the velocity vector is, for example, as shown in Fig. 13.21, ϕ_u can be determined with an interpolation between nodal points ϕ_2 and ϕ_3, and the value of ϕ_{ip2}, which is not a grid point. If we have an even steeper slope of the velocity vector, ϕ_u can be determined by an interpolation between ϕ_{ip2} and ϕ_{ip4}. When the values of ϕ_{ip} at the four integration points of the element are equated, there may be situations in which all integration points may be function of each other, creating a 4×4 linear system that must be solved to determine all values ϕ_{ip} for the element as function of nodal values. The value of ϕ at the integration point must be function of grid points only.

Fig. 13.21 Alternatives for determining ϕ_u [9]

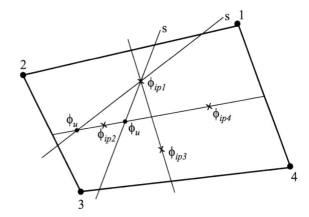

This 4 × 4 system is easily inverted, and this coupling between integration points is beneficial as it has a result equivalent to mesh refinement within the element [9]. It is always possible to relate the value at the integration point to nodal points only, avoiding solving the 4 × 4 system for each element. In Fig. 13.21, in the first case, ϕ_u, lying on the left boundary of the element would be calculated with an interpolation between nodes 2 and 3, and in the second case, lying on the southern boundary, with an interpolation between nodes 3 and 4. However, there is some losses in robustness with this procedure.

If $\phi_{ip} = \phi_u$, a pure upwind scheme is implied, while it is possible to create higher order schemes adding to the value of ϕ_u the term $(\partial \phi / \partial s)$ obtained from the full differential equation governing the phenomenon under study. For example, for the components of the momentum equations, this would be

$$\frac{\Delta u_i}{\Delta s} = \frac{1}{\rho |V|} \left(-\frac{\partial P}{\partial x_i} + \frac{\partial}{\partial x_j} \left(\mu \left(\frac{\partial u_i}{\partial x_j} + \frac{\partial u_j}{\partial x_i} \right) \right) + S^{u_i} \right) \qquad (13.93)$$

in which $|V|$ is the magnitude of the velocity vector along s and u_i represent the 3 components of the velocity vector. Calculating u_{ip} using Eq. (13.90) and adding a second order correction to the upstream scheme using Eq. (13.93), one is introducing in the interpolation function terms which embodies the local physics, including pressure, important for the pressure–velocity coupling. To avoid misunderstanding in variable sub-indexes, just remember that the convention used herein states that $\phi_1, \phi_2, \ldots, \phi_N$ identifies variables at grid nodes, while $\phi_{ip1}, \phi_{ip2}, \ldots, \phi_{ipN}$ identifies variables at the integration points. For a quadrangular element, as in Fig. 13.20, there will be 4 nodes and 4 integration points. The number of integration points will be always equal to the number of grid points, irrespective of the geometry.

Still related to Eq. (13.93), the term on the RHS needs to be approximated numerically. This approximation can be done employing the shape functions, whenever possible. Pressure and diffusive terms exhibit elliptic behavior, but only pressure can be approximated using the shape functions. The second order derivatives of the diffusive term can't, because these derivatives cancel each other out. Central differencing approximations must then be used, properly defining the diffusion length, procedure that can be found in [9, 10, 26].

Although the determination of ϕ_{ip} in this manner is a superior approach than simply using upwind or central differences, the determination of ϕ_u can still produce negative coefficients. As described in [9], these negative coefficients are of small magnitude and much less detrimental than the coefficients caused when central difference schemes are employed. There are situations, however, in which negative coefficients are not permitted under any circumstances, and one of such situation is the solution of the equations for $(k - \varepsilon)$ or $(k - \omega)$ for turbulent flows [9].

13.3.5 Family of Positive Advection Schemes

Positiveness is a desired property in numerical schemes since it prevents the development of non-physical spatial oscillations and the introduction of new extrema into the numerical solution. This is always the goal when developing advecting schemes in finite volume formulations. Since the pioneering work of [25] it is recognized the importance of considering the actual flow direction in developing advection schemes. Although the skewed upwind scheme proposed in [25] does not satisfy positivity conditions, for various problems it yields more accurate results than the simpler upwind scheme. Several schemes have been proposed since then in the pursuit of an optimal numerical approximation for the advection term in fluid flow equations.

In [9] it is presented a scheme which guarantees positive coefficients, later extended in [26] for a family of schemes satisfying this property, also in the context of EbFVM. These schemes belong to the class of methods called Mass Weighted Upstream Schemes (MWUS), as it is recognized that the determination of the function at the integration points strongly rely on a correct identification of where the information to the integration point comes from. In other words, the correct determination of ϕ_u is what matters in order to obtain positive coefficients, for avoiding oscillations.

We begin revisiting the scheme presented in [9] one of the most well-behaved schemes proposed in the element-based finite volume framework described in [9, 10]. This scheme considers the skewness of the flow by measuring local mass flow rates required for finding the correct donor cell, to obtain the value of the variable at the integration point. The element-based construction of the finite volume method allows to develop schemes of this nature.

Figure 13.22a shows a quadrangular element with the grid nodes (1, 2, 3, 4) on the vertices, and the integration points ($ip1, ip2, ip3, ip4$) on the centroid of the faces. These integration points are on the boundary surfaces of the control volumes connected with this element. Figure 13.22b reports the convention for the fluxes calculation, which will be helpful when assembling the equations, since the flux

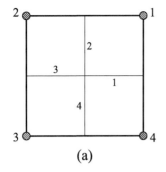
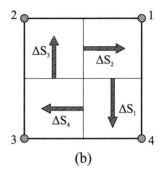

Fig. 13.22 Integration points (**a**) and fluxes convention (**b**)

which enters in a control volume leaves its neighbor one. For example, in the integration point $ip2$, the flux entering/leaving the control volume 1, leaves/enters the control volume 2.

Inspecting the sub-control volume in the upper right quadrant of Fig. 13.23, reveals that there are two internal faces in whose centroids lie the integration points $ip1$ and $ip2$, and the grid node, ϕ_1, related to this sub-control volume. If the interest is to find the value of ϕ_{ip1}, the mass flow influence on it can come from ϕ_1, from ϕ_{ip2} or from both. As it is possible to evaluate the mass flow which crosses the surfaces, the ratio of the mass flows will tell us the donor cell, or, in other words, what region donate the information to ϕ_{ip1}. Therefore, is easy to equate ϕ_{ip1} as a blend among these two variables based on the mass flow ratio, as

$$\phi_{ip1} = (1 - \lambda_{ip1})\phi_1 + \lambda_{ip1}\phi_{ip2}, \qquad (13.94)$$

in which λ_{ip1} is a weighting factor whose value depends on the local flow direction.

Figure 13.24 depicts three typical flow configurations accounted for in the configuration considered in [9]. Each configuration corresponds to some possible flow arrangement and is related to the mass flow ratio $\dot{m}_{ip2}/\dot{m}_{ip1}$.

Case (a) corresponds to the condition $(\dot{m}_{ip2}/\dot{m}_{ip1}) \geq 1$, which means that all fluid flowing through face $ip1$ came from surface $ip2$, and nothing came from the grid node ϕ_1. According to what we have intensively discussed about the donor of the

Fig. 13.23 Mass flux in $ip1$

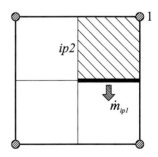

13.3 EbFVM—Element-based Finite Volume Method

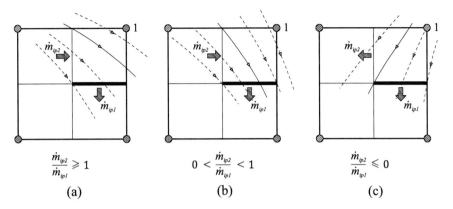

Fig. 13.24 Flow direction alternatives

information, in this case the donor is solely ϕ_{ip2}. Therefore the ϕ_u of ϕ_{ip1} is ϕ_{ip2} in this case. Therefore, the value of λ_{ip1} in Eq. (13.94) must be 1.

Case (c) is the opposite and happens when $(\dot{m}_{ip2}/\dot{m}_{ip1}) \leq 0$, what requires in Eq. (13.94) $\lambda_{ip1} = 0$. Pay attention that according to the convention shown in Fig. 13.22b, $\dot{m}_{ip2}$ is negative. It is easy to realize observing Fig. 13.24 that the whole flow leaving the sub-control volume through both faces comes integrally from grid node 1. So, the donor of the information in this case is ϕ_1, what means that the upstream value ϕ_u of ϕ_{ip1} is ϕ_1.

Case (b) is the intermediate case, and happens when $0 \leq (\dot{m}_{ip2}/\dot{m}_{ip1}) \leq 1$, what means that the information to ϕ_{ip1} comes from ϕ_{ip2} and from ϕ_1. Equation (13.94) represents this situation with $\lambda_{ip1} = (\dot{m}_{ip2}/\dot{m}_{ip1})$. The three flows configuration can be recovered defining

$$\lambda_{ip1} = \max[0, \min(1, \omega_{ip1})], \quad (13.95)$$

in which

$$\omega_{ip1} = \dot{m}_{ip2}/\dot{m}_{ip1} \quad (13.96)$$

In this scheme, as mentioned, the value at the integration point can be function of another integration point, as in cases (a) and (b), what would require the solution of a 4×4 matrix for every element of the grid. In [9] it is argued that this 4×4 linear system can be easily inverted without penalizing the computational efforts.

The same analysis can be generalized for the whole element using the 4-cycle $\{1, 2, 3, 4\}$ convention [27], in which $k + 1 = 1$ when $k = 4$, whereas $k - 1 = 4$ when $k = 1$. To avoid excessive sub-index usage, let's define the values of ϕ at the grid nodes by $\hat{\phi}$, keeping ϕ for the integration points. Therefore, Eq. (13.94) can be written as,

$$\phi_k = \begin{cases} (1-\lambda_k)\hat{\phi}_k + \lambda_k \phi_{k+1}, & \text{if } \dot{m}_k > 0 \\ (1-\lambda_k)\hat{\phi}_{k-1} + \lambda_k \phi_{k-1}, & \text{if } \dot{m}_k < 0 \end{cases}, \quad (13.97)$$

and ω_k by

$$\omega_k = \begin{cases} \dfrac{\dot{m}_{k+1}}{\dot{m}_k}, & \text{if } \dot{m}_k > 0 \\ \dfrac{\dot{m}_{k-1}}{\dot{m}_k}, & \text{if } \dot{m}_k < 0 \end{cases} \quad (13.98)$$

The parameters ω and λ are always related to the integration points, therefore, the sub-index k defines them properly.

A family of positive advection schemes can be defined also considering the same framework previously described in [9, 10]. Advantages of that framework are twofold. First, it can be applied similarly to structured and unstructured grids, and second, it can handle properly the possible local variation of the flow velocity at element level. Every member of the family of schemes has a functional relationship between the weighting factor used in Eq. (13.97) and the mass flux ratio defined in Eq. (13.98). For instance, the functional relationship in [10] is given by Eq. (13.95).

In order to a given scheme of the family be positive, it is enough that the weighting factor remains in the positivity region [28], defined by

$$0 \le \lambda_k(\omega_k) \le \max[0, \min(1, \omega_k)], \quad \text{for } -\infty < \omega_k < \infty \quad (13.99)$$

The upper limit of the positivity region, as depicted in Fig. 13.25, corresponds to scheme described in [10]. The lower limit, $\lambda_k = 0$, corresponds to the common upwind scheme, as we can see in Eq. (13.94) or Eq. (13.97), for that particular weighting factor, the face value takes the value of the upstream node, $\phi_{ip1} = \phi_1$, regardless of the actual mass flux ratio. It is worth noting that all schemes in the family must have $\lambda_k = 0$ for $\omega_k < 0$ (See Fig. 13.26).

For the problems solved, it is demonstrated that the best scheme of the family is the one which uses $\lambda_k = \omega_k/(1+\omega_k)$ Some numerical properties of the scheme and recommendations for the assemble of the discretized equations can be also found in [27].

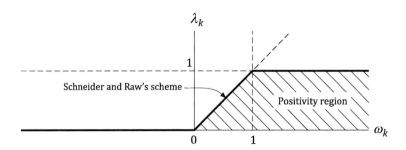

Fig. 13.25 Positive region—Schneider and Raw scheme

13.3 EbFVM—Element-based Finite Volume Method

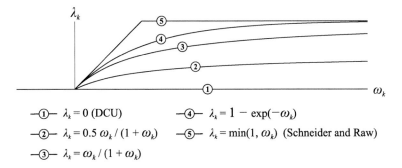

— ① — $\lambda_k = 0$ (DCU)   — ④ — $\lambda_k = 1 - \exp(-\omega_k)$
— ② — $\lambda_k = 0.5\,\omega_k/(1+\omega_k)$ — ⑤ — $\lambda_k = \min(1, \omega_k)$ (Schneider and Raw)
— ③ — $\lambda_k = \omega_k/(1+\omega_k)$

Fig. 13.26 Positive region. All schemes

There are several others advecting schemes available in the literature which works avoiding undershoots and overshoots in the numerical solution. In [23] the high-resolution scheme developed for triangular unstructured grids [29] is used. This methodology involves computing ϕ_{min} and ϕ_{max} at each node using a stencil involving adjacent nodes, and including the node itself. Following, for each integration point around the node, Eq. (3.100) is solved for β to ensure that it does not undershoot or overshoot,

$$\phi_{ip} = \phi_P + \beta(\nabla\phi)_P \cdot r \tag{13.100}$$

The nodal value for β is taken to be the minimum value of all integration point values surrounding the node. The value of β is also not permitted to exceed 1. This algorithm can be shown to be Total Variation Diminishing (TVD) when applied to one-dimensional situations [23]. Having all ingredients in hands, the integration of the conservation equations can be done.

13.3.6 Integration of the Conservation Equations

The system of equations of interest is the mass conservation equation, the three components of the momentum conservation equations and possible scalars, as temperature, concentration of species, saturation, kinetic energy of turbulence and its dissipation, among others. The important set of equations which requires careful treatment is composed of mass and momentum conservations, due to it intricated coupling. It was deserved considerable time already discussing the solution strategies in previous chapters, whether segregated or simultaneous, and no more efforts on this topic is needed, since all what was learnt applies directly for the EbFVM. The partial differential equations, written in compact form, are

$$\frac{\partial \rho}{\partial t} + \frac{\partial}{\partial x_j}(\rho u_j) = 0 \tag{13.101}$$

$$\frac{\partial(\rho u_i)}{\partial t} + \frac{\partial}{\partial x_j}(\rho u_j u_i) = -\frac{\partial p}{\partial x_i} + \frac{\partial}{\partial x_j}\left(\mu\left(\frac{\partial u_i}{\partial x_j} + \overline{\frac{\partial u_j}{\partial x_i}}\right)\right) + S^{u_i} \quad (13.102)$$

$$\frac{\partial(\rho\phi)}{\partial t} + \frac{\partial}{\partial x_j}(\rho u_j \phi) = \frac{\partial}{\partial x_j}\left(\Gamma^\phi \frac{\partial \phi}{\partial x_j}\right) + S^\phi \quad (13.103)$$

In Eq. (13.102) u_i represents the three Cartesian components of the velocity vector, reproducing the three equations of motion. Normally, the terms denoted by the overbar are amalgamated in the source term, since they do not involve the principal variable of this equation. This is plausible when solving the equation system in a segregated manner, but in the simultaneous approach it is wise to have all variables present in all equations. Therefore, leaving this term implicitly in the integration of the equations helps the conditioning of the matrix, since all velocity components will appear in all momentum conservation equations. When μ is constant and $\nabla \cdot V = 0$, the overbar term is zero, but even in this situation this portion of the stress tensor should be retained in the equations. The term which can be put in the source term, independent of the solution procedure, is $-(\partial/\partial x_i)(2/3)\mu\nabla \cdot V \delta_{ii}$.

There are no changes in essence when using EbFVM compared to a conventional finite volume technique. All techniques integrate the partial differential equation in conservative form. It differs greatly in the construction of the final linear system, which is done in an element-by-element basis, supported by the local matrices, the approach learnt from finite element procedures. All calculations are done for the elements, creating a local matrix for each element, and the global linear system is constructed while sweeping the domain element-by-element.

Considering the control volume centered at P and the corresponding elements used to construct the control volume, depicted in Fig. 13.27, the integration of Eqs. (13.101–13.103) reads,

$$\frac{M_P - M_P^o}{\Delta t} + \sum_{ip}(\rho V \cdot n \Delta S)_{ip} = 0 \quad (13.104)$$

$$\frac{M_P(u_i)_P - M_P^o(u_i)_P^o}{\Delta t} + \sum_{ip}(\rho V \cdot n \Delta S)_{ip}(u_i)_{ip}$$

$$= -\sum_{ip}(p_i \cdot n \Delta S)_{ip} + \sum_{ip}(\tau_i^* \cdot n \Delta S)_{ip}$$

$$+ (S_P(u_i)_P + S_C)\Delta V \quad (13.105)$$

$$\frac{M_P \phi_P - M_P^o \phi_P^o}{\Delta t} + \sum_{ip}(\rho V \cdot n \Delta S)_{ip} \phi_{ip}$$

$$= \sum_{ip}\left(\Gamma^\phi \nabla \phi \cdot n \Delta S\right)_{ip} + \left(S_P^\phi \phi_P + S_C^\phi\right)\Delta V \quad (13.106)$$

13.3 EbFVM—Element-based Finite Volume Method

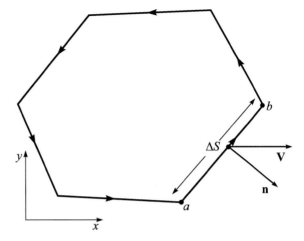

Fig. 13.27 Generic control volume

In Eq. (13.105), τ_i^* is part of the stress tensor, given by,

$$\tau_i^* = \mu \begin{bmatrix} \left(\frac{\partial u}{\partial x} + \frac{\partial u}{\partial x}\right) & \left(\frac{\partial u}{\partial y} + \frac{\partial v}{\partial x}\right) & \left(\frac{\partial u}{\partial z} + \frac{\partial w}{\partial x}\right) \\ \left(\frac{\partial u}{\partial y} + \frac{\partial v}{\partial x}\right) & \left(\frac{\partial v}{\partial y} + \frac{\partial v}{\partial y}\right) & \left(\frac{\partial v}{\partial z} + \frac{\partial w}{\partial y}\right) \\ \left(\frac{\partial u}{\partial z} + \frac{\partial w}{\partial x}\right) & \left(\frac{\partial w}{\partial y} + \frac{\partial v}{\partial z}\right) & \left(\frac{\partial w}{\partial z} + \frac{\partial w}{\partial z}\right) \end{bmatrix}, \quad (13.107)$$

The scalar pressure, to be integrated following the usual procedure, can be written as,

$$p = \begin{bmatrix} p & 0 & 0 \\ 0 & p & 0 \\ 0 & 0 & p \end{bmatrix} \quad (13.108)$$

such that each line of Eqs. (3.107) and (13.108) enters in one of the components of the momentum conservation equations.

Inspecting the integrated equations, there are several terms which need to be evaluated. To this end, consider Fig. 13.27, in which a general control volume, its integration surfaces, the normal and velocity vectors are shown. By convention, the normal is always directed outward, and the direction of integration is counterclockwise. The normal unit vector is,

$$\mathbf{n} = \left(\frac{\Delta y}{\Delta S}\right)\mathbf{i} - \left(\frac{\Delta x}{\Delta S}\right)\mathbf{j}, \quad (13.109)$$

with the scalar product among the normal and velocity vector resulting in,

$$\mathbf{V} \cdot \mathbf{n} \Delta S = u\Delta y - v\Delta x, \quad (13.110)$$

or, applying Eq. (13.110) to all integration points in the surface of the control volume, it becomes,

$$\sum_{ip} (V \cdot n\Delta S)_{ip} = \sum_{ip} (u\Delta y - v\Delta x)_{ip} = \sum_{ip} \left(u_j \Delta n_j\right)_{ip} \quad (13.111)$$

in which Δx and Δy must always be obtained by sweeping the surface of the control volume counterclockwise, i.e., $\Delta x = x_b - x_a$ and $\Delta y = y_b - y_a$, maintaining the resulting sign. In the second equality in Eq. (13.111), in the summation process in j, at each integration point, one has, for $j = 1$, $\Delta n_1 = \Delta y$ and $u_1 = u$, and for $j = 2, \Delta n_2 = -\Delta x$ and $u_2 = v$, reproducing the first equality. This notation [23] helps in programming the code, and this is the only reason for using it here. See that all terms present in the conservation equations, except for the source term, are scalar products. The others scalar products needed, besides the one given by Eq. (13.111), are $\sum_{ip} \left(\mu \tau_i^* \cdot n\Delta S\right)_{ip}$ and $(p_i.n)\Delta S$ for the momentum equations, and $\sum_{ip} \left(\Gamma^\phi \nabla \phi \cdot n\Delta S\right)_{ip}$ for the general scalar ϕ. They are, considering a 2D condition for simplicity,

$$\sum_{ip} \left(\mu \tau_i^* \cdot n\Delta S\right)_{ip} = \sum_{ip} \left(\mu \left(\frac{\partial u}{\partial x} + \frac{\partial u}{\partial x}\right)\Delta y - \mu \left(\frac{\partial u}{\partial y} + \frac{\partial v}{\partial x}\right)\Delta x\right)_{ip}$$

$$= \sum_{ip} \left(\mu \left(\frac{\partial u_i}{\partial x_j} + \frac{\partial u_j}{\partial x_i}\right)\Delta n_j\right)_{ip}, \quad (13.112)$$

and for the pressure term,

$$\sum_{ip} (p_i \cdot n\Delta S)_{ip} = (p\Delta n_i)_{ip}, \quad (13.113)$$

which can be applied, for $i = 1$ and $i = 2$, for the x and y components of the momentum conservation equation, as

$$\sum_{ip} (p_i \cdot n\Delta S)_{ip} = \sum_{ip} (p\Delta y)_{ip} \quad (13.114)$$

$$\sum_{ip} (p_i \cdot n\Delta S) = -\sum_{ip} (p\Delta x)_{ip} \quad (13.115)$$

To find the pressure at the integration points the shape functions are employed, since pressure is essentially elliptic, and a bi-linear interpolation function suffices, as

$$p_{ip} = \sum_i N_i \left(\xi_{ip}, \eta_{ip}\right) p_i$$

13.3 EbFVM—Element-based Finite Volume Method

The discrete equations, written in compact form, are, for the mass conservation equation,

$$\frac{M_P - M_P^o}{\Delta t} + \sum_{ip}(\rho u_j \Delta n_j)_{ip} = 0 \tag{13.116}$$

for the momentum conservation equations,

$$\underbrace{\frac{M_P(u_i)_P - M_P^o(u_i)_P^o}{\Delta t}}_{A} + \underbrace{\sum_{ip}(\rho u_j \Delta n_j)_{ip}^o}_{B}\underbrace{(u_i)_{ip}}_{C} =$$

$$-\underbrace{\sum_{ip}(p \Delta n_i)_{ip}}_{D} + \sum_{pi}\left(\mu \underbrace{\left(\frac{\partial u_i}{\partial x_j} + \frac{\partial u_j}{\partial x_i}\right)}_{E} \Delta n_j\right)_{ip} + \underbrace{\left(S_P^{u_i}(u_i)_P + S_C^{u_i}\right)\Delta V}_{F},$$

$$\tag{13.117}$$

and, for a general scalar ϕ,

$$\frac{M_P \phi_P - M_P^o \phi_P^o}{\Delta t} + \sum_{ip}(\rho u_j \Delta n_j)^o \phi_{ip} = \sum_{i}\left(\Gamma^\phi \frac{\partial \phi}{\partial x_j}\Delta n_j\right)_{ip} + \left(S_P^\phi \phi_P + S_C^\phi\right)\Delta V \tag{13.118}$$

To complete the discretization, let's recap term by term of Eq. (13.117). Term A represents the changing in time of the momentum quantity inside the control volume. A first order approximation was used for discretization. Term B is the mass flow at all faces of the control volume, identified by the integration point (ip) which lie on the centroid of the faces.

Term C, as learned in several sections, is the most important step in obtaining the discrete equations, since it may introduce the pathologies known as numerical diffusion and numerical oscillation. In this term the velocities (advected velocities) at the integration points should be replaced by velocities at nodal points. The choice of an interpolation function which generates positive coefficients is recommended.

Term D, due to its elliptic nature can be easily determined using the shape functions. The derivatives appearing in term E need to be represented by the derivatives of the shape functions using the coordinate transformation. As they are diffusive terms, they can be obtained using the shape functions. Term F is the source term in which $S_P^{u_i}$ goes to the diagonal of the matrix and S_C to the independent term. When all representations of the variables at integration points are replaced by their respective schemes as function of the grid nodes, the linear system is obtained.

The strategy of the element-base method is to calculate all terms of Eq. (13.117) for each element. For example, for a rectangular element the values of each sub-control volume of one element will contribute to four different control volumes.

Therefore, when sweeping element-by-element the control volumes which shares this element will receive the corresponding contribution to create the coefficients of the linear system. An assembling procedure should, then, be created, which is not unique, and the efficiency depends on the code programming. Same comments hold for Eq. (13.118) or any other.

13.3.7 Assembling Strategies

To obtain the global linear system, the discretized equations for the control volumes could be obtained by sweeping by control volumes, gathering the data from the sub-control volumes (sub-element) which are required for the control volume in consideration. This would require, when dealing with a specified control volume, to get information from different elements. This is not a recommended strategy, since there will be a lot of duplication in the calculations.

To exemplify, Fig. 13.28 shows the global numbering for 8 grid nodes. The flux at the position $ip1$ of the element 7654 (see Fig. 13.29) is the same flux which enters in the equation for the control volumes 7 and 4. If the sweeping is done via control volumes, this flux may be calculated twice, once when in the control volume 4, and later for the control volume 7. The code programming may be more cumbersome for avoiding double calculations without using the elements.

Therefore, the best strategy is to sweep by elements and, what is more important, filling the entries of the global linear system as the sweeping goes on. Even following

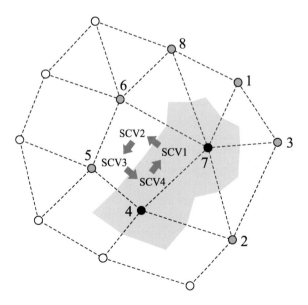

Fig. 13.28 Contribution of element 7654 in control volumes 7 and 4

13.3 EbFVM—Element-based Finite Volume Method

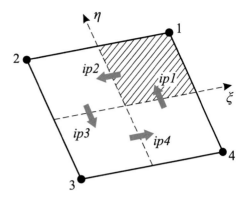

Fig. 13.29 General element and its integration points

this strategy, good programming practices should be followed to take advantage of the element-by-element nature of the sweeping.

According to Fig. 13.28 and using the proper partial differential equation to be solved, all data is calculated for each element. Advective and diffusive fluxes, represented by the arrows, volumes, source terms, including the transient term from the previous time level, are calculated for each sub-control volumes of the element. This process creates a N × N local matrix (being N the number of grid nodes of the element), whose elements feed the global matrix as the sweeping is processed. Considering 8 unknowns (Fig. 13.28), the line number 7, 6, 5 and 4 of the global matrix will receive contributions from the 4 × 4 local matrix of element 7654.

To explore a bit more the use of the elements, consider Fig. 13.29, which is a general element, in which case the local number 1 correspond to the global number 7 of Fig. 13.28. Let's consider the integration of Eq. (13.118) in the sub-control volume 1(SCV1) of the element 7654, region which is part of the control volume centered at 7. The integration gives,

$$\frac{M_{SCV1}\phi_1 - M^o_{SCV1}\phi^o_1}{\Delta t} + \dot{M}_{ip1}\phi_{ip1} + \dot{M}_{ip4}\phi_{ip2} - \left(\Gamma^\phi \sum_{k=1}^{4} \frac{\partial N_k}{\partial x}\phi_k \Delta y\right)_{ip1}$$
$$+ \left(\Gamma^\phi \sum_{k=1}^{4} \frac{\partial N_k}{\partial y}\phi_k \Delta x\right)_{ip1} - \left(\Gamma^\phi \sum_{k=1}^{4} \frac{\partial N_k}{\partial x}\phi_k \Delta y\right)_{ip2}$$
$$+ \left(\Gamma^\phi \sum_{k=1}^{4} \frac{\partial N_k}{\partial y}\phi_k \Delta x\right)_{ip2} - S_P\phi_1 \Delta V_{SVC1} - S_C \Delta V_{SVC1}$$

(13.119)

Expression (13.119) is part of the full equation for the control volume 1 (or 7 in the physical domain). In Eq. (13.119) ϕ_{ip1} and ϕ_{ip2}, will be written in terms of the nodal values, $(\phi_7, \phi_6, \phi_5, \phi_4)$ depending on the advection scheme employed. The diffusion terms, after substituting the partial derivatives with respect to x and y by the derivatives with respect to ξ and η using the coordinate transformation, will be also function of the nodal values (ϕ_7, ϕ_6, ϕ_5, ϕ_4). The same will happens with the sub-control volumes SCV2, SCV3 and SCV4, entering in the balances for the nodal point 6, 5 and 4. Those values are inserted in the global matrix when the element 7654 is visited.

When building the global matrix, in the line 7, there will be contribution of grid nodes (ϕ_7, ϕ_6, ϕ_5, ϕ_4) because of the quadrangular element 7654, a contribution of grid nodes 768 because of SCV1 of the neighbor triangular element, and so on. Therefore, the discrete equation for a control volume, which represents a line of the global matrix, goes being constructed adding the information of each sub-control volume which forms that control volume when the element-by-element sweep is realized.

Observe that if the construction of the discrete equation is done in a control volume sweeping, each line of the matrix would be constructed at once.

To exercise a bit more the assemble strategy, Fig. 13.30 depicts three quadrangular elements and the respective local matrices. Symbols are given for each element, representing in which entry of the global matrix the element contributes. As the sweep goes on in an element-by-element basis, it is shown in each rows and columns the contribution of each element in the global matrix is. Different symbols mean that different elements have contributed with the equation for that node. See control volume 5, for example, which have received contribution of three elements represented by the symbol of a triangle upside down, which is the addition of the contributions of elements represented by symbols circle, diamond and star (elements 1, 2 and 3). In this configuration it is the only one control volume which receives contributions of all elements.

13.3.8 Boundary Conditions

To conclude this chapter, boundary conditions are now discussed. By construction, cell-vertex finite volume method has grid nodes at the boundaries and the procedures do not differ to what was discussed in Chaps. 3 and 7 for Cartesian grids, and in Chap. 12 for boundary-fitted coordinates. Figure 13.31 depicts a boundary control volume which involves two elements sharing the nodal point P. It is equivalent, of course, to a half control volume for the Cartesian grid. What should be done in order to apply the boundary condition is to realize a balance at the boundary control volume, respecting the existing boundary condition.

If the condition is of Neumann type, that is, flux is prescribed at the boundary faces of the control volume, the value of the flux just enter the balance directly, since this type o boundary condition is the natural one for finite volume methods. Just for refreshing, when it is said that a balance is performed at the control volume level, it is equivalent to integrate the differential equation in conservative form.

If the condition is of Dirichlet type, this grid node can be removed from the linear system but, as already stated in the previous chapters, conservation will not be obeyed

13.3 EbFVM—Element-based Finite Volume Method

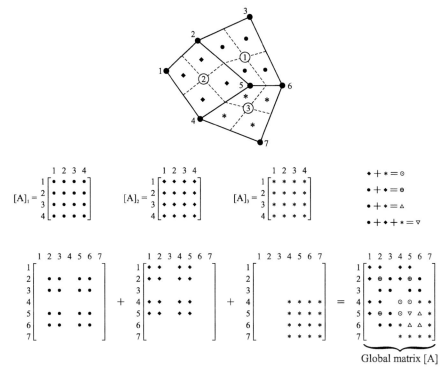

Fig. 13.30 Elements local matrices and assembling

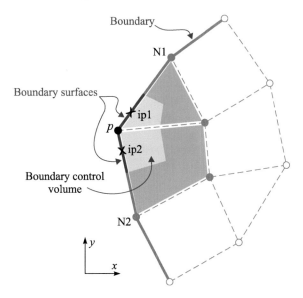

Fig. 13.31 Boundary control volume

in all control volumes lying on the boundaries. The EbFVM construction allows to escape from this non-satisfaction of the conservation principles by performing a balance at the boundary control volume respecting the prescribed value of the variable.

Consider it is known the ϕ distribution at the boundary, therefore, the values of ϕ_{N1}, ϕ_{N2} and ϕ_P are known. This allows to find the values of ϕ at the integration points $ip1$ and $ip2$ via an average of the known values at grid nodes P, N1 and N2. Using these two values, a flux at these integration points can be approximated using the other interior nodes, and a balance can be done for the boundary control volume, leaving the value of the variable at node P as unknown, which will be found through the solution of the linear system. It is not difficult recognize that the result will be as if the node P was at the center of the boundary control volume in which the balance was done. It is equivalent of applying a boundary condition in a cell-center method in which the control volume would be half. Depending on the grid resolution at the boundaries, the calculated value will be slightly different from the one prescribed at the point P. This doesn't harm the solution and conservation will be obeyed for the entire domain.

Inflow and outflow boundary conditions follows the same rationale already seen in previous chapters.

13.4 Conclusions

The flexibility of using unstructured grids for the discretization of fluid flow equations was shown in the context of cell-center and cell vertex methods. Both methods are widely used in commercial applications. Cell-center methods, also called the conventional finite volume method, are better known by the engineering community, as it was used since the beginning of the major developments in computational fluid dynamics. The cell-vertex method that received our attention was the EbFVM—Element-base Finite Volume Method, also known in the literature as CVFEM—Control volume Finite Element Method. The EbFVM borrows from finite element technologies the concept of the element, which allows all calculations to be done in the element and the construction of local matrices used in an element-by-element sweep to build the global linear system. It is a very attractive methodology, as it is prone to an elegant and neat programming of the code because of the element-by-element approach. This strategy favors the use of object-oriented programming in the context of C++. The method is, as any other finite volume method, full conservative, what respects the physics at discrete level and enhances robustness.

As a learning aid, the author has recorded a series of 12 video-lessons about cell-center and cell-vertex methods in unstructured grids which may help to follow this chapter. This material can be seen at the link provided in [30].

13.5 Exercises

13.1 Consider two elements, one of which is shown in Fig. 13.32, and the other one is a rectangular element defined by the points $P_1(5, 3)$, $P_2(0, 3) P_3(0, 0)$, and $P_4(5, 0)$. Given the function $\phi(x, y) = (1 + x)(1 + y)$ defined on these elements, draw the transformed plane of these elements and calculate $\partial \phi / \partial x$ and $\partial \phi / \partial y$ at the point $((\xi, \eta) = (0, 0))$, analytically and through the shape functions and compare the results. Why for the element in Fig. 13.32 the approximations by the shape functions do not agree with the analytical values?

13.2 Find the area of the element of Fig. 13.32 through the areas of its sub-control volumes.

13.3 To exercise the use of EbFVM, solve the 2D steady state heat diffusion problem given by Eq. (10.4) and Fig. 13.33, with a and b of your choice, using the element-based volume method, finite volume method with the element as the control volume (cell-center) and by finite differences using the mesh formed by the elements. The elements are formed by a Cartesian grid. Use central differences in the interpolation functions for the derivatives. This problem has an analytical solution given by Eq. (10.5). With a mesh with 4×4 elements, compare the analytical solution with the three numerical solutions. Note that for the EbFVM and finite differences the points in which the temperature is calculated are the same, while for the conventional finite volume method the calculation point is in the centroid of the element (control volume) and to compare an interpolation of values is required. Comment about the comparisons. What kind of error is present? Successively, refine the mesh and observe what happens to the solutions. Calculate the heat flux at some point on the north boundary using all methods and see what happens as the grid is refined.

13.4 Our goal in this exercise is to derive the coefficients of the approximate equations for the EbFVM formulation and compare them to the conventional finite volume method. Figure 13.34a shows an equally spaced mesh in x and y (Δx

Fig. 13.32 Figure for Problem 13.1

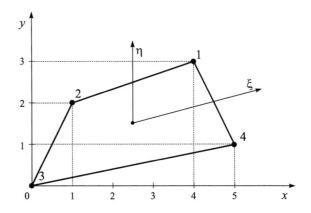

Fig. 13.33 Figure for Problem 13.3

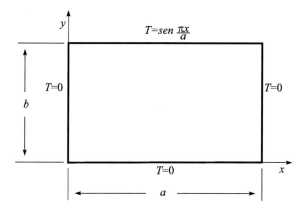

can be different from Δy), showing the element 1234 and the neighboring elements contributing to the control volume centered at 1. Consider the two-dimensional transient equation with no source term for a generic variable ϕ given by,

$$\frac{\partial}{\partial t}(\rho\phi) + \frac{\partial}{\partial x}(\rho u\phi) + \frac{\partial}{\partial y}(\rho v\phi) = \frac{\partial}{\partial x}\left(\Gamma^\phi \frac{\partial\phi}{\partial x}\right) + \frac{\partial}{\partial y}\left(\Gamma^\phi \frac{\partial\phi}{\partial y}\right) \quad (13.120)$$

(a) Using for the approximations of the diffusive fluxes the shape functions and for the advective ones, the WUDS interpolation scheme, given by,

$$\phi_{ip1} = \left(\frac{1}{2} + \alpha\right)_{ip1} \phi_{23} + \left(\frac{1}{2} - \alpha\right)_{ip1} \phi_{14} \quad (13.121)$$

with a similar equation for ϕ_{ip4}, in which α is positive for $u > 0$ and negative for $u < 0$. For the equation for ϕ_{ip4} the signal of v is considered. Obtain Eq. (13.122), which corresponds to the integration for the sub-control volume 1.

$$\frac{M_1\phi_1 - M_1^o\phi_1^o}{\Delta t} + \dot{M}_{ip1}\left(\left(\frac{1}{2}+\alpha\right)_{ip1}\phi_{23} + \left(\frac{1}{2}-\alpha\right)_{ip1}\phi_{14}\right)$$

$$+ \dot{M}_{pi4}\left(\left(\frac{1}{2}+\alpha\right)_{ip4}\phi_{34} + \left(\frac{1}{2}-\alpha\right)_{ip4}\phi_{12}\right)$$

$$+ \left(\Gamma^\phi \sum_{k=1}^{4} \frac{\partial N_k}{\partial x}\phi_k \frac{\Delta y}{2}\right)_{ip1} + \left(\Gamma^\phi \sum_{k=1}^{4} \frac{\partial N_k}{\partial y}\phi_k \frac{\Delta x}{2}\right)_{ip4} \quad (13.122)$$

Substitute the values of ϕ_{23}, ϕ_{14}, ϕ_{34} and ϕ_{12} by an linear average of the nodal values and write the equation in the form,

13.5 Exercises

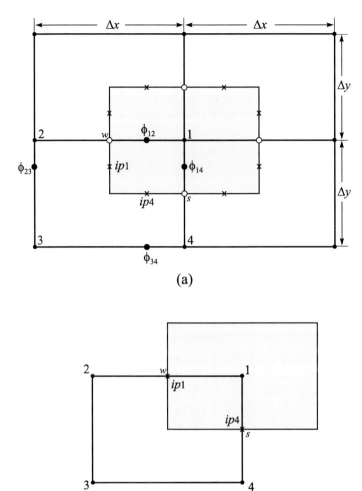

Fig. 13.34 Figure for Problem 13.4

$$A_1\phi_1 = A_2\phi_2 + A_3\phi_3 + A_4\phi_4 + B_1, \tag{13.123}$$

and calculate the coefficients of Eq. (13.123)

(b) Consider now the integration points in the element located as shown in Fig. 13.34b, which in Fig. 13.34a correspond to the open circles, creating the well-known situation of a single integration point in the middle of each face, typical of the conventional finite volume method. For the volume centered at 1, and considering only the integration points s ($ip4$) and w ($ip1$), and using WUDS, obtain, integrating Eq. (13.120), using the conventional finite volume method, the following equation,

$$\left[\frac{M_1}{\Delta t} - \dot{M}_w\left(\frac{1}{2}-\alpha\right)_w - \dot{M}_s\left(\frac{1}{2}-\alpha\right)_s + \Gamma^\phi_w\frac{\Delta y}{\Delta x} + \Gamma^\phi_s\frac{\Delta x}{\Delta y}\right]\phi_1$$
$$= \left[\dot{M}_w\left(\frac{1}{2}+\alpha\right)_w + \Gamma^\phi_w\frac{\Delta y}{\Delta x}\right]\phi_2 + \left[\dot{M}_s\left(\frac{1}{2}+\alpha\right)_s + \Gamma^\phi_s\frac{\Delta x}{\Delta y}\right]\phi_4 + \frac{M^o_1\phi^o_1}{\Delta t}$$
(13.124)

in which, one is calling ϕ_W of ϕ_2, ϕ_S of ϕ_4 and ϕ_P of ϕ_1.

(c) For this new location of the integration points, starting from Eq. (13.122) and writing in the form of Eq. (13.123), find the following coefficients,

$$A_1 = \frac{M_1}{\Delta t} + \dot{M}_{ip1}\left(\frac{1}{2}-\alpha\right)_{ip1} + \dot{M}_{pi4}\left(\frac{1}{2}-\alpha\right)_{ip4}$$
$$+ \left(\Gamma^\phi\frac{\Delta y}{2\Delta x}\right)_{ip1} + \left(\Gamma^\phi\frac{\Delta x}{2\Delta y}\right)_{ip4} \quad (13.125)$$

$$A_2 = -\dot{M}_{ip1}\left(\frac{1}{2}+\alpha\right)_{ip1} + \left(\Gamma^\phi\frac{\Delta y}{2\Delta x}\right)_{ip1} \quad (13.126)$$

$$A_4 = -\dot{M}_{pi4}\left(\frac{1}{2}+\alpha\right)_{ip4} + \left(\Gamma^\phi\frac{\Delta x}{2\Delta y}\right)_{ip4} \quad (13.127)$$

$$A_3 = 0 \quad (13.128)$$

$$B_1 = M^o_1\phi^o_1/\Delta t, \quad (13.129)$$

See that the coefficients are identical to the coefficients of Eq. (13.124), except the corresponding area, which now is Δy and no longer $\Delta y/2$. See also that the sign of the advective part of the coefficients A_2 and A_4, from Eqs. (13.126) and (13.127), are opposite to their counterparts in Eq. (13.124).

(d) Take a time to look at the coefficients of Eq. (13.122) and observe that the upwind approximation at integration point $ip1$, for example, is being split between points 2 and 3, when u is positive, what changes the diagonal of the matrix coefficients. What influence does this have on the possible stability of the method compared to when only point 2 is used? The same happens with the upwind approximation in y, which is split between points 3 and 4 for positive v.

References

1. Mavriplis DJ (2000) Mesh generation and adaptivity for complex geometries and flows. In: Peyret R (ed) Handbook of computational fluid mechanics. Academic Press
2. Morgan K, Peraire J, Peiró J (1994) Unstructured grid methods for the simulation of 3d transient flows. Final Report, NASA-CR-196139, Department of Civil Engineering, University of Wales, Swansea, UK
3. Jameson A, Baker TJ, Weatherhill NP (1986) Calculation of inviscid flow over a complete aircraft. AIAA Paper 86-0103
4. Mavriplis DJ, Jameson A, Martinelli L (1989) Multigrid solution of the Navier–Stokes equations on triangular meshes. AIAA Paper 89-0120
5. Mavriplis DJ (2003) Revisiting the least-squares procedure for gradient reconstruction on unstructured meshes. NIA Report 2003-06, National Institute of Aerospace, Hampton, USA
6. Baliga BR, Patankar SV (1980) A new finite element formulation for convection-diffusion problems. Numer Heat Transfer 3:393–409
7. Baliga BR, Phan TT, Patankar SV (1983) Solution of some two-dimensional incompressible fluid flow and heat transfer problems using a control-volume finite-element method. Numer Heat Transfer 6:263–282
8. Prakash, C, Patankar SV (1986) A control volume-based finite-element method for solving the Navier-Stokes equations using equal-order velocity-pressure interpolation. Numer Heat Transfer 9:253–276
9. Raw MJ (1985) A new control-volume-based finite element procedure for numerical solution of the fluid flow and scalar transport equations, Ph.D. thesis, University of Waterloo, Canada
10. Schneider GE, Raw MJ (1986) A skewed positive influence coefficient upwinding procedure for control-volume-based finite-element convection-diffusion computation. Numer Heat Transfer 9:1–26
11. Minkowycz WJ, Sparrow EM, Schneider GE, Pletcher RH (eds) (1988) Handbook of numerical heat transfer. Wiley
12. Park MA, Krakos TM, Loseille A, Alonso JJ (2016) Unstructured Grid Adaptation: status, potential impacts and recommended investments towards CFD vision 2030, AIAA Aviation. https://doi.org/10.2514/6.2016-3323
13. Maliska CR, Raithby GD (1984) A method for computing three dimensional flows using boundary-fitted co-ordinates. Int J Numer Meth Fluids 4:519–537
14. Peters S, Maliska CR (2017) A staggered grid arrangement for solving incompressible flows with hybrid unstructured meshes. Numer Heat Transfer, Part B-Fundam 71:50–65
15. Golub GH, Loan CEV (1996) Matrix computations. John Hopkins University Press
16. Ansys Fluent Theory Guide (2021) Ansys, Inc. Canonsburg, PA
17. Cerbato G, Hurtado FSV, Silva, AFC, Maliska CR (2014) Analysis of gradient reconstruction methods on polygonal grids applied to petroleum reservoir simulation. In: Proceedings of the ENCIT/2014
18. Van Leer B (1979) Toward the ultimate conservative difference scheme. V. A second order sequel to Godunov's method. J Comp Phys 32:101–136
19. Raithby GD, Torrance KE (1974) Upstream-weighted differencing schemes and their application to elliptic problems involving fluid flow. Comput Fluids 2:191–206
20. Van Doormaal, JP, Raithby GD (1984) Enhancements of the simple method for predicting incompressible fluid flow. Numer Heat Transfer 7:147–163
21. Maliska CR, Vasconcellos JFV (2000) An unstructured finite volume procedure for simulating flows with moving fronts. Comp Meth Appl Mech Eng 182:401–420
22. Marchi CH, Maliska CR (1994) A nonorthogonal finite volume method for the solution of all speed flows using co-located variables. Numer Heat Transfer 26(3):293–311
23. Ansys CFX Solver Theory Guide (2021) Ansys, Inc. Canonsburg, PA
24. Zienkiewicz OC, Taylor RL (2000) The finite element method, vol 1: the basis
25. Raithby GD (1976) Skew upstream differencing schemes for problems involving fluid flow. Comp Meth Appl Mech Eng 9:153–164

26. Hurtado FSV, Maliska CR (2012) A family of positive flow-weighted advection schemes for element-based finite volume methods. Numer Heat Transfer, Part B, Fundam 62(2–3):113–140
27. Roe PL, Sidilkover D (1992) Optimum positive linear schemes for advection in two and three dimensions. SIAM J Numer Anal 29:1542–1568
28. Hurtado FSV (2005) An element-based finite volume formulation for two-phase immiscible flow in porous media. M.Sc. dissertation, Mech. Eng. Department, UFSC (in Portuguese). http://repositorio.ufsc.br/handle/123456789/101678
29. Barth TJ, Jespersen DC (1989) The design and application of upwind schemes on unstructured meshes. AIAA-89-0366, 27th Aerospace Sciences Meeting, Reno, Nevada
30. Maliska CR (2019) Video-lessons on unstructured finite volume methods. https://youtube.com/playlist?list=PLIPfGy5ZylmPaTq6634hKIDhckK3LgMOE

Chapter 14
Pressure Instabilities: From Navier–Stokes to Poroelasticity

14.1 Introduction

The scope of this chapter is beyond the topic covered so far in this textbook. The reason for having this chapter in this book is due to the similarities between the pressure–velocity coupling in the Navier–Stokes equations and the pressure–displacement coupling in poroelasticity. It will be seen that the pathologies present in poroelasticity with respect to instabilities in the pressure field can be solved with the same remedies applied to the Navier–Stokes equations five decades ago. It becomes relevant to discuss this topic because the instabilities in poroelasticity are still a difficulty which challenges the numerical analysts in this area.

Besides the instabilities, another issue present in the day-to-day activities of the aquifer management engineer or of the petroleum engineer, to name two important areas of application, is that this coupling is usually solved with two different methods. The classical finite element method, largely used in solid mechanics computations is used for solving the rock mechanics problem, and the finite volume method, widely employed in fluid flow calculations, due to its conservative behavior, is used to solve the flow in the porous media. This strategy employs, thus, two different numerical methods, two different grids and creates a cumbersome engineering workflow. Pressure needs to feed the finite element algorithm for calculating the displacement and, in the way back, displacements should feed the flow simulator to find pressure. This is an iterative procedure which requires interpolation of variables if the grids are not the same, as usually is, for example, in petroleum engineering. It is desirable to have a method which solves a single linear system having as unknowns the pressure and the displacements, avoiding this back-and-forth among two different methodologies.

This chapter highlights a methodology using EbFVM, the method analyzed in the previous chapter, for solving both physics, the fluid flow and the rock mechanics in a single numerical method. In doing so, it allows the prescription of the remedies used in Navier–Stokes to eliminate pressure instabilities in poromechanics. Only enough detail for explaining the scheme for suppressing oscillations will be shown. Simple one-dimensional problems for the Biot's equation will be used to demonstrate the feasibility of using finite volumes for both physics. It is not in the scope of this book to deep the discussion on the numerical treatment of poroelasticity problems. This can be seen in [1–10]. Moreover, it is assumed that the reader has enough background in poroelasticity to follow the developments. There are excellent books on the topic in the literature [11–15], in which detailed theoretical basis can be found. An interesting piece of history involving the pioneering researchers in the field of poroelasticity can be enjoyed in [16].

14.2 Pressure Instabilities

To establish the connection between the instabilities in the pressure field arising when solving the mass conservation equation in the Navier–Stokes equations and the mass conservation equation in poroelasticiy, it is advisable to briefly recapitulate this problem, already seen in Chap. 7. The fundamental issue that causes the decoupling of the pressure and velocity, and thus, instabilities, is linked to the conservation of mass and the co-located arrangement, in which the velocities are not available on the faces of the control volume to apply mass conservation. When solving a Navier–Stokes problem, the mass conservation equation taking part on the equation system is

$$\frac{\partial \rho}{\partial t} + \nabla \cdot (\rho \mathbf{V}) = 0, \tag{14.1}$$

in which $\mathbf{V}$ is the fluid velocity and ρ the density, or specific mass.

Considering a control volume in a porous media with rock compaction, there is a solid mass flow entering the control volume carrying fluid due to this compaction. A balance of mass fluid in this control volume gives,

$$\frac{\partial}{\partial t}(\rho \phi) + \nabla \cdot [\rho(\mathbf{V} + \phi \mathbf{V}_S)] = 0 \tag{14.2}$$

in which ϕ is the porosity of the medium, $\mathbf{V}$ is the Darcy's velocity, which is the fluid velocity relative to the solid, and $\mathbf{V}_s$ is the solid velocity due compaction. The Darcy's velocity is replaced by its expression related to the pressure gradient and, therefore, it is not an unknown of the problem. The solid velocity, $\mathbf{V}_s$, by its turn, is given by

$$\mathbf{V}_S = \frac{\partial \mathbf{u}}{\partial t}, \tag{14.3}$$

in which $\mathbf{u}$ is the displacement vector, the unknown of the problem. Indirectly, the solid velocity is an unknown in a poroelasticity problem.

14.2 Pressure Instabilities

What is known from the simulation of fluid flows (Navier–Stokes) is that it is mandatory to have a consistent[1] advecting velocity at the interfaces of the control volume in which mass balance is performed. This is the co-located pressure–velocity coupling (CPVC) discussed in Chap. 7. This suggests, inspecting Eq. 14.3, that the displacement **u**, which represents the velocity of the fluid carried by the solid due to the compaction, must be available at the interfaces of the control volume too, following what happens in fluid mechanics.[2] In the solution of a poroelasticity problem the unknowns are pressure and displacement and, consequently, a pressure–displacement coupling, exactly as in Navier–Stokes takes place. In poroelasticity one could name it co-located pressure–displacement coupling (CPDC). This means that all strategies used for calculating the velocity at the interfaces of a control volume for mass conservation in Navier–Stokes, can be applied to displacement in poroelasticity. Since the pathology is the same, the known remedies used in fluid mechanics apply.

In certain physical situation the consolidation takes place in a much smaller time scale, such that $V \ll V_s$, what implies that the solid velocity is the main actor in the mass conservation equation, Eq. 14.2, making the difficulties even worse. This condition happens in the undrained consolidation, at the beginning of the transient and at the interfaces of different materials, conditions in which oscillatory pressure fields are observed. The two remedies for the pressure oscillatory pathology are now described. The reader is referred to Chap. 7 for details.

14.2.1 Remedy 1

The first remedy is an old recipe motivated by instabilities in the pressure field when using co-located grid arrangement in fluid mechanics. Developed in the 60s [17] (see Chap. 7), the solution is the use of the staggered grid arrangement. Placing the velocities at the interface of the control volume makes them available for the mass balances. This is what promotes a tight coupling among pressure and velocity. In this variable arrangement, the velocity which enters in the mass conservation is the proper unknown velocity of the problem. Therefore, the coupling between pressure and velocity is treated in the best way. Recognizing the similarity [1, 2, 4–9] among the couplings in fluid mechanics and poroelasticity, the staggered grid approach, as depicted in Fig. 14.1, is the first remedy for avoiding pressure instabilities in the solution of the Biot's consolidation problem.

For the purposes of this chapter the Biot's consolidation model is used. The mass and momentum conservation equations, without gravity effects, governing the coupled fluid flow/geomechanics problem are, respectively,

---

[1] By consistent velocity it is meant a velocity which promotes the coupling between pressure and velocity. In finding this velocity, pressure should enter the scheme.

[2] Fluid mechanics in this section refers to the solution of the Navier–Stokes equations, a non-porous media flow.

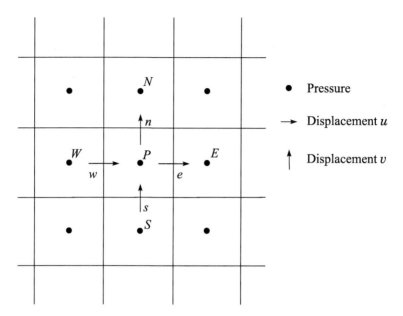

Fig. 14.1 Staggered grid for pressure and displacement

$$\frac{1}{M}\frac{\partial p}{\partial t} + \nabla \cdot (V + \alpha V_s) = q \tag{14.4}$$

$$\nabla \cdot \sigma - \alpha \nabla p = b \tag{14.5}$$

in which M is the Biot's module, q is a possible source term representing a well, for example, V is the Darcy's velocity and V_s the solid velocity due to the compaction. For the equilibrium equation (momentum equation), σ is the effective stress tensor [18], α is the Biot's coefficient, p is the pressure and b stands for a possible force by unit of volume acting on the system. The Biot's module and the Darcy's equation are, respectively, given by

$$M = \left[\phi c_f + (\alpha - \phi)c_s\right]^{-1} \tag{14.6}$$

$$V = -\frac{k}{\mu}\nabla p, \tag{14.7}$$

in which k is the permeability tensor, and c_f and c_s are the compressibility of the fluid and solid, respectively.

To demonstrate the non-appearance of instabilities when staggered grid is used, it is enough to work with a 1D problem [2, 4, 5]. Using Eq. (14.4) an replacing the divergence of the solid velocity by the time variation of displacements, and using Darcy's equation, one has, for the mass conservation equation

14.2 Pressure Instabilities

Fig. 14.2 Staggered grid for the 1D problem

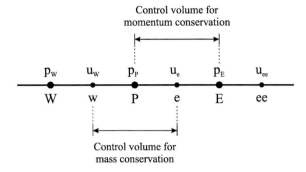

$$\left(\frac{1}{M}\right)\frac{\partial p}{\partial t} - \nabla \cdot \left(\frac{k}{\mu}\nabla p\right) = q - \alpha\frac{\partial \varepsilon}{\partial t}, \tag{14.8}$$

in which ε is the volumetric deformation. The integration of Eq. (14.8) in the control volume shown in Fig. 14.2 and in time, gives

$$\frac{\Delta x}{M}\frac{p_P}{\Delta t} - \frac{k}{\mu}\left[\left.\frac{\partial p}{\partial x}\right|_e - \left.\frac{\partial p}{\partial x}\right|_w\right] + \frac{\alpha}{\Delta t}(u_e - u_w) = q_P\Delta x + \frac{\Delta x}{M}\frac{p_P^o}{\Delta t} + \frac{\alpha}{\Delta t}(u_e - u_w)^o \tag{14.9}$$

Inspecting Eq. (14.9) against the grid layout of Fig. 14.2, the displacements are available where they are required for the calculation of the solid velocity, the issue of utmost importance for pressure stability. The pressure gradients are calculated by

$$\left.\frac{\partial p}{\partial x}\right|_e = \frac{p_E - p_P}{\Delta x} \tag{14.10}$$

$$\left.\frac{\partial p}{\partial x}\right|_w = \frac{p_P - p_W}{\Delta x} \tag{14.11}$$

The pressure gradients in the above equations are the responsible for driving the Darcy's velocity. For unstructured grids using cell-center methods, a gradient recovery method should be employed, since they are not easily available as in Cartesian grids. In all equations reported, a constant Δx is used, and when non-uniform grids are employed, as in one of the results to be presented, the proper local dimension should be used in the gradient approximation. Substituting Eqs. (14.10) and (14.11) into Eq. (14.9), the linear system to be solved for the pressure determination is obtained.

To complete the integration procedure, the momentum conservation equation, Eq. (14.5), should be integrated in space, using the grid shown in Fig. 14.2. Now, due to the staggered grid, the control volume for displacement is centered in "e". Neglecting the source term, the integration gives,

$$\sigma_E - \sigma_P = \alpha(p_E - p_P) \tag{14.12}$$

in which the driving force for the displacement is correctly located on the grid, another factor contributing for the stability of the scheme, now present in the momentum equation. For a 1D problem the stress tensor is written as

$$\sigma_E = (\lambda + 2G)\frac{\partial u}{\partial x}\bigg|_E \approx (\lambda + 2G)\frac{(u_{ee} - u_e)}{\Delta x} \tag{14.13}$$

$$\sigma_P = (\lambda + 2G)\frac{\partial u}{\partial x}\bigg|_P \approx (\lambda + 2G)\frac{(u_e - u_w)}{\Delta x} \tag{14.14}$$

Introducing Eqs. (14.13) and (14.14) into Eq. (14.12), the linear system for the determination of the u displacement with the fluid pore pressure present in the equation is obtained. Therefore, one has two linear systems, one for pressure and one for displacement, which can be solved using several strategies, which can be classified generally in segregated solution and simultaneous solution. There are a considerable amount of literature dealing with the segregated solution in poroelasticity, specially using fixed-stress methods.

To demonstrate the appearance of instabilities, the 1D problem will be solved using a co-located grid arrangement, according to Fig. 14.3. The mass conservation equation is repeated here for completeness,

$$\frac{\Delta x}{M}\frac{p_P}{\Delta t} - \frac{k}{\mu}\left[\frac{\partial p}{\partial x}\bigg|_e - \frac{\partial p}{\partial x}\bigg|_w\right] + \frac{\alpha}{\Delta t}(u_e - u_w) = q_P \Delta x + \frac{\Delta x}{M}\frac{p_P^o}{\Delta t} + \frac{\alpha}{\Delta t}(u_e - u_w)^o \tag{14.15}$$

with the marked difference that, now, the displacements required at points e and w, at the interfaces of the control volume for mass conservation, are not available, requiring, therefore, some stabilization scheme, as commented and fully discussed in Chap. 7, when the co-located pressure–velocity coupling was addressed.

The 1D problem considered is the Terzaghi's column subject to a load on the top under undrained conditions. The top is, then, opened to atmosphere with zero effective pressure, while the load is kept acting. This problem has an analytical solution widely used for checking numerical methods in poroelasticity. At the surface, at $t = 0$ there is a discontinuity on the pressure variable, and the time step and the spatial scheme employed will dictate the behavior of the solution near the open

Fig. 14.3 Co-located grid for the 1D problem

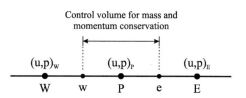

14.2 Pressure Instabilities

surface. This condition also happens at the interface of different materials, as will be seen in the two-layers Terzaghi's column. This problem was solved with different number of grids and with slightly and highly non-uniform grids.

The data used for this problem can be found in [5, 6] and the geometry of the problem is displayed in Fig. 14.4.

The solution at the beginning of the transient, as shown Fig. 14.5, demonstrates that the model using co-located variables presents strong oscillations with completely wrong solution for the displacement. These results get even worse if highly non-uniform grids are employed, as can be seen in [6]. In the same figure, using staggered grids no oscillations appear at all, confirming that the same remedy, that is, staggered grids, used in fluid mechanics, are applicable in poroelasticity. The reason for the oscillation is the absence of a consistent solid velocity (represented by the displacements) at the boundaries of the control volume for the mass balance. Instead, a simple velocity averaging is used, which is known not to promote the required coupling. A convergence analysis, demonstrating a second order convergence rate obtained with the EbFVM can be found in [2, 4, 6, 10, 19].

It looks like that these instabilities are not related with incompressibility in solids or fluids, but just because the pressure–displacement coupling is not well treated. Perhaps, what is still lacking in the literature, is the recognition that what is required for avoiding instabilities in poroelasticity is to have the solid velocities located where it is needed for mass conservation. Since solid velocities are variation of displacement in time, therefore, what is required, in other words, is to have displacements at the boundaries of a mass conservation control volume. This is the perfect analogy with pressure–velocity coupling in Navier–Stokes.

When other methods than finite volume, in general finite element, are used, both pressure and displacements are located at the same point, at the vertices of the element. Since there is no concept of control volume, there is not even the concept

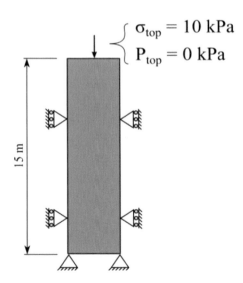

Fig. 14.4 Terzaghi's column

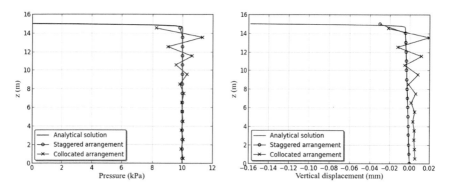

Fig. 14.5 Terzaghi's column. Staggered versus co-located

of balances, and, therefore, it can't even be said that the displacement is not stored where it is needed for mass balance. The consequence is that the stabilization techniques used in these methods are purely mathematics, without physical reasoning for its construction. Perhaps this is the reason why this issue of instabilities is still a crucial point when solving poroelasticity using other methods than finite volumes.

If the displacements are not stored displaced in relation to pressure, stabilization schemes are needed. These stabilization schemes can be physically based, as PIS, or mathematically constructed.

In [10] a detailed analysis of the error approximation using Taylor series was performed when staggered and co-located meshes are employed. Fundamental works on the mathematical reasons for the appearance of instabilities in this type of problems can be found in [20–23], among several others.

Some interesting findings are shown in [10, 19], in which the errors are decomposed into errors due to approximations in pressure and displacement, as functions of space and time, when applicable. For example, using co-located variables, the error in the momentum equation is composed of two terms, both related to the space coordinate, since time is absent in this equation. One part of the error is connected to the displacement field and the other one to the pressure field, both due to a central differencing scheme used. The errors are, then, coupled.

For the mass conservation equation, the time coordinate also enters in the determination of the errors and, for co-located variables, the errors are composed of three terms, one related to the displacement and two related to pressure.

For a staggered grid, the error in the mass conservation is related to pressure only, corresponding to the approximation in time and space. There are no errors connected with displacement. For the momentum conservation in a staggered grid, errors are related to displacement only. This allows to conclude that the staggered grid eliminates the errors linked with displacements in the mass conservation equation.

It was shown in [10] that when solving a poroelasticity problem using co-located variables the errors are coupled, which may result in instabilities, depending on the coupling strength. When using staggered grids, the errors are uncoupled, promoting

14.2 Pressure Instabilities

stability. A detailed convergence analysis was undertaken in [10] and the interested reader on the topic is invited to visit this document.

Figure 14.6 brings a sample of the convergence analysis of the Terzaghi's problem solved with staggered grids [6], depicting a second order convergence rate for both pressure and displacement. These results were obtained with a highly non-uniform grid.

To complete the analysis, staggered grids are used in unstructured grids solving the Terzaghi's problem [24, 25]. In this case, as in the Cartesian system, there will be one control volume for pressure (mass conservation) and another one for displacements (momentum conservation). A sketch of the mesh and these control volumes are shown in Fig. 14.7. The purpose of this study is to demonstrate that the staggered grid is efficient in eliminating instabilities in the pressure field in any type of grid. In the other hand, it is known that its use is not feasible in practice due to coding problems, seriously aggravated in 3D.

Figure 14.8 shows the results for the two-layered Terzaghi's column using unstructured grids and staggered arrangement of variables, demonstrating that no oscillations appear at the interface of the two materials. The conclusion is that the use of staggered grid suppresses the oscillations in all situations.

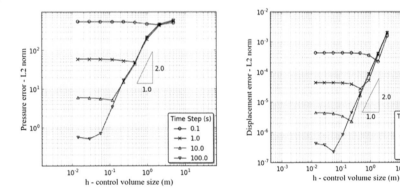

Fig. 14.6 Terzaghi's problem—convergence rate

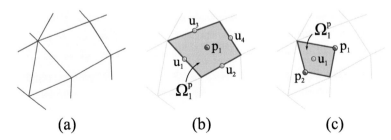

Fig. 14.7 Staggered unstructured grid. Control volumes

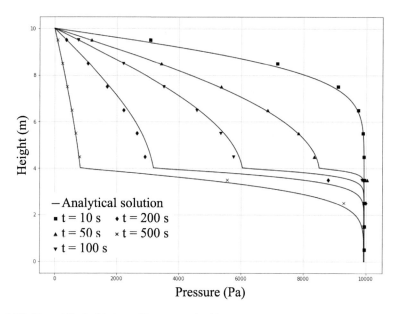

Fig. 14.8 Terzaghi's double layer. Unstructured grids

The results obtained so far demonstrated that the solution of the coupled problem using a finite volume method for both physics and staggered grids would be the recommended numerical recipe for this class of problems. However, as commented, for real problems staggered grids is impractical, and the instabilities should be managed with a new approach. This is the second remedy which will be under analysis now.

14.2.2 Remedy 2

In engineering practice all finite volume simulation packages uses co-located variables. This simplifies considerably the code by having just one control volume for all variables, with the same fluxes calculated at the boundaries applying for all variables. The key issue, already deeply discussed in Chap. 7 and highlighted in the previous section, is the calculation of the advecting velocity at the control volume boundaries. In poroelasticity, the counterpart of the advecting velocity of fluid mechanics is the solid velocity, which is calculated by the time derivative of the displacement. Therefore, displacements need to be calculated at the boundaries of the control volumes. The remedy is, again, an old strategy applied in fluid mechanics long ago, which uses physical reasoning to create the stabilizing scheme. In fluid mechanics the most known used schemes are the Rhie-and-Chow-like [26, 27] and PIS-like schemes [28, 29]

14.2 Pressure Instabilities

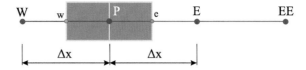

Fig. 14.9 One-dimensional co-located grid

Again, a 1D analysis suffices for demonstrating the applicability of the PIS scheme [2, 4]. Consider Fig. 14.9 in which a 1D grid is show with the same control volume for both pressure and displacement.

The idea behind PIS is to apply the governing equation at the integration point of interest and extract the variable at the interface from this equation. In the present case, the integration points are the east and west face of the control volume. Displacement is located at the point P, and should be calculated at the interfaces, required for mass conservation. Using a Taylor series expansion around point e one can write,

$$u_E \approx u_e + \frac{\Delta x}{2} \frac{\partial u}{\partial x}\bigg|_e + \frac{\Delta x}{8} \frac{\partial^2 u}{\partial x^2}\bigg|_e \qquad (14.16)$$

$$u_P \approx u_e - \frac{\Delta x}{2} \frac{\partial u}{\partial x}\bigg|_e + \frac{\Delta x}{8} \frac{\partial^2 u}{\partial x^2}\bigg|_e \qquad (14.17)$$

Adding these equations, the variable at the interface e, reads

$$u_e \approx \frac{u_E + u_P}{2} - \frac{\Delta x}{8} \frac{\partial^2 u}{\partial x^2}\bigg|_e \qquad (14.18)$$

If the second derivative is neglected, a second order central differencing scheme is recovered. This would be equivalent of using the shape functions for finding u_e, which is known, it doesn't promote the required coupling among pressure and displacement. It is also not helpful to find the second derivative using the neighbor points, since it would become a second order central differencing. The idea is to obtain a second order derivative of the displacement from the governing equation. This equation, for the 1D problem and applied to point e, is

$$\frac{\partial^2 u}{\partial x^2}\bigg|_e = \frac{\alpha}{(\lambda + 2G)} \frac{\partial p}{\partial x}\bigg|_e \qquad (14.19)$$

The interesting point in Eq. (14.19) is that the second derivative of the displacements is replaced by material properties and by the first order pressure derivative. This derivative is easily available using the two neighbor points of pressure in this 1D case, or through a pressure gradient reconstruction in a general unstructured control volume. Equation (14.19), replacing the first derivative of pressure reads,

$$\frac{\partial^2 u}{\partial x^2}\bigg|_e \approx \frac{\alpha}{(\lambda + 2G)} \frac{(p_E - p_P)}{\Delta x} \qquad (14.20)$$

Substituting Eq. (14.20) into Eq. (14.18), the expression for the displacement at the face e of the control volume is

$$u_e \approx \frac{u_E + u_P}{2} - \frac{\alpha \Delta x}{8(\lambda + 2G)}(p_E - p_P) \qquad (14.21)$$

The same procedure is done for finding u_w to complete the mass conservation equation. A physical interpretation of this equation can be seen in [2], in which the 3D formulation is also derived using the PIS scheme. Rhie and Chow-like methods can also be applied for finding the displacements at the interfaces of the control volume for mass conservation. Those methods can be seen in Chap. 7.

Few 3D results from [2] and available also in [4] are now addressed, just for the sake of illustrating the capability of the method in suppressing pressure oscillations when the PIS method is employed with co-located grids in three-dimensional situations.

The problem of underground water withdrawal is of interest in the water supply of cities and for several applications in environmental engineering, among others. This geometry was first described in [30] and used in [2, 4] for validation of the numerical schemes, and consists of a soil with four different rocks, as shown in Fig. 14.10, totaling 50 m thickness and 250 m radius.

The three-dimensional unstructured grid and a zoom identifying the region of the well connecting with the sand layer and the grid used for the well are shown in Fig. 14.11. The unstructured grid is composed of around 20.000 elements and 4000 nodes. The well model employed was of a Peaceman type, and all numerical and physical data can be found in [2, 4, 30]. In [30], solving a surface load problem, both physics were solved in the framework of finite elements, with a mixed finite element for solving the fluid flow. This approach is required in order to avoid instabilities, since the classical Galerkin used for the rock mechanics is not adequate for the solution of fluid flows. The disadvantage of this alternative is that the mixed finite element method requires more complex elements and computer implementation and, certainly, additional computer time to obtain the solution.

The strategy of solving both problems with a finite volume method and a PIS stabilization scheme, as proposed in [2–6] has no drawbacks at all. The alternative

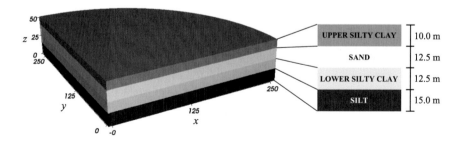

Fig. 14.10 Four layers rock. Water withdrawal problem [2]

14.2 Pressure Instabilities

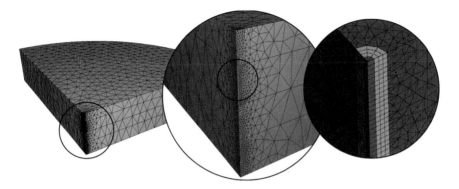

Fig. 14.11 Unstructured grid. Water withdrawal problem [2]

of using finite volume methods for both physics can also be seen in [31, 32] in the context of cell-center unstructured grids.

Figure 14.12 shows a plot of the pressure along the vertical well, in which very interesting physical behavior is seen. As expected, the pressure in the sand layer decreases due to the water production, while it remains constant in all other layers. On the numerical aspects the results put clear the numerical difficulties for the methods in capturing the effects occurring in the interface of two different materials. The element-based finite volume method non-stabilized produces results damaged by the instabilities, while using the PIS scheme the results follow the physics. In a 3D plot of the region, Fig. 14.13 shows the oscillations near the well transmitted inside the domain represented by non-continuous and blurred gray color map.

In order to verify the behavior of the PIS scheme in different grid topologies, the Terzaghi's column is solved for hexahedra, tetrahedra, prims and pyramids elements [2]. Recall that the EbFVM construct its control volumes based on the elements. Despite of Terzaghi's problem be 1D it is solved as a 3D problem, what helps in checking the algorithm, since all results should be constant in a z-constant plane. Figure 14.14 shows the geometry and the four grids employed.

Figure 14.15 depict the pressure after a certain time level for the four different grids, revealing the same difficulty for the scheme without stabilization to face the initial transient when a pressure discontinuity exists.

The appearance of instabilities in the Terzaghi's problem with one material was already well documented in this section. To complete the cases in which instabilities appear, a two-layered Terzaghi's column is solved, as shown in Fig. 14.16, using hexas and tetras as grid elements.

Figure 14.17 shows the pressure distribution along the depth of the column revealing strong instability at the interface of the sand and clay materials for the non-stabilized scheme. By its turn, the solution which used PIS as stabilization scheme shows excellent results with no oscillations at all. Following the open circles in the figure it is seen that the smooth behavior could be even improved if a more refined grid were used.

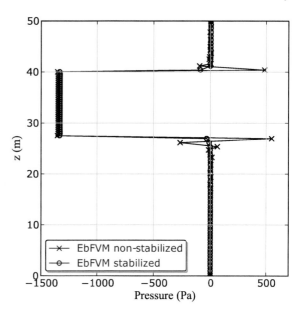

Fig. 14.12 Pressure instabilities near well. Water withdrawal [2]

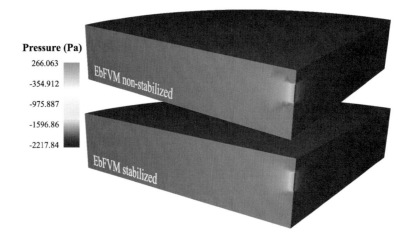

Fig. 14.13 Pressure instabilities near well. 3D visualization [2]

To end this chapter, Fig. 14.18 depicts the convergence rate for the Terzaghi's problem with the four grids employed. It can be appreciated that both pressure and displacement run with a second order convergence for all grids.

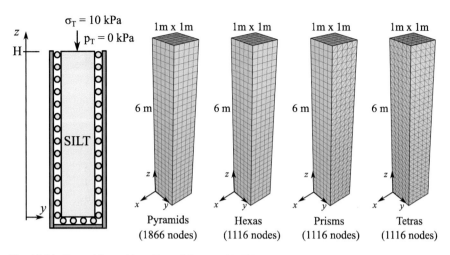

Fig. 14.14 Terzaghi's problem. Four different grids [2]

14.3 Conclusions

This brief chapter presented and advanced a strategy using an element-based finite volume method (EbFVM) for the solution of the fluid flow coupled with geomechanics in porous media. In the scope, it was demonstrated the origin of the oscillations in fluid pressure in certain situations in compacting porous media flows. The remedies to suppress these oscillations were also presented. Validation using 1D problems were done in Cartesian grids as well as 3D problems using element-based finite volume method. Due to the physical background of the stabilization method, no mathematical artifacts are required to obtain solutions free of oscillations.

It was demonstrated that in a single algorithm using finite volumes for both physics using the same grid, it is possible to solve pressure and displacement simultaneously. In contrast, if other non-conservative method is used for solving both physics, it will require more expensive techniques to treat the fluid flow. The PIS method, a stabilization scheme developed on physically based principles, has demonstrated good performance for suppressing oscillations in fluid mechanics as well as in poroelasticity.

The results obtained so far solving compacting porous media flows with diverse boundary conditions in transient 1D, 2D and 3D situations, with grids of different topologies, encourage to follow this route seeking the development of a tool for the day-to-day of the engineers in several areas of applications.

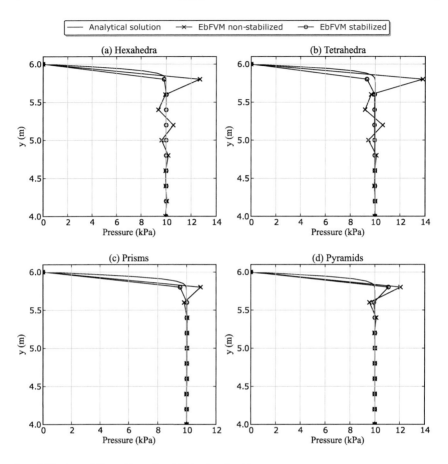

Fig. 14.15 Terzaghi's problem. Instabilities in all grids

14.3 Conclusions

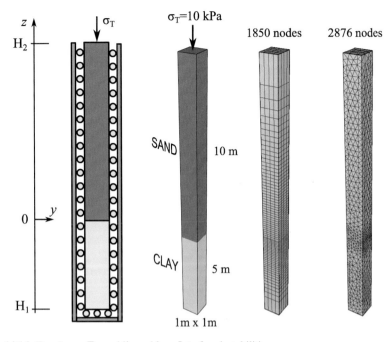

Fig. 14.16 Two layers Terzaghi's problem. Interface instabilities

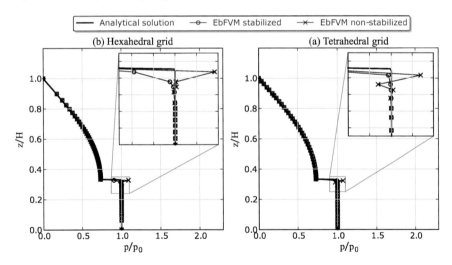

Fig. 14.17 Terzaghi's two-layers. EbFVM stabilized

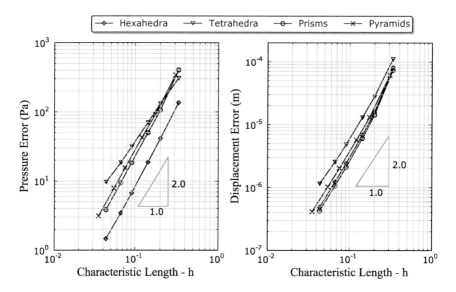

Fig. 14.18 Terzaghi's problem. Convergence rate

References

1. dal Pizzol A, Maliska CR (2012) A finite volume method for the solution of fluid flows coupled with the mechanical behavior of compacting porous media. AIP Conf Proc 1453:205–212. https://doi.org/10.1063/1.4711176
2. Honorio HT (2018) A stabilization technique for treating numerical instabilities in three-dimensional poroelasticity, Ph.D. thesis, Mech. Eng. Dept. UFSC. https://repositorio.ufsc.br/handle/123456789/198088
3. Ribeiro GG (2016) Volumes finitos baseado em elementos para problemas de poroelasticidade. Ph.D. Thesis, Mech. Eng. Dept. UFSC. https://repositorio.ufsc.br/xmlui/handle/123456789/174709
4. Honorio HT, Maliska CR, Ferronato M, Janna C (2018) A stabilized element-based finite volume method for poroelastic problems. J Comp Phys 364:49–72
5. Maliska CR, Honorio HT (2019) Pressure-displacement coupling in poroelasticity. Further details of a stable finite volume formulation. In: Oñate E, Papadrakakis M, Schrefler B (eds) VIII International conference on computational methods for coupled problems in science and engineering, pp 574–585. http://congress.cimne.com/coupled2019/frontal/doc/EbookCoupled2019.pdf
6. Maliska CR, Honorio HT (2018) An oscillation-free finite volume method with staggered grids for solving problems of poroelasticity. ICCM Proc 5:372–385, ISSN 2374-3948 (online). https://www.sci-en-tech.com/ICCM2018/ICCM2018-Proceedings.pdf
7. Gao C, Grey KE (2017) The development of a coupled geomechanics and reservoir simulator using a staggered grid finite difference, SPE-187186-MS
8. Asadi R, Ataie-Ashtiani B and Simmons CT (2014) Finite volume coupling strategies for the solution of a Biot consolidation model. Comput Geotech 55:494–505
9. Gaspar FJ, Lisbona FJ, Vabishchevich PN (2003) A finite difference analysis of Biot's consolidation model. Appl Numer Math 44(4):487–506

10. Ferreira, CAS (2019) Analysis of numerical schemes in collocated and staggered grids for problems of poroelasticity. M.Sc. dissertation, Mech. Eng. Dept., UFSC. https://repositorio.ufsc.br/handle/123456789/215259
11. Wang HF (2000) Theory of linear poroelasticity with applications to geomechanics and hydrogeology. Princeton University Press
12. Rice JR, Cleary MP (1976) Some basic stress diffusion solutions for fluid saturated elastic porous media with compressible constituents. Rev Geophys Space Phys 14:227–241
13. Verruijt A (2016) Theory and problems of poroelasticity. Delft University of Technology, Netherlands
14. Vermeer P, Verruijt A (1981) An accuracy condition for consolidation by finite elements. Int J Numer Anal Methods Geomech 5:1–14
15. de Boer R (2000) Theory of porous media—Highlights in historical development and current state. Springer-Verlag
16. de Boer R (2005) The engineer and the scandal. Springer-Verlag
17. Harlow FH, Welch JE (1965) Numerical calculation of time-dependent viscous incompressible flow of fluid with free surface. Phys Fluids 8:2182–2189
18. Terzaghi K (1923) Die berechnung der durchlassigkeitsziffer des tones aus dem verlauf der hydrodynamischen spannungsercheinungen. Sitz Akad Wissen Wien Math Naturwis 2a:125–128
19. Ferreira CS, Honório HT and Maliska CR (2019) Finite volume stabilizing schemes for undrained condition in poroelasticity. In: CILAMCE 2019 Proceedings of the XL Ibero-Latin-American congress on computational methods in engineering, ABMEC. Natal/RN, Brazil, 11–14 Nov 2019
20. Babuska I (1971) Error bounds for finite-element method. Numer Math 16(322–333):1971
21. Brezzi F (1974) On the existence of uniqueness and approximation of saddle-point problems arising from lagrangian multipliers. Rev Fr d'Automatique Informatique Rech Operationnelle 8:129–151
22. Zienkiewicz OC, Taylor RL (2000) The finite element method, vol 1: the basis, 5th edn. Butterworth-Heinemann, Oxford, UK
23. Murad M, Loula L (1994) On stability and convergence of finite element approximation of Biot's consolidation problem. Int J Numer Eng 37:645–668
24. Quintino dos Santos GW (2022) Unstructured grids with staggered arrangement of variables for the solution of poroelastic problems, M.Sc. dissertation, Mech. Eng. Dept., Federal University of Santa Catarina (in Portuguese)
25. Quintino dos Santos GW (2022) Unstructured meshes with staggered grids for the solution of poroelasticity problems. National Congress in Mechanical Engineering, Braz. Soc. of Mech. Eng-Events, August 2022 (in Portuguese)
26. Rhie CM (1981) A numerical study of the flow past an isolated airfoil with separation, Ph.D. thesis, University of Illinois, Urbana-Champaign
27. Peric M, Kessler R, Scheuerer G (1988) Comparison of finite volume numerical methods with staggered and colocated grids. Comput Fluids 16:389–403
28. Schneider GE, Raw MJ (1987) Control volume finite-element method for heat transfer and fluid flow using co-located variables-computational procedure. Numer Heat Transfer 11:363–390
29. Raw MJ (1985) A new control-volume-based finite element procedure for numerical solution of the fluid flow and scalar transport equations, Ph.D. thesis, University of Waterloo, Canada
30. Ferronato M, Casteletto N, Gambolati G (2010) A fully coupled 3D mixed finite element model of Biot consolidation. J Comp Phys 229:4813–4830
31. Nordbotten J (2014) Cell-centered finite volume discretizations for deformable porous media. Int J Numer Meth Eng 100:399–418
32. Nordbotten J (2016) Stable cell-centered finite volume discretization for Biot equations. SIAM J Numer Anal 54:942–968

Chapter 15
Applications

15.1 Introduction

This chapter presents some results obtained with the methodologies described in this text to illustrate the applications of the finite volume method for both structured and unstructured meshes. In the structured class, generalized curvilinear coordinates, or boundary-fitted grids are used, while for unstructured meshes, problems with triangular and quadrangular elements in 2D and 3D hybrid grids employing the element-based finite volume method (EbFVM) are presented. Voronoi diagram are also used with triangular meshes with the traditional finite volume formulation (cell centered method) for solving porous media flow.

The problems were chosen due to the availability of results still in raw data files, allowing the generation of figures via computer and their transfer to a text editor. In some problems, the results presented are only qualitative, for the sole purpose of illustration. In others, they are mesh size dependent, and for this reason care should be taken if they are used for comparisons. It is recommended, if this is the objective, to look for the works that originated them. The few problems reported are with the solely purpose of demonstrating the generality and the applicability of the finite volume method for solving engineering problems.

15.2 Aerodynamics

15.2.1 All Speed Flow Over a Blunt Body

Many problems in aerodynamics have been solved with the methodology for all-speed flows described in Chap. 8 [1–3]. Among them we can mention two-dimensional flows over arbitrary bodies, flows inside nozzles with different inlet and outlet boundary conditions, and three-dimensional flows over blunt bodies in all flow regimes. From all those, few results of the last mentioned problem will be presented.

Figure 15.1 shows the frontal part of a blunt body showing one plane of the 3D grid. This grid is very easy to generate, just rotating a 2D mesh around the axis of the body generating the 3D mesh. The flow is symmetric in relation to a vertical plane, what means that a 180°mesh is used. The all speed flow methodology employed is described in Chap. 8 and the flow is governed by the Navier–Stokes equations.

Figure 15.2 shows the pressure coefficient (c_p) over the body surface for a flow with free stream Mach number for the subsonic region equal to 0.50 and angle of attack of 6°.

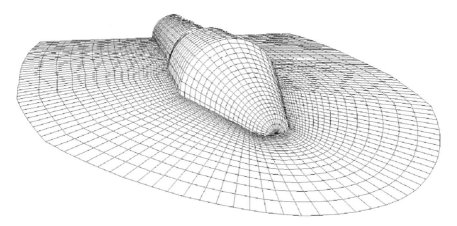

Fig. 15.1 Boundary-fitted grid. 3D flow over a blunt body

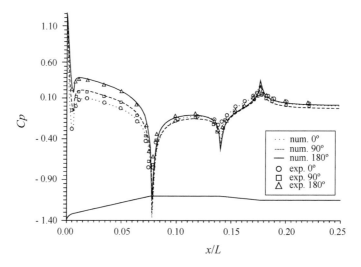

Fig. 15.2 Flow over a blunt body. Free stream Ma = 0.50

15.2 Aerodynamics

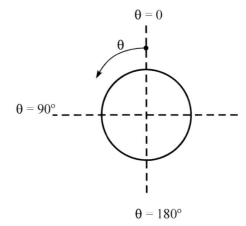

Fig. 15.3 Planes of flow calculations and measurements

Results are shown for θ equal to 0°, 90° and 180°, with θ counting according to Fig. 15.3. It can be seen from the figure that the results are in well agreement with the experimental ones obtained in [4]. All expansion and compression zones on the surface of the body are well captured.

It is worth remembering that the all-speed methodology employed doesn't use any special schemes to deal with transonic flows, which is the most difficult flow regime to be solved. It is, in fact, a unique scheme applied indistinctively to all flow regimes. The scheme, according to the linearization of the mass flow used (see Chap. 8) adapts to the flow regime.

Figure 15.4 shows the results for the free stream Mach of 0.90, again for angle of attack of 6°. In this flow regime, i.e., transonic, the agreement between numerical and experimental data is more difficult in the regions of expansions and compressions, as usual. It was verified through a large number of tests that by refining the mesh, the numerical and experimental results get closer, what is expected. The difficulty in accurately capturing the phenomena in this flow regime is well known by the researchers due to the undefined behavior of the partial differential equations in this range of Mach. Shocks are weak and more difficult to be captured. Even though, the results are considered good.

According to this reasoning, the numerical and experimental results for high Mach number flows should agree much better, because in this flow the shocks are strong and perfectly defined. Figure 15.5 shows precisely that, and presents the results for a free stream Mach number equal to 3.0. One can see from this figure that the results compare very well, showing only a small discrepancy for θ equal to 180°.

To complete the illustrations, Fig. 15.6 shows the spatial Mach number distribution in the blunt body of the previous figures for Mach $= 1.15$. The expansions and compressions suffered by the flow in the regions in which there are geometry changes from conical to cylindrical and vice-versa are clearly captured.

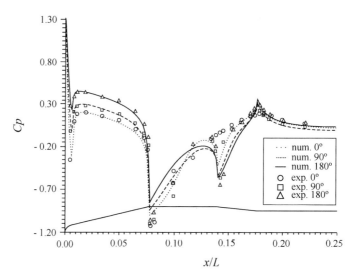

Fig. 15.4 Flow over a blunt body. Free stream Ma = 0.90

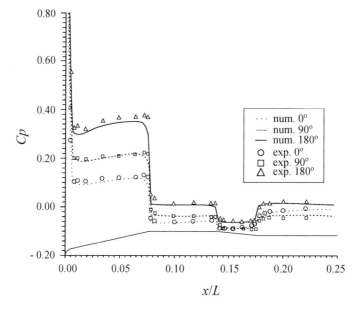

Fig. 15.5 Flow over a blunt body. Free stream Ma = 3.0

15.2 Aerodynamics

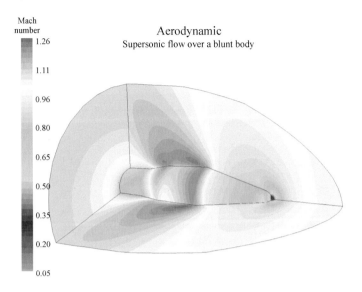

Fig. 15.6 Spatial Mach number distribution. Free stream Ma = 1.15

15.2.2 Ice Accretion on Aerodynamic Profiles

Another important application in aerodynamics is the calculation of ice accretion on aerodynamic devices, being wing ice accretion of special interest because lift can be compromised depending on the thickness of the ice layer formed. This occurs under certain flight conditions, when water particles in the clouds crash against the aircraft, creating a layer of ice with different shapes and structures. The solution to this problem involves first calculating the flow over the profile of interest followed by the application of a thermodynamic model that calculates the deposited ice layer. The new geometry of the airfoil is fed back to the simulator to calculate a new three-dimensional flow, followed by the calculation of the new deposited ice distribution, and so on, until reaching the desired simulation time. In this way it is possible to verify the influence of icing on lift, and designing the anti-icing system, a mandatory item in aircrafts.

Figure 15.7 shows an airfoil subjected to a flow whose thermodynamic conditions are favorable to icing. The ice growth profiles obtained experimentally and by the applications Dera[1] and Lewice[2] are compared against the results of Aeroicing[3] [5–7]. The thermodynamic model that predicts ice formation can be coupled to any software that calculates the 3D turbulent velocity field over the aerodynamic profile.

---

[1] DERA Defense Evaluation and Research Agency (United Kingdom).

[2] LEWICE (LEWis ICE accretion program). Developed by the Icing Branch at NASA Glenn Research Center.

[3] Aeroicing is the application developed in the CFD Lab – SINMEC/UFSC for icing calculation in aerodynamic profiles in a joint project with EMBRAER S/A.

In the specific case of these results, the flow is determined by the panel method [5]. Figure 15.8 shows the grid used in the icing simulation, in which a layer of Cartesian grids adapts to the unstructured triangular grids used to construct the control volumes for the EbFVM.

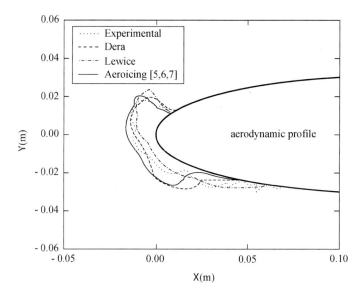

Fig. 15.7 Ice accretion comparisons. Numerical versus experimental

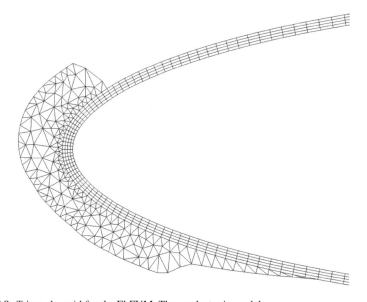

Fig. 15.8 Triangular grid for the EbFVM. Thermodynamic model

15.3 Porous Media Flows   413

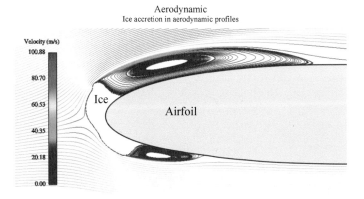

Fig. 15.9 Ice layer influence on the flow

To end the description of this problem, Fig. 15.9 reproduces the fluid flow streamlines, the ice collected on the wing and the recirculation zones caused by the perturbation of the ice on the flow. The aerodynamic performance of the airfoil can be severely prejudiced by the ice in these situations.

15.3 Porous Media Flows

Although porous media flows aren't Navier–Stokes problems, to demonstrate the use of Voronoi grids in 2D two problems will be shown in the area of petroleum reservoir simulation. Unstructured Voronoi meshes were also the subject of study in the field of petroleum reservoir simulation in the 1990s [8–10]. As seen in Chap. 13, Voronoi meshes respect local orthogonality, and therefore all numerical schemes can be developed as an extension of the models for Cartesian meshes.

In the following, some examples of simulating two-phase and single-phase flows with tracer, using Voronoi meshes, will be presented. In petroleum reservoir engineering, the main goal is to determine the amount of oil recovered in the production wells due to the oil displacement caused by the flow of the water injected. It is a two-phase oil/water flow. In this 2D problem the unknowns are the pressure field and the saturation of water, solved using a Newton-like method. The oil saturation comes from the global mass conservation, while the velocities are found using the Darcy's equation.

Figure 15.10 shows a hypothetic triangular reservoir [9] with two producers and one injector wells. The Voronoi mesh [10] is made polar around the wells and hexagonal in the rest of the domain. This is an advantage of this meshing process, because near the wells the flow is practically radial, and the polar mesh is the most appropriate.

The results obtained from the simulation of the two-phase flow in the triangular reservoir are compared with those of [11] in Fig. 15.11. The results, for mobility ratios 10 and 50, compare very well.

In Fig. 15.12 it is shown a petroleum reservoir with two injection wells and six production wells, in which one can see again the suitability of the Voronoi grids around the wells [12]. The polar grid is now placed in a large interior portion of

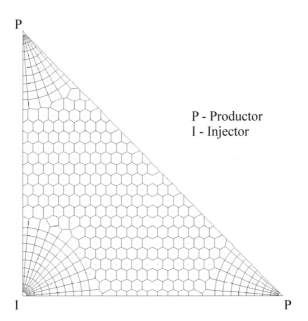

Fig. 15.10 Voronoi discretization for a hypothetical reservoir

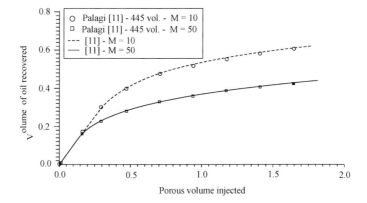

Fig. 15.11 Comparisons for the hypothetical reservoir

15.3 Porous Media Flows

the reservoir, not just in the corner of the domain, as in the previous problem. The physics is the same.

In this problem a grid refinement study was performed and is shown in Fig. 15.13 for the production wells P_1, P_2 and P_5, using 804 and 1062 control volumes.

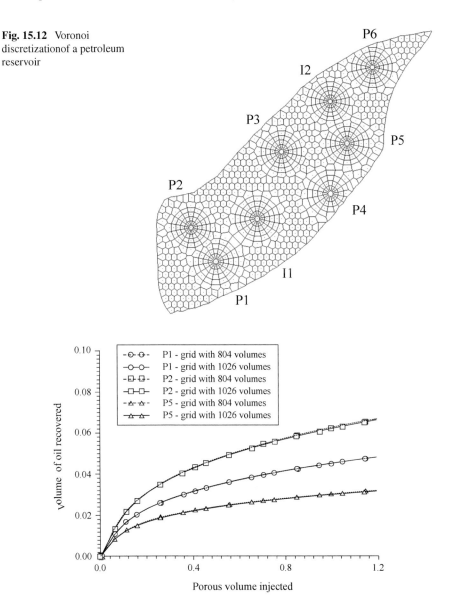

Fig. 15.12 Voronoi discretization of a petroleum reservoir

Fig. 15.13 Grid refinement results. Problem of Fig. 15.12

The next problem is a single-phase flow with a tracer [8]. A tracer is a substance injected along with the water that can be identified in the flow reaching the wells and, therefore, serves to study the behavior of the reservoir by observing (measured) the concentration of tracer in the well and inferring the path of the tracer. Of course, other techniques like seismic 4D and analysis of rock samples, among others, are used for characterizing the reservoir. The tracer doesn't affect the flow and, therefore it is solved as an advection/diffusion problem in which the unknown is the concentration of the tracer. The validation uses a 5-spot configuration in which all boundaries are impermeable with the injection well containing the tracer at the one corner, and the production well at the opposite corner on the diagonal.

To illustrate the tracer problem, consider the geometry shown in Fig. 15.14, where a hexagonal Voronoi mesh is used to discretize the geometry. A steady-state single-phase flow is injected into well I and produced in the well P. Suddenly, a tracer is injected into well I. One wants to know the concentration distribution of this tracer in the reservoir over time. The advection/diffusion equations of the tracer in the porous medium must be solved. Two situations are analyzed. The first one, where the tracer injection is continuous, and the second, where a pulse of 0.4 *PVI* of the tracer is injected. Knowing the injection flow rate (m^3/s, it is possible to find the time it takes for injecting some amount of PV (m^3). In petroleum reservoir engineering PVI, the porous volume injected is used as a time coordinate.

Figure 15.15 shows the results for the two situations compared with the experimental data from [13]. The agreement is very good. This problem has also been solved using structured meshes. Figure 15.16 shows the results obtained with the Voronoi mesh and a curvilinear structured mesh. The contamination of the solution

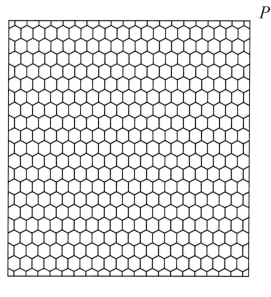

Fig. 15.14 Tracer problem. Five-spot configuration

15.3 Porous Media Flows

with numerical diffusion is much lower with the Voronoi mesh, as can be seen by the smearing of the pulse, probably because of the involvement of more neighboring volumes in the formulation for calculating the fluxes.

To end this brief section showing applications using the finite volume algorithms discussed along the book, the use of tridimensional hybrid grids in petroleum reservoir simulation is related [14]. The only purpose is to demonstrate the applicability and flexibility of the EbFVM for solving 3D problems with different types of grids and combinations of them, as shown by the illustration in Fig. 15.17. This illustrative grid shows, for example, a well surrounded by a cylindrical grid connected to a polyhedral transitional grid, which connects to a Cartesian-like discretization. Near-well flows can be solved accurately with this kind of grid, task not easily accomplished with the usual commercial petroleum reservoir simulators due to the size of the control volumes near the wells. Local grid refinement is not an easy matter in those applications.

To demonstrate the applicability of the transitional and hybrid grids, two problems of a two-phase flow of oil/water in a 3D reservoir are solved. Grids consisting of Cartesian, pure tetrahedral (unstructured) and hybrid are used. Hybrid in this case is the mixture of a Cartesian grid with an unstructured grid inserted in a portion of the domain. The idea was to compare the software against each other due to their different grid topologies and to check the algorithms at the interface of the grids, since tetrahedra must conform with hexahedra, needing special treatment for fluxes calculation to keep conservation at the interface. The problem is the simulation of oil/water flow in a cubic reservoir.

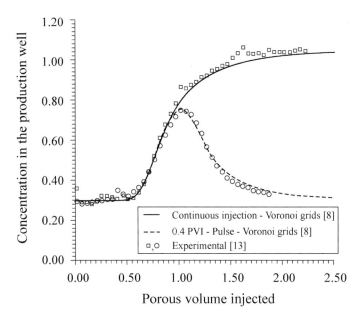

Fig. 15.15 Numerical versus experimental. Problem of Fig. 15.14

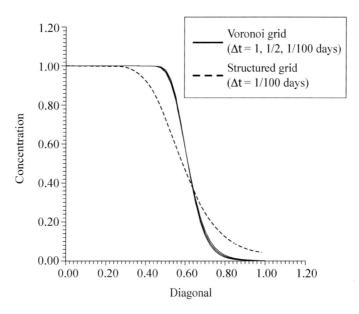

Fig. 15.16 Tracer concentration along diagonal. Structured and Voronoi

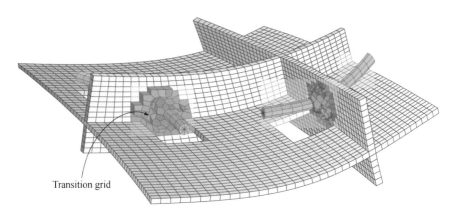

Fig. 15.17 A illustrative 3D hybrid grid with transitions

Figure 15.18 shows a cut of the cube in half showing the interior of the domain, allowing the observation of the tetras in both grids, the pure tetrahedral and the one with an insertion of a tetrahedral into a Cartesian grid. The boundary conditions are shown in Fig. 15.19 for the faces identified. At the other three faces of the domain zero pressure and zero saturation of water is set. For the simulator using pure Cartesian meshes (not shown in the figure), 42,875 volumes were used, while for the tetrahedral mesh simulator 46,507 control volumes were employed. The number of tetrahedra required for this number of control volumes is much larger since, as this is not a

15.3 Porous Media Flows

cell-center method, the tetrahedra are agglomerated resulting in a smaller number of unknowns.

The grid for the hybrid simulator involves an unstructured grid of 70,946 tetrahedra elements resulting in 11,885 vertices (number of unknowns) inserted centralized in the cube with Cartesian meshes, as shown in Fig. 15.18b, totalizing 49,847 control volumes. The unknowns of the problem are pressure and water saturation and solved simultaneously using a GMRES solver with a SOR preconditioner.

Figure 15.20 shows the saturation along the line shown in Fig. 15.21c, demonstrating the good agreement among the results obtained with the simulators. One of the key numerical issues is this problem is the development of conservative fluxes at

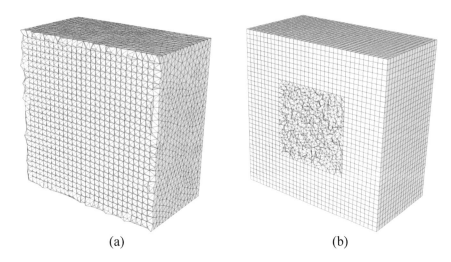

Fig. 15.18 Fully unstructured and hybrid grids

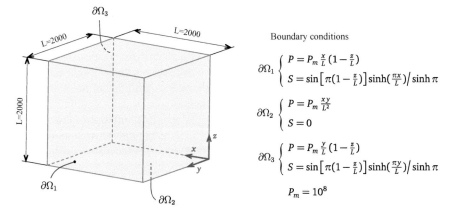

Boundary conditions

$$\partial\Omega_1 \begin{cases} P = P_m \frac{x}{L}\left(1-\frac{z}{L}\right) \\ S = \sin\left[\pi\left(1-\frac{z}{L}\right)\right]\sinh\left(\frac{\pi x}{L}\right)\Big/\sinh\pi \end{cases}$$

$$\partial\Omega_2 \begin{cases} P = P_m \frac{xy}{L^2} \\ S = 0 \end{cases}$$

$$\partial\Omega_3 \begin{cases} P = P_m \frac{y}{L}\left(1-\frac{z}{L}\right) \\ S = \sin\left[\pi\left(1-\frac{z}{L}\right)\right]\sinh\left(\frac{\pi y}{L}\right)\Big/\sinh\pi \end{cases}$$

$$P_m = 10^8$$

Fig. 15.19 Boundary conditions for the cubic reservoir

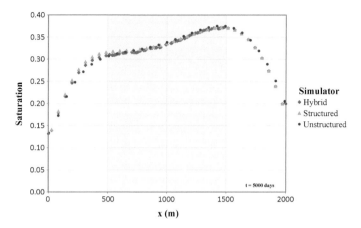

Fig. 15.20 Saturation along x for the three simulators. See Fig. 15.21c

the interface of control volumes, as seen in Fig. 15.22. In this case at the interface of the unstructured and the Cartesian meshes with conforming and non-conforming grids. This problem gives a real idea of the advanced numerical technologies required for simulating three-dimensional engineering problems with this kind of hybrid grids.

15.4 Conclusions

The goal of this chapter was to give a flavor of the capacity of the numerical techniques for simulating engineering problems, displaying some developments realized by the author and his colleague using the tools described in the book. Discretization using curvilinear coordinate system, locally orthogonal Voronoi grids and 3D unstructured grids for simulating several physics were presented. Navier–Stokes equation with ice accretion and aerodynamic flows over blunt bodies in all flow regimes were reported. The 3D problems using hybrid grids were not in the field of CFD, but of porous media flow in petroleum reservoir simulation, in which the discretization of irregular domains with wells is critical. The techniques presented along the textbook and these few demonstrative problems gives an overview of the capabilities of the numerical simulation using finite volume methods with unstructured grids. As pictured in the Introduction, the world of simulation has no boundaries. All engineering problems will be simulated in a near future. And finite volume techniques have an important role to play in this scenario.

15.4 Conclusions

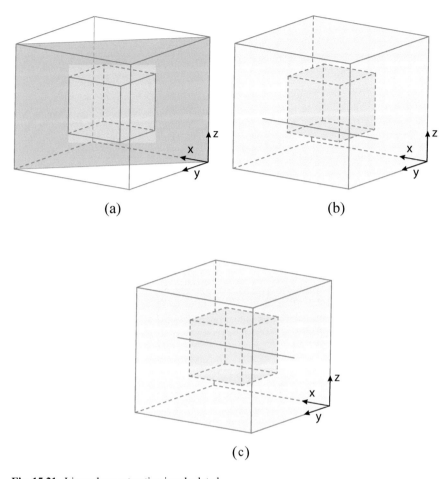

Fig. 15.21 Lines along saturation is calculated

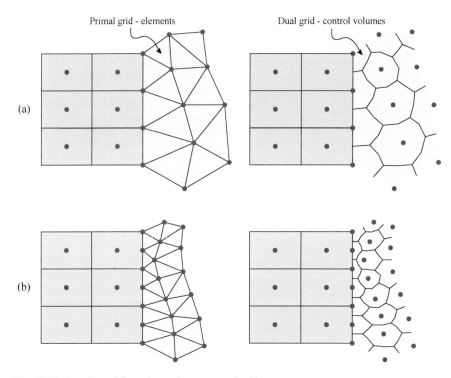

Fig. 15.22 Interface of Cartesian and unstructured grids

References

1. Marchi CH, Maliska CR (1993) Numerical computations of the 3D flow over the VLA-Brazilian launch vehicle. XII Congresso Brasileiro de Engenharia Mecânica. Brasília, Brasil, pp 805–808
2. Marchi CH, Maliska CR (1994) A nonorthogonal finite volume method for the solution of all speed flows using co-located variables. Numer Heat Transfer Part B Fundamentals 26:293–311
3. Azevedo JLF, Moraes Jr P, Maliska CR, Marchi, CH and Silva AFC (1996) Code validation for high-speed flow simulation over satellite launch vehicle. J Spacecraft Rockets 33–1 January-February
4. Moraes P Jr, Neto AA (1990) Aerodynamic experimental investigation of the Brazilian satellite launch vehicle-VLS. Brazilian Society of Mechanical Engineering, III National Thermal Science Meeting, pp 211–215
5. Silveira RA, Maliska CR (2001) Numerical simulation of ice accretion on the leading edge of aerodynamic profiles. In: J. Pontes (ed) Computational heat and mass transfer—ICHMT 2001. E-Papers Publishing House Ltd., Rio de Janeiro, pp 33–40
6. Silveira RA (2006) Numerical simulation of 3D ice accretion on aerodynamic profiles (in Portuguese). Ph.D. Thesis, EMC/UFSC, https://repositorio.ufsc.br/handle/123456789/88472
7. Donatti CN, Silveira RA, Maliska CR, Bridi G (2007) Ice accretion simulation in presence of a hot air anti-icing system November conference: 19th international congress of mechanical engineering, Brasília, DF, Brazil. https://doi.org/10.13140/2.1.4922.0802
8. Maliska CR, Maliska CR Jr (1994) A finite volume method using voronoi grids for the solution of miscible displacement in porous media. J. of the Braz. Soc Mech Sci 16:415–422

9. Marcondes F, Zambaldi MC, Maliska CR (1995) Comparação de Métodos Estacionários e GMRES em Simulação de Reservatório de Petróleo Utilizando Malhas Não Estruturadas de Voronoi. Revista Brasileira de Ciências Mecânicas XVI I(4):360–372
10. Maliska Jr CR (1993) A robust Voronoi diagrams generator for discretization of irregular domains. XIV Ibero-Latin_American Congress on Comp. Meth. for Eng, São Paulo, pp 753–762
11. Palagi C (1992) Generation and Application of Voronoi Grid to Model Flow in Heterogeneous Reservoir. Ph.D. Thesis, Stanford University, California, EUA
12. Marcondes F, Maliska CR and Zambaldi MC (2009) A comparative study of implicit and explicit methods using unstructured Voronoi meshes in petroleum reservoir simulation. J Braz Soc Mech Sci Eng
13. Santos RLA, Pedrosa Jr. OA and Corrêa ACF (1992) An efficient finite-volume approach for modelling miscible displacement. Latin-American Petroleum Engineering Conference, Caracas, Venezuela
14. Maliska CR et al. (2016) Development of a 3D oil/water simulator using hybrid grids. Internal report, Computational Fluid Dynamics Laboratory. Mech Eng Dept., Federal University of Santa Catarina

Correction to: Fundamentals of Computational Fluid Dynamics

Correction to:
C. R. Maliska, *Fundamentals of Computational Fluid Dynamics*, **Fluid Mechanics and Its Applications 135, https://doi.org/10.1007/978-3-031-18235-8**

The original version of the chapter was published with incorrect symbol for the v-velocity in Chapters 2, 5, 9, 10, 12 and 13, which has now been corrected. The book has been updated with the changes.

The updated versions of these chapters can be found at
https://doi.org/10.1007/978-3-031-18235-8_2
https://doi.org/10.1007/978-3-031-18235-8_5
https://doi.org/10.1007/978-3-031-18235-8_9
https://doi.org/10.1007/978-3-031-18235-8_10
https://doi.org/10.1007/978-3-031-18235-8_12
https://doi.org/10.1007/978-3-031-18235-8_13

© The Author(s), under exclusive license to Springer Nature Switzerland AG 2023
C. R. Maliska, *Fundamentals of Computational Fluid Dynamics*, Fluid Mechanics and Its Applications 135, https://doi.org/10.1007/978-3-031-18235-8_16

Index

A
ACM Multigrid, 103, 107
Advected velocity, 280
Advecting solid velocity, 396
Advecting velocity, 30, 166, 168, 280, 319, 343, 389, 396
Advection, 6, 7, 18, 19, 24, 25, 29, 30, 33, 41, 113, 115, 118, 120, 121, 123, 124, 126, 131, 134–136, 139, 145, 153, 154, 157, 171, 200, 221, 242, 244, 278, 304, 309, 313, 314, 346, 347, 367, 370, 377, 416
Advective dominant, 115, 124, 135, 141
Aerodynamics
    all speed flows, 207
    ice accretion, 411, 420
    supersonic flow, transonic flow, 207
Anisotropy, 95, 96, 105
Assembling, 359, 363, 367, 376, 379

B
Basis vector, 271–278, 284, 288, 336
Biot's model, 389
Block-by-block, 41
Blunt body, 34–36, 217, 218, 300, 407–410
Boundary conditions
    elliptic, 31–33, 38, 120
    hyperbolic, 31–33
    inflow, 200, 344, 380
    outflow, 200, 344, 380
    parabolic, 31–33, 38, 120, 200–202, 221, 230, 232, 246
Boundary element method (BEM), 8
Boundary-fitted, 326, 344, 378, 407, 408
Boundary Layer, 33, 35, 38, 223, 224, 228, 233, 292

C
C++, 236–238, 380
Cell-based method, 338
Central coefficient, 59, 73, 76, 172, 188, 248, 316
Checkerboard, 164, 166
Code, 9, 10, 41, 48, 79, 86, 92, 113, 137, 153–155, 164, 217, 235–239, 242, 243, 246–248, 250, 251, 256, 263, 284, 292, 299, 325–327, 344, 345, 374, 376, 380, 396
Coefficients anisotropy, 105
Co-located grid, 160, 163, 164, 166–168, 186, 189, 191, 209, 318, 319, 351, 389, 392, 397, 398
Compiling, 240
Compressible flow, 31, 190, 202, 207, 208
Compressible formulation, 161, 177, 207
Computational domain, 81, 158, 256, 264, 270, 289, 291–295, 297, 298, 300, 301, 306, 308, 311, 324, 325, 327, 333, 359
Concurrent Version System (CVS), 238
Connectivity, 48, 256, 363
Conservative, 7–9, 11, 20, 24, 28, 43–45, 47, 51, 74, 75, 78, 98, 100, 114, 256, 301, 303, 304, 307, 310, 322, 355, 356, 372, 378, 380, 387, 401, 419
Consistency, 41, 78, 79, 118
Control volume, 7–9, 14, 16–21, 23–25, 28, 29, 37, 39, 43–49, 51, 52, 54, 57, 59, 64–69, 73–75, 79, 81, 86, 88, 90,

100–103, 110, 113, 114, 116, 119, 123, 140, 145, 163–166, 168, 169, 171–173, 176, 180, 182, 184, 186, 189, 192, 193, 198, 199, 203, 204, 207, 209–211, 215, 216, 218, 236, 247, 253, 256, 258–261, 271, 278, 280–282, 290, 299, 300, 306, 307, 309, 311–314, 317, 319–327, 333, 335–338, 340, 341, 343–346, 348–351, 353–357, 359, 360, 363, 367, 368, 372–382, 388, 389, 391–393, 395–399, 412, 415, 417–420
Control Volume-based Finite Element Method (CVFEM), 8, 354, 380
Convergence, 10, 36, 41, 43, 51, 58, 63, 76, 78, 79, 86, 87, 90, 92–97, 102 106, 128, 141, 150, 161, 162, 176, 177, 179, 182, 183, 185–189, 192, 193, 201, 203, 215–217, 227, 232, 239–241, 250, 296, 317, 348, 352, 393, 395, 400, 404
Convergence criteria, 240, 241
Coordinate systems
  Cartesian, 6, 10, 28, 47, 210, 253, 271, 288, 301, 306, 317, 336
  contravariant components, 276, 278
  covariant components, 277, 280, 288
  nonorthogonal, 314, 322, 337
  orthogonal, 6, 47, 253, 269, 290
  unitary vectors, 276
Coordinate transformation, 10, 256, 263, 282, 294, 302, 325, 329, 357, 359, 362, 363, 375, 378
Coupled flow/rock mechanics, 387, 398
CPU time, 31, 42, 48, 58, 107, 186, 257
Crank Nicolson, 60
Cross-derivatives, 306, 314, 325
Curvilinear grid, 9, 192, 257, 259, 260, 263, 278, 288, 292, 317, 318, 344

D

Darcy's equation, 390, 413
Deformation, 21, 391
Density relations, 213, 215
Differential equations, 2–4, 6, 7, 9, 14, 30–32, 38, 41–47, 58, 63, 72, 76, 78, 85, 92, 120, 122, 131–134, 136, 137, 154, 157–159, 161, 168, 173, 176, 195, 242–244, 250, 253, 263, 301, 322, 342, 349, 364, 366, 371, 372, 377, 378, 409

Diffusion, 10, 23–25, 32, 33, 50, 55, 58, 64, 65, 67, 68, 70, 75, 76, 78, 79, 113, 115, 119–121, 123, 124, 126, 127, 129, 131, 133–136, 139, 141, 142, 145, 146, 153, 154, 165, 166, 198, 221, 227, 244, 245, 259, 304, 309, 312–314, 336, 341, 346–348, 367, 377, 381, 416
Digital Twins, 1, 3
Dirichlet, 54, 64, 67, 90, 94, 198, 335, 378
Displacement, 39, 308, 387–398, 400, 401, 413
Distorted transient, 30, 35, 36, 58, 61, 85, 150, 153, 161, 216
Divergence theorem, 16, 20, 25, 114, 346, 348
Donor cell, 121, 128, 139, 140, 343, 349, 363, 367, 368
Double column, 396
Duct entrance, 231

E

Element-based, 8, 10, 11, 47, 139, 140, 256, 259, 261, 333, 355, 367, 381, 399, 401, 407
Element-based Finite Volume Method (EbFVM), 8, 10, 11, 47–49, 67, 139, 256, 333, 334, 344, 354, 355, 359, 363, 364, 367, 371, 372, 380, 381, 387, 393, 399, 401, 403, 412, 417
Element-by-element sweep, 8, 355, 359, 378, 380
Elements
  hexahedron, 48, 257, 356
  prism, 49
  pyramid, 49, 257, 333, 338, 399
  rectangle, 49
  tetrahedron, 48, 257
  triangle, 49, 256–258, 338, 378
Element sweep, 8
Energy conservation equation, 23, 28, 37, 162
Enthalpy, 28, 30
Entities, 30, 47, 273, 282, 305, 344, 355, 356
Error wavelength, 110
Exact, 56, 81, 85, 92, 97, 109, 115, 123, 124, 134, 135, 141, 142, 154, 193, 244, 292, 339, 346
Exact solution, 42, 56, 58, 76, 79, 92, 94, 97, 115, 121, 123, 131–133, 135, 154, 155

Index 427

Exercises, 17, 20, 39, 56, 60, 78, 82, 92, 94, 107, 109, 203, 216, 217, 233, 273, 275, 284, 288, 306, 327, 346, 378, 381
External flows, 218, 224, 227, 228, 301

F

Fictitious volumes, 65
Finite differences, 6, 7, 9, 42, 47, 52, 76, 78, 80, 81, 95, 139, 171, 290, 292, 307, 308, 342, 381
Finite element
  Galerkin, 7, 47, 48, 398
Finite volume methods
  cell-center, 8, 9, 11, 41, 47–50, 64, 67, 79, 139, 201, 258, 259, 261, 311, 333, 359, 363, 399
  cell-vertex, 8, 10, 11, 41, 47, 48, 50, 51, 67, 79, 139, 260, 333, 378
  Voronoi diagrams, 8, 10, 11, 256, 334, 407
First order, 7, 48, 52, 53, 77, 120, 129, 131, 133, 171, 187, 200, 227, 244, 257, 336, 357, 358, 375, 397
Flux continuity, 41, 73, 74
Force balance, 20, 48
Formulations
  explicit, 52–56, 81, 82, 149–151, 227
  fully implicit, 41, 52, 57, 58, 61, 64, 68, 79, 82, 113, 116, 141, 146, 153, 244, 312, 346
  implicit, 52, 60, 149, 153
Fortran, 237

G

Gauss elimination, 60, 92
Generic variable, 19, 24, 329, 382
Geometric conservation law, 307
Global coordinate systems, 254, 256, 263, 264, 289, 363
Global mass flow, 224
Global matrices, 47, 261, 355, 377, 378
GMRES, 419
Gradient, 35, 104, 115, 117, 118, 123, 129, 131, 133, 134, 138, 142, 147, 162, 170, 171, 173, 192, 193, 224, 225, 228–230, 232, 258, 274, 299, 336–341, 344, 361, 363, 391
Gradient reconstruction, 339, 363
Graphical Interface (GUI), 236, 237

H

Heat balance, 309
Heat conduction, 10, 35, 38, 39, 50, 72, 81, 82, 104, 108–110, 115, 121, 123, 135, 242, 243, 327, 346
Heat generation, 38, 81, 109, 309

I

Impermeable, 199, 416
Inclined grid, 133–135
Incompressible flow, 160, 175, 177, 192, 201, 203, 207–209, 211, 222, 233, 329, 350
Incompressible formulation, 208
Industry 4.0, 3
Inflow, 20, 200, 344, 380
Initial conditions, 37, 55, 153, 161, 216, 243, 354
Instabilities, 52, 79, 117, 162, 163, 208, 387–390, 392–396, 398–400, 402, 403
Integration points, 28, 114, 115, 123, 127, 128, 130, 138–141, 145, 146, 148, 166, 173–175, 184, 186, 189–191, 199, 209–211, 258–261, 314, 315, 337, 338, 341, 343, 346, 350, 351, 356, 361, 363–371, 374, 375, 377, 380, 383, 384, 397
Interface, 28, 41, 52, 70, 73–75, 113, 115, 123, 124, 145, 147, 158, 164, 166, 168, 170–173, 176, 182, 188–191, 197, 203, 207–211, 215, 216, 235–237, 244, 247, 278, 314, 319, 320, 341, 343, 351–354, 359, 389, 392, 393, 395, 397–399, 403, 417, 420, 422
Internal energy, 27, 30
Internal flows, 218, 223, 224, 230
Internet of Things (IoT), 1, 3
Interpolation functions
  central differencing scheme (CDS), 115, 119–121, 123, 341, 365
  deferred correction, 128
  exponential, 56, 122
  general interpolation, 139–141, 209, 342, 343
  high resolution scheme, 140
  montonic upstream centered scheme (MUSCL), 140, 342
  quick scheme, 128
  seond order upwind (SOU), 140
  skew upstream differencing scheme (SUDS), 137–139, 364

upwind differencing scheme (UDS), 119, 121, 124, 128, 342
weighted upstream differencing scheme (WUDS), 142, 154, 244, 314, 348
Inverse, 267, 285, 286, 292, 326, 327, 340

J
Jacobian, 10, 72, 267, 270, 271, 282, 286, 288, 302, 303, 327, 360, 362
Jet, 221, 224, 225, 233

L
Languages, 4, 235, 237, 238
Least square, 48, 339, 340
Linearization, 41, 50, 62, 63, 76, 147, 161, 196, 208–210, 217
Linearization of mass flow, 209, 217, 409
Linear system, 10, 30, 36, 41, 47, 51, 54, 57–60, 62, 63, 65–67, 70–73, 79, 82, 85, 86, 88, 92–96, 98–100, 102, 103, 107, 117, 128, 150, 152–154, 160, 168, 176, 181–183, 185, 186, 190, 192, 193, 195, 197, 208, 215, 216, 226, 227, 230–232, 241, 242, 246–248, 250, 255, 257, 268, 312, 313, 318, 320, 333, 338, 340, 342, 343, 353, 354, 359, 363, 365, 369, 372, 375, 376, 378, 380, 387, 391, 392
   diagonal dominance, 69, 94, 95
   direct solution, 60, 71, 88, 93, 102
   iterative solution, 10, 153, 160
   matrix of coefficients, 41, 70, 160
   matrix structure, 70
   sparsity, 71, 93
Local coordinate systems, 11, 256, 263, 357
Local matrices, 359, 372, 377–380
Lower triangular, 91

M
Mach, 34, 190, 408, 409
Mach number, 190, 202, 207, 217, 408, 409, 411
Mapping, 264, 291–300, 360
Mass conservation, 18, 25, 31, 44, 79, 148, 157, 158, 168, 171, 173, 178, 182, 184, 186, 188, 189, 192, 194, 196, 203, 208–210, 215, 216, 224, 225, 228, 229, 231, 290, 316, 320, 321, 343, 351, 353, 388, 389, 392–395, 397, 398, 413

Mass conservation equation, 4, 16, 18, 22, 24, 37, 43, 44, 126, 148, 161, 162, 164, 171, 173, 175–178, 180–184, 191, 192, 195, 196, 203, 208–212, 214, 216, 217, 306, 318–321, 349, 352, 371, 375, 388–390, 392, 394, 398
Mass flow ratio, 368
Mathematical model, 2–4, 13, 57, 222, 224, 227, 228, 246, 250, 251, 317
Mesh, 6–8, 11, 32, 50, 54, 56, 59, 65, 71, 72, 78, 79, 81, 82, 88, 91, 94, 96–98, 100, 102, 105, 110, 116, 117, 119, 123, 124, 129, 133–135, 137, 141, 142, 150, 160, 164, 173, 232, 236, 240, 244, 245, 248, 250, 254–256, 258, 260, 263, 264, 284, 285, 288, 292, 295, 296, 299, 300, 306–308, 321, 329, 333, 334, 336, 339, 341, 345, 349, 350, 366, 381, 394, 395, 407–409, 413, 416, 418–420
Mesh generator, 48, 256
Methods
   analytical, 2, 42
   experimental, 3, 4
   numerical, 1–4, 6–9, 14, 37, 42, 43, 46, 59, 60, 73, 78, 85, 90, 113, 117, 207, 250, 254, 258, 280, 281, 284, 292, 299, 325, 387, 392
Metrics of the transformation, 10, 264, 266, 282, 284, 285, 288, 289, 292, 302, 325, 361
Metric tensor, 10, 268, 269, 286, 288, 306, 309, 310, 326, 337
Minimization, 339–341
Modified Strongly Implicit (MSIP), 93
Momentum equations, 15, 19, 20, 24, 30, 32, 37, 38, 124, 165, 166, 168–175, 179, 182–184, 186, 188–191, 196, 210, 215, 216, 224, 226, 227, 229, 231, 311, 316, 319, 320, 344, 366, 374, 390, 392, 394
Moving grids, 308
Multigrid, 10, 42, 86, 90, 94–96, 98–101, 103–105, 107, 160, 197, 333, 344

N
Navier-Stokes equations, 4, 6, 10, 11, 14, 18, 22, 30, 32, 39, 47, 72, 129, 176, 221, 290, 318, 387–389, 408, 420
Newmann, 56, 64, 67, 80, 82, 192, 193, 198
Newton's Second Law, 17

Index   429

Non linearities, 6, 10, 13, 15, 18, 23, 30,
    35–37, 39, 47, 72, 79, 92, 157,
    159–161, 183, 197, 217, 232, 244,
    250, 290
Nonorthogonal grid, 66, 318
Non-uniform grids, 391, 393, 395
Nonzeros, 60, 197
[L] Numerical approximation, 42, 70, 72,
    73, 76–78, 80, 118, 129, 147, 149,
    154, 159, 178, 230, 244, 246, 248,
    313, 345, 367
Numerical diffusion, 2, 7, 113, 117, 118,
    123, 127, 129–131, 133–138, 141,
    244, 314, 341, 342, 349, 364, 375,
    417
Numerical oscillation, 2, 10, 113, 117, 119,
    124, 128, 129, 131, 134–136, 141,
    142, 363, 375
Numerical validation, 3, 4

O

Object-Oriented, 236, 238, 380
One-sided approximation, 121
Outflow, 20, 200
Overshoots, 117, 140, 171, 371

P

Parabolic flows, 222–225, 227
Parallel plates, 38, 109, 233, 246–249
Peclet, 38, 116, 121–125, 136, 138, 139,
    174, 244, 317, 348
Petroleum engineering, 387
Physical Influence Scheme (PIS), 168, 173
Physical validation, 4
Physical velocity, 10
Phyton, 236–238
Plumes, 224, 233
Point-by-point, 57, 82, 86, 87, 94, 95, 107,
    150, 186
Point level, 7, 9
Polygonal grid, 338–340
Poroelasticity, 11, 387–389, 392–394, 396,
    401
Porous media flow, 401, 407, 413, 420
Positiveness, 367
Positiveness of the coefficients, 73, 149,
    211, 342
Positive region, 370, 371
Power-Law, 123
Prescribed pressure, 193, 194, 201
Pressure decoupling, 171, 222, 225
Pressure-displacement, 11, 387, 389, 393

Pressure gradient, 32, 37, 38, 109, 147,
    159, 164, 166, 168, 170, 173–175,
    178, 183, 190, 192, 194, 201, 216,
    223, 224, 228–231, 320, 351, 353,
    388, 391, 397
Pressure oscillation, 163, 164, 166, 175,
    398
Pressure-velocity couplings
  CPVC - co-located
    physical influence scheme (PIS),
      176, 189, 208, 343
    Rhie ad Chow, 168, 319, 321
  elliptic, 221, 225, 227, 228, 230, 231,
    233
  parabolic direction, 221–224, 226, 228,
    230, 233
  SPVC – segregated
    Chorin, 177–179
    pressure implicit with split operator
      (PISO), 188
    pressure implicit momentum explicit
      (PRIME), 176, 186, 188, 189, 194
    simple consistent (SIMPLEC), 176,
      189, 190, 213, 321
    simple-revisited (SIMPLER), 183,
      192
Pulse, 117, 118, 131–133, 244, 245, 416,
    417

R

Recommendations, 10, 236, 239, 241, 289,
    307, 344, 370
Relative velocity, 16, 20, 30, 280, 306, 307,
    325
Residue, 46, 48, 93, 99, 100, 103, 177, 197,
    247
Reynolds, 14, 44, 95, 116, 123, 136, 248
Rhie and Chow method, 168
Robin, 65, 198

S

Second order, 32, 33, 48, 52–54, 77, 78, 80,
    120, 128, 129, 140, 171, 187, 223,
    306, 341, 364, 366, 367, 393, 395,
    397, 400
Shape functions, 8, 47, 258, 334, 355,
    357–363, 365–367, 374, 375, 381,
    382, 397
Shock waves, 34, 129, 208
Simply connected domains, 296
Simulator, 299, 301, 387, 411, 417–420
SKS skyline storage, 72

Solid mechanics, 5, 11, 47, 387
Solution
    segregated, 10, 153, 160, 168, 194, 195, 201, 203, 210, 227, 321, 343, 392
    simultaneously, 160, 173, 194, 217, 238, 250
Solution of linear systems
    direct solution, 102
    Gauss-Seidel, 82, 107
    incomplete LU decomposition, 91, 92
    iterative solution, 10
    Jacobi, 82
    line-by-line, 57
    LU decomposition, 91, 92
    point-by-point methods, 57
    successive over relaxation (SOR), 82
    tridiagonal, 88
    Tri Diagonal Matrix Algorithm (TDMA), 88
Solution symmetry, 247
Source term, 24, 25, 30, 37, 41, 50, 51, 54, 57, 62, 63, 72, 76, 95, 113–115, 136, 137, 155, 165, 166, 179, 188–191, 216, 247, 248, 309, 317, 329, 330, 346, 351, 354, 372, 374, 375, 377, 382, 390, 391
Stability, 41, 54, 56, 60, 70, 78–80, 113, 163, 188, 192, 227, 232, 244, 248, 384, 391, 392, 395
Stabilization schemes, 166, 392, 394, 398, 399, 401
Staggered grid, 8, 163, 166–168, 182, 189, 201, 203, 207, 208, 319, 389–391, 393–396
State equation, 31, 157, 161, 213, 216
Steady state, 30, 36, 37, 55, 56, 58, 60, 61, 70, 81, 108, 110, 123, 151, 153, 161, 162, 164, 177, 183, 188, 215, 222, 381
Strong Implicit Procedure (SIP), 93
Subsonic flows, 34, 192, 202, 207, 217, 218
Supersonic flows, 3, 34, 35, 192, 202, 208, 217
System, 2, 4, 6, 8–11, 15–18, 22–25, 28–31, 35–38, 42, 45, 51, 57–60, 65, 70, 72, 78, 79, 82, 85, 86, 92–94, 98–100, 103, 107, 117, 128, 133, 153, 154, 157–160, 162, 163, 167, 173, 176, 182, 183, 185, 186, 188, 189, 192–197, 207, 210, 217, 223–225, 227, 228, 230, 238, 247, 248, 250, 253, 256, 258, 261, 263–265, 267, 269, 272–278, 280–282, 284, 286–288, 290–294, 301, 302, 304–306, 309, 311, 314, 318, 319, 321, 325–328, 337, 340, 342, 345, 350, 366, 371, 372, 388, 390, 395, 411, 420

T

Temperature, 24, 28, 31, 33, 36, 52, 53, 55–60, 64–66, 73, 76, 81, 87, 107–110, 116, 120, 121, 151, 154, 155, 158, 161, 162, 176, 182, 185, 186, 199, 200, 202, 207, 211, 215–217, 232, 242, 243, 309, 327, 329, 344, 371, 381
Terzaghi, 392–396, 399–404
Test problems, 242
Thermal conductivity, 33, 35, 36, 82, 109
Time step, 30, 35, 36, 53, 56–58, 60, 70, 81, 149–151, 153, 183, 186, 188, 192, 215, 354, 392
Tools, 2, 4, 5, 9, 14, 48, 78, 107, 113, 167, 176, 235, 237–239, 299, 301, 401, 420
Tracer, 413, 416, 418
Transformed equations, 263, 303, 304, 306, 311, 325, 329
Transient heat conduction, 35, 41, 107, 109, 243
Transition grids, 418
Transport coefficient, 302, 346
Tridimensional geometries, 299
True transient, 14, 53, 57, 150, 161
Turbulence
    direct numerical simulation (DNS), 14
    k-epsilon, 14
    k-omega, 14
    Reynolds averaged Navier–Stokes equations (RANS), 14, 158
Two-phase flow, 414, 417

U

Undershoots, 140, 171, 371
Undrained, 389, 392
Unstructured grids, 8–11, 47, 48, 67, 136, 141, 163, 167, 209, 236, 253, 257, 259, 261, 263, 282, 284, 292, 296, 326, 333, 334, 336, 341, 342, 344, 347, 356, 357, 364, 370, 371, 380, 391, 395, 396, 398, 399, 417, 419, 420, 422
Upper triangular, 91

V
Validation, 3, 4, 235, 237, 242, 398, 401, 416
Verification, 3, 4
Vertex-based method, 338
Viscosity, 22, 31, 109, 196, 350

W
Weighted residual approach, 7, 46–48
Weighting function, 7, 8, 47, 48
Withdrawal, 398–400

Printed by Printforce, the Netherlands